Oxford Graduate Texts i

Series Editors
R. Cohen   S. K. Donaldson   S. Hildebrandt
T. J. Lyons   M. J. Taylor

OXFORD GRADUATE TEXTS IN MATHEMATICS

Books in the series

1. Keith Hannabuss: *An introduction to quantum theory*
2. Reinhold Meise and Dietmar Vogt: *Introduction to functional analysis*
3. James G. Oxley: *Matroid theory*
4. N. J. Hitchin, G. B. Segal, and R. S. Ward: *Integrable systems: twistors, loop groups, and Rieman surfaces*
5. Wulf Rossmann: *Lie groups: An introduction through linear groups*
6. Qing Liu: *Algebraic geometry and arithmetic curves*
7. Martin R. Bridson and Simon M. Salamon (eds): *Invitations to geometry and topology*
8. Shmuel Kantorovitz: *Introduction to modern analysis*
9. Terry Lawson: *Topology: A geometric approach*
10. Meinolf Geck: *An introduction to algebraic geometry and algebraic groups*
11. Alastair Fletcher and Vladimir Markovic: *Quasiconformal maps and Teichmüller theory*
12. Dominic Joyce: *Riemannian holonomy groups and calibrated geometry*
13. Fernando Villegas: *Experimental Number Theory*
14. Péter Medvegyev: *Stochastic Integration Theory*
15. Martin Guest: *From Quantum Cohomology to Integrable Systems*
16. Alan D. Rendall: *Partial Differential Equations in General Relativity*
17. Yves Félix, John Oprea and Daniel Tanré: *Algebraic Models in Geometry*

Partial Differential Equations in General Relativity

Alan D. Rendall
Max Planck Institute for Gravitational Physics,
Albert Einstein Institute, Am Mühlenberg 1,
14476 Potsdam, Germany

OXFORD
UNIVERSITY PRESS

Great Clarendon Street, Oxford OX2 6DP

Oxford University Press is a department of the University of Oxford.
It furthers the University's objective of excellence in research, scholarship,
and education by publishing worldwide in

Oxford New York

Auckland Cape Town Dar es Salaam Hong Kong Karachi
Kuala Lumpur Madrid Melbourne Mexico City Nairobi
New Delhi Shanghai Taipei Toronto

With offices in

Argentina Austria Brazil Chile Czech Republic France Greece
Guatemala Hungary Italy Japan Poland Portugal Singapore
South Korea Switzerland Thailand Turkey Ukraine Vietnam

Oxford is a registered trade mark of Oxford University Press
in the UK and in certain other countries

Published in the United States
by Oxford University Press Inc., New York

© Alan Rendall 2008

The moral rights of the author have been asserted
Database right Oxford University Press (maker)

First published 2008

All rights reserved. No part of this publication may be reproduced,
stored in a retrieval system, or transmitted, in any form or by any means,
without the prior permission in writing of Oxford University Press,
or as expressly permitted by law, or under terms agreed with the appropriate
reprographics rights organization. Enquiries concerning reproduction
outside the scope of the above should be sent to the Rights Department,
Oxford University Press, at the address above

You must not circulate this book in any other binding or cover
and you must impose the same condition on any acquirer

British Library Cataloguing in Publication Data

Data available

Library of Congress Cataloging in Publication Data

Data available

Typeset by Newgen Imaging Systems (P) Ltd., Chennai, India
Printed in Great Britain
on acid-free paper by
Biddles Ltd., King's Lynn, Norfolk

ISBN 978–0–19–921540–9 (hbk.)
ISBN 978–0–19–921541–6 (pbk.)

10 9 8 7 6 5 4 3 2 1

To Eva

Preface

General relativity is a physical theory which nowadays plays a key role in astrophysics and in this way is important for a number of ambitious experiments and space missions. At the same time it is central to our conceptual framework for understanding such basic concepts as space, time and gravitation. This theory requires a lot of mathematical input. It is formulated in the language of differential geometry and there are a number of excellent books which provide students and researchers with the relevant background knowledge on that topic. Much less is available concerning the necessary background on partial differential equations (PDE). The Einstein equations, which are the centrepiece of general relativity, are a system of partial differential equations which are essentially hyperbolic in nature. For this reason it is clear that the theory of partial differential equations is of central importance for general relativity. The aim of this book is to give an exposition of the parts of the theory of partial differential equations that are needed in this subject and to present a variety of applications.

The book is written with several types of readers in mind. It is directed to graduate students interested in learning about general relativity, to researchers in general relativity who feel the need to learn more about partial differential equations and to graduate students and researchers working on partial differential equations who are tempted by the prospect of learning something about the Einstein equations. This is a very special system of PDE, a fact which is in part due to its links with physics and geometry. Experience shows that many people in the area of PDE would like to learn about the Einstein equations but are intimidated by the unfamiliar notation and approach. One of the main aims of this book is to give them a helpful point of entry to the subject. On the other hand it seems that the education in PDE of the community of researchers in general relativity is much poorer than that which they have in differential geometry and the book is planned to help to change this.

A book of this kind must start by putting the material to be covered into context, from the point of view of both physics and mathematics. It is not the plan here to compete with standard textbooks on general relativity. The aim is to give just enough background to reach the point where the role of differential equations can be discussed. In practice the Einstein equations, which describe gravitational fields, are coupled with equations describing the matter generating those fields. It is thus important to understand the mathematical description of matter in detail and this is treated thoroughly in the book. Many of the mathematical problems involving PDE that arise in general relativity are very hard and so to obtain manageable problems it is often necessarily to make simplifications. A frequent way of doing this is to make symmetry assumptions. The book contains a systematic discussion of symmetry classes of solutions.

The phrase 'partial differential equations' in the title is understood to include ordinary differential equations (ODE). Ordinary differential equations are important in applications in general relativity and they also serve to introduce important concepts in a relatively simple setting. The reader is not assumed to have a wide knowledge of functional analysis and some of the central ideas of that subject needed are explained in a down-to-earth way. Hyperbolic equations are central to the subject and so are naturally focused on. However elliptic and kinetic equations are also discussed.

There seems to be a widespread belief among researchers in general relativity that if differential equations cannot be solved explicitly then it is necessary to resort to approximations, whether analytical or numerical. While I recognize the immense practical importance of approximations even when they are not backed up by mathematical theorems, I have tried to demonstrate in the book that this is not the only way to go. There is a lot of theory out there which can be used to obtain rigorous information on the qualitative behaviour of solutions. The book attempts to explain how this can work and presents examples of how it has worked. My hope is that in this way graduate students can develop a more sophisticated relationship to PDE theory than they would otherwise have had and that this can benefit research in general relativity as well as the scientific development of the individual student.

The material in this book is a mixture of rigorous mathematics, heuristic arguments and physics. It is clear to me that all of these elements are important in general relativity and should be applied together to achieve the best results. At the same time I am convinced that it is of paramount importance to distinguish clearly between these different areas of intellectual activity: what is a rigorous proof, what is a persuasive heuristic argument and what is a piece of physical intuition? I trust that the reader will have no difficulty recognizing what is what in the following.

The choice of topics in this book, as in any book, is influenced by the personal tastes of the author and no attempt has been made to give a balanced survey of the literature on the subject. The reader is referred to the author's article in the online journal *Living Reviews in Relativity* [176] for a broader picture with more references. This complements the book, where the main emphasis has been to provide material which will be of practical use for those wanting to tackle problems involving the application of partial differential equations in general relativity.

Acknowledgements

I am grateful to the many people who have shared their knowledge and insights concerning general relativity and partial differential equations with me over the last twenty years and who have inspired me by their enthusiasm and vision. It would be impracticable to thank all those who deserve it here and I confine myself to mentioning the following individuals to whom I feel particularly indebted: Håkan Andréasson, Demetrios Christodoulou, Mihalis Dafermos, Jürgen Ehlers, Sergiu Klainerman, Vincent Moncrief, Gerhard Rein and Bernd Schmidt. I also want to thank these people and others in the research community for their lasting friendship and practical help.

I thank Roger Bieli, Tobias Lamm and Blaise Tchapnda for a careful reading of parts of the manuscript. Of course all remaining errors and failings in the book are entirely the responsibility of the author.

I also thank the following people for discussions which contributed to the development of the book: Roger Bieli, Sergio Dain, Michael Munzert, Hermann Nicolai, Todd Oliynyk, Hans Ringström, Bernd Schmidt, Blaise Tchapnda and Juan Velázquez. Finally, I thank Roger Bieli for preparing the figures.

Contents

1	Introduction	1
1.1	Physical background	1
1.2	Mathematical background	4
1.3	Structure of the book	6
2	General relativity	8
2.1	Basic concepts	8
2.1.1	Lorentzian algebra	8
2.1.2	Lorentzian geometry	11
2.1.3	Geodesic deviation and singularity theorems	16
2.1.4	Volume and integration	20
2.2	The Einstein equations	21
2.3	The $3+1$ decomposition	22
2.4	Conformal rescalings	28
2.5	Covering spaces and foliations	29
2.6	Further reading	29
3	Matter models	30
3.1	Scalar fields	33
3.2	The Maxwell and Yang–Mills equations	39
3.3	Continuum mechanics	41
3.4	Kinetic theory	43
3.5	Other matter models	46
3.6	Further reading	49
4	Symmetry classes	50
4.1	Static and stationary models	52
4.2	Spatially homogeneous models	56

	4.3	Surface symmetry	61
	4.4	T^2 symmetry	63
	4.5	$U(1)$ symmetry	68
	4.6	Further reading	69
5	Ordinary differential equations		71
	5.1	Existence and uniqueness	73
	5.2	Dynamical systems	75
	5.3	Formal power series solutions and asymptotic expansions	76
	5.4	Linearization and the Hartman–Grobman theorem	79
	5.5	Examples (Bianchi models)	80
	5.5.1	The Wainwright–Hsu system	80
	5.5.2	Models of Bianchi types II and VI_0	83
	5.6	Centre manifolds and the reduction theorem	85
	5.7	Further examples	86
	5.7.1	Bianchi types II and VI_0 revisited	86
	5.7.2	The massive scalar field	88
	5.7.3	Bianchi type III Einstein–Vlasov	90
	5.8	Bifurcation theory	93
	5.9	Global existence for homogeneous spacetimes	94
	5.10	An application to surface symmetry	98
	5.11	Further reading	102
6	Functional analysis		103
	6.1	Abstract function spaces	103
	6.2	Distributions	106
	6.3	Concrete function spaces	108
	6.4	Littlewood–Paley theory	115
	6.5	Pseudodifferential operators	116
	6.6	Further reading	118
7	Elliptic equations		119
	7.1	The concept of ellipticity	119
	7.2	Boundary value problems	122
	7.3	Douglis–Nirenberg ellipticity	123
	7.4	Fredholm operators	123
	7.5	The Einstein constraints	127
	7.6	Further reading	131

8	Hyperbolic equations		132
	8.1	The Cauchy problem	132
	8.2	Examples of ill-posed problems	134
	8.3	Symmetric hyperbolic systems	136
	8.4	Strong hyperbolicity	150
	8.5	Leray hyperbolicity	152
	8.6	The analytic Cauchy problem	154
	8.7	Initial boundary value problems	155
	8.8	The null condition	158
	8.9	Global difficulties	160
	8.10	Comparison with parabolic equations	162
	8.11	Fuchsian methods	164
	8.12	Further reading	169

9	The Cauchy problem for the Einstein equations		170
	9.1	Coordinate conditions	170
	9.2	The local Cauchy problem	171
	9.3	Inclusion of matter	177
	9.4	Cosmic censorship	179
	9.5	The BKL picture	182
	9.6	Further reading	184

10	Global results		186
	10.1	Gowdy spacetimes	186
	10.2	Stability of de Sitter space	193
	10.3	Stability of Minkowski space	199
	10.4	Stability of the Milne model	203
	10.5	Stability of the flat Bianchi type III model	203
	10.6	The Newtonian limit	207
	10.7	Further reading	211

11	The Einstein–Vlasov system		213
	11.1	Other kinetic equations	213
	11.2	Small data global existence	214
		11.2.1 Schwarzschild coordinates	214
		11.2.2 Maximal-isotropic and double null coordinates	220

		11.3	Cosmological solutions	223
			11.3.1 Einstein–Vlasov solutions with T^2 symmetry	224
			11.3.2 T^2 symmetry and CMC time	232
			11.3.3 Einstein–Vlasov solutions with surface symmetry	241
			11.3.4 Spherical symmetry and CMC time	245
			11.3.5 Strong cosmic censorship without full asymptotics	245
		11.4	Isotropic singularities	248
		11.5	Weak cosmic censorship and internal structure of black holes	249
		11.6	Further reading	250
12	The Einstein–scalar field system			252
		12.1	Asymptotically flat solutions	252
		12.2	Weak null singularities	255
		12.3	Price's law	258
		12.4	Cosmological solutions	260
		12.5	Further reading	262
References				263
Index				277

1 Introduction

1.1 Physical background

In the seventeenth century Isaac Newton showed that the way objects fall in everyday life and the motion of the moon and other celestial bodies can be explained in the context of a single theory. In this theory, the Newtonian theory of gravitation, there is a force of attraction between all material bodies. For most physical phenomena in which gravitation plays a role the Newtonian theory is sufficient for all practical purposes even today. There is, however, another description of gravitation which is more accurate in some situations. This is general relativity which was introduced by Albert Einstein in 1915.

The description of gravity is closely related to the concepts of space and time. In Newtonian physics there is an absolute time while space has a fixed geometry which is that of Euclid. Einstein discovered that to have a correct description of space and time which is valid for objects moving with very high velocity a new framework is necessary. This is the special theory of relativity, introduced by Einstein in 1905. In special relativity time loses its absolute character and space and time merge into the concept of spacetime. It was the effort to reconcile special relativity and gravity which led Einstein to general relativity. There spacetime is curved and indeed the curvature of spacetime is intimately related to gravity itself.

The physical situations in which general relativity becomes important are those in which both high velocities and large masses are involved, matter is concentrated into a region small compared to its mass or, in the absence of these factors, very high precision is required. There is only one case in everyday life where general relativity is of practical importance and it is connected to the last of the three circumstances just mentioned. This is the navigation system GPS (Global Positioning System). It requires comparing high precision measurements made by clocks which are in relative motion and experience different gravitational fields. If general relativistic effects were ignored then GPS would not work.

The applications of general relativity other than GPS are all related to astrophysical phenomena. In recent decades it was discovered that there are many violent processes occurring in the universe where the high velocities and large masses requiring the use of general relativity are frequently encountered. In special and general relativity the speed of light is an upper limit for the velocity of any object and to be precise the phrase 'high velocities' in the above discussion means velocities comparable to the speed of light. A particularly interesting phenomenon is the collapse of a large concentration of matter such as a star under its own weight to form a black hole. The description of this collapse process is beyond the range of the Newtonian gravitational theory and black holes are one of the characteristic features of Einstein's theory. General relativity is also very important in cosmology, which is the attempt to describe the structure and evolution of our universe on the largest scales that we can observe.

It has been known since the 1930s that our universe is expanding. This means that sufficiently distant objects such as galaxies are all moving away from us. From a Newtonian point of view these objects attract each other and so their relative velocities should decrease with time. Under plausible assumptions this conclusion remains true in general relativity. Unfortunately for this theoretical consideration, since the mid 1990s observational evidence has been accumulating that the relative velocities are increasing – the expansion of our universe is accelerated. Obtaining a better understanding of the cosmic acceleration is a challenge which now drives a lot of research in cosmology, both theoretical and observational.

One of the most important sources of information about cosmology comes from the observation of supernovae of type 1a, a class of exploding stars. The explosion produces so much energy that it can be seen at cosmological distances. Moreover the events have essentially constant intrinsic brightness which means that their observed brightness can be used to determine their distance. The spectra of these stars are shifted towards the red as a consequence of the cosmological expansion. The relation between redshift and distance gives a record of the cosmological evolution and this was what led to the realization that the expansion of the universe is accelerated.

General relativity leads to the conclusion that there was a time in the past when matter densities and temperatures were very high. Intuitively this comes from following the cosmic expansion backwards in time. In fact general relativity implies under certain general conditions that in the approach to a certain time in the past quantities like the matter density diverge and the model breaks down. This is the initial singularity or big bang, a subject discussed in more detail later. The earliest time that we can see directly by means of electromagnetic radiation corresponds to a time about three hundred thousand years after the big bang. Before that time free electrons scattered photons so effectively that the universe was opaque. Since that time, known as decoupling, photons have been travelling almost freely. These photons can be observed today and constitute what is known as the cosmic microwave background radiation (CMB). The temperature of the CMB in different directions in the sky has been measured very accurately by various instruments, notably those of the satellites COBE and WMAP. The first result is that the temperature is very uniform and the second is that the variations encode a wealth of information about the evolution of our universe. In 2006 the Nobel Prize in Physics was awarded to John Mather and George Smoot for their contributions to the COBE project.

Another characteristic feature of general relativity which has given rise to an active research field are gravitational waves. Even in the absence of material bodies variations in the gravitational field can propagate through space. This is analogous to the propagation of electromagnetic waves including light and the speed of propagation of gravitational waves is also the speed of light. Gravitational waves can be produced in situations such as collapse to a black hole and carry information about the processes going on to us here on Earth. The only problem is that detecting these waves requires devices of extremely high sensitivity. It is necessary to measure lengths to an accuracy of one part in 10^{24}. Several detectors such as GEO600 and LIGO have now been built and are already making these amazing measurements. Despite this it is at the moment (2007) the case that they have not yet seen any gravitational waves. Much remains to be done in improving the detectors and our understanding of the sources of gravitational waves. A great deal of effort (and money – this is a billion dollar project) is being invested into the programme to detect these waves on Earth. It should be noted that there is strong indirect evidence for gravitational waves. This comes from astronomical observations starting with those of the binary pulsar 1913+16 for which Russell Hulse and Joseph Taylor won the Nobel Prize in Physics in 1993. This is a system of two objects orbiting around each other, one of which is a neutron star. The latter sends out pulses of radio waves whose distribution in time carries information about the dynamics of the system. It is observed that the system is losing energy in a way which is in excellent agreement with the amount predicted to be carried away due to the emission of gravitational waves.

Apart from the relations to observations just described, general relativity raises interesting conceptual questions which lead to fascinating mathematics. A prominent example is provided by the concept of cosmic censorship introduced by Roger Penrose. The dynamics of general relativity often leads to singularities where the theory breaks down. The example of the big bang singularity has already been mentioned. Singularities are also predicted to occur in the interior of black holes. It might happen that the occurrence of singularities causes a breakdown of predictability, a fundamental property which a physical theory should have. Cosmic censorship is a possible mechanism for saving predictability within general relativity. Whether cosmic censorship holds is an open question which is being investigated intensively. This problem is one of the sources of inspiration in the development of mathematical general relativity today and is considered from many points of view in this book.

The centrepiece of general relativity is a system of partial differential equations, the Einstein equations. The main task of mathematical general relativity is to understand as much as possible about the solutions of these equations. The questions of physical interest described briefly in this section give indications which properties of solutions deserve particular attention. Experience shows that the questions which are physically relevant are often those which are most interesting from a mathematical point of view.

1.2 Mathematical background

The theory of partial differential equations (PDE), including that of ordinary differential equations (ODE), is of immense importance in modern science. It often happens that one mathematical technique is important in very different applications and that one area of scientific application requires many different mathematical techniques. The latter is the case for the application which is the focus of interest in this book, general relativity.

Partial differential equations represent restrictions on the way in which their solutions vary as a function of the independent variables, which often represent space and/or time. They encode information about the actual dependence of the solutions on these variables and in order to use PDE it is necessary to extract some of this information. One way of doing so is to obtain explicit solutions in terms of elementary functions. In practice this is only possible for very simple PDE or very simple solutions of more general PDE. In general other techniques must be used. This applies to the Einstein equations which play a central role in this book.

In the absence of explicit solutions there are two main ways of obtaining information about solutions of PDE. One is to use some kind of approximation. If there is reason to believe that in an application of interest certain quantities in the solutions of a PDE are small they may be set to zero to obtain a simplified system. The simplified system approximates the original system. In favourable cases more is true and solutions of the simplified system approximate solutions of the original system. It is, however, a non-trivial question whether this is true in a given application and this must be investigated on a case by case basis. The procedure just described includes as an example the use of numerical methods to calculate solutions. In that case the small quantities are quantities of the order of the scale of discretization and the simplified system is a system of algebraic equations which can be solved by a computer.

In the physics literature statements can be found of the type, 'These equations cannot be solved explicitly and so we must resort to approximations (or numerical calculations)'. In this book it is demonstrated that there is another possibility which can be used and which has been used successfully in the past in general relativity.

The alternative approach to PDE which cannot be solved explicitly is to do a rigorous qualitative analysis. This means that certain properties of the solutions are proved directly. A common tool for doing this are estimates of solutions. In comparison with the derivation of explicit solutions this means dealing with inequalities instead of equations. It is important to be able to specify which solutions are being dealt with and a typical way of doing so is to prescribe certain initial or boundary conditions. This is often very natural in the context of the applications. Consider for instance the case of a system whose development in time is to be modelled. Then a deterministic evolution is to be expected and it is natural to specify initial data. After deciding on a way of characterizing solutions by data it is necessary to check that a specification of this kind is really possible for the given system. This creates the need to prove theorems on existence and uniqueness of solutions with prescribed data before proceeding to study the properties of these solutions.

To fix ideas, consider the case of an evolution equation describing the behaviour of a system as a function of time t. The first step in obtaining a solid mathematical understanding is to prove the existence of solutions for a short time corresponding to suitable initial data. Next it can be investigated whether these solutions can be extended to all times. This is not always the case since singularities may occur. Once the maximal interval of existence of a solution has been determined the question of asymptotic behaviour arises. The aim is to obtain a qualitative description of the dynamics as $t \to \infty$ or as a singularity is approached. If there are time independent solutions their stability may be investigated. If explicit classes of dynamical solutions are known it can be investigated whether perturbing their initial data leads to a qualitative change in the dynamics. These steps are illustrated by examples coming from general relativity in later chapters. An analysis of this kind is sometimes a more effective way of obtaining information about the behaviour of a solution than using an explicit formula for it, even when one is available.

An issue which plays an important role in practice in studying these questions is that of regularity properties of functions such as continuity and differentiability. There are many more kinds of regularity which were invented specially to treat PDE. A reasonable mathematical theory of the solutions of a particular PDE can only be obtained if the right kind of regularity has been discovered. Consideration of the regularity properties of functions leads to the concept of function spaces which are sets of functions (typically vector spaces) with a given regularity endowed with a suitable concept of convergence of functions. The proofs in the theory of PDE often involve proving the convergence of sequences of functions in an appropriate function space. The theory of function spaces in general also has an important role to play. A certain knowledge of these matters is a necessary prerequisite for a real understanding of PDE. This book endeavours to include a kind of 'user's guide' to the relevant theory.

There are many difficult problems which arise in the study of ODE but there is no doubt that PDE are much more complicated than ODE. At the same time a good understanding of ODE brings many insights into PDE and so it is natural that ODE are treated in this book before coming to PDE. Another motivation for doing so comes from the fact that general relativity directly leads to interesting ODE problems.

1.3 Structure of the book

Chapters 2–4 introduce a number of basic concepts which are frequently applied in general relativity. In Chapter 2 the significance of some of these concepts is illustrated by a brief introduction to a classical topic in mathematical general relativity, the singularity theorems. Chapters 5–8 are devoted to explaining basic concepts of importance in the theory of ordinary and partial differential equations. Chapter 5 also contains some examples coming from general relativity which show a number of the ODE concepts in action. In Chapters 9–12 knowledge from these two sources is combined and applied to a variety of problems in general relativity.

In more detail the contents of the different chapters are as follows. Chapter 2 introduces the basic geometrical concepts needed in general relativity and uses them to define the Einstein equations. It also contains a collection of useful equations applicable to general solutions of the Einstein equations. In Chapter 3 the most important matter models used in general relativity are introduced. These include both matter models of practical use in astrophysics and more speculative ones arising from high energy physics and cosmology. Some alternatives to general relativity are also mentioned.

Chapter 4 explains the different types of symmetry assumptions which play a role in general relativity, proceeding from the strongest to the weakest assumptions. Along the way some of the most important explicit solutions of the Einstein equations are introduced.

Ordinary differential equations are the subject of Chapter 5. General concepts are introduced which are also important for partial differential equations: existence and uniqueness, the conceptual framework of dynamical systems, asymptotic expansions, linearization, centre manifolds and bifurcation theory. These concepts are illustrated by some applications to spatially homogeneous spacetimes and surface symmetric spacetimes. Chapter 6 covers both abstract functional analysis and the concrete function spaces which are essential for PDE theory. The treatment of PDE begins with that of elliptic equations in Chapter 7. Apart from the general theory the applications to the Einstein constraint equations are described. Chapter 8 is concerned with the class of PDE which is most important in general relativity, the hyperbolic equations. In discusses the local initial value problem and different possible definitions of hyperbolicity. It then goes on to questions of global existence and non-existence. The important concept of Fuchsian equations is introduced.

In Chapter 9 the material on hyperbolic equations is applied to the local Cauchy problem for the Einstein–matter equations. There is a discussion of cosmic censorship and the BKL picture, concepts which come up in the study of the global Cauchy problem in general relativity. Chapter 10 presents those results on the global properties of solutions of the Einstein equations which require little or no symmetry to be assumed. There are stability results for a number of basic explicit solutions and results on the relations between general relativity and the Newtonian theory of gravity. There is also a discussion of the Gowdy spacetimes which do have a lot of symmetry but deserve particular attention due to their flagship role in the evolution of our understanding of the global dynamics of solutions of the vacuum Einstein equations.

There are a variety of global results on solutions of the Einstein–matter equations with symmetry and certain boundary conditions. The two matter models for which most results are known are given by the Vlasov equation (treated in Chapter 11) and the scalar field (treated in Chapter 12). They illustrate key features of cosmic censorship and singularities, both in the cosmological and asymptotically flat cases. There are discussions of weak null singularities, which play a role in the internal structure of black holes, and Price's law, which is related to the question of stability of the Schwarzschild solution.

2 General relativity

2.1 Basic concepts

2.1.1 Lorentzian algebra

In this section some of the basic concepts of general relativity are introduced. First some necessary linear algebra is presented. Let V be a finite-dimensional real vector space. A vector in V can be represented by its components v^α in a given basis. The dual space V^* consists of the linear functionals on V, i.e. the linear mappings from V to the real numbers $\mathbb{R}$. Corresponding to a basis e_μ of V there is a unique basis θ^μ of V^*, the dual basis, satisfying the relation $\theta^\mu(e_\nu) = \delta^\mu_\nu$ where δ^μ_ν is the Kronecker delta. Indices from the beginning of the Greek alphabet are used to label components while indices from the middle of the alphabet are used to label the elements of a basis. An element of V^*, a covector, can be represented by its components η_α in the basis dual to the given basis of V. Then the relation $\theta^\mu_\alpha e^\alpha_\nu = \delta^\mu_\nu$ holds. Here and in the following the Einstein summation convention is used, which says that a sum is understood over any repeated index. Thus for instance $v^\alpha \eta_\alpha$ really means $\sum_{\alpha=1}^n v^\alpha \eta_\alpha$, where n is the dimension of the vector space. In the literature on general relativity, and in this book, the summation convention is generally used without this being mentioned explicitly. In exceptional cases where it is not used this is specified. It should be noted that the use of upper and lower indices has a very specific meaning in many formulae of general relativity. The purpose of this notational convention is to make complicated computations as efficient as possible. The terminology 'the vector v^α' is often used instead of 'the vector v'. In this case the indices are called 'abstract indices' and there is a mathematical framework which guarantees that this use of language is logically consistent. For details see [206]. Let T be a multilinear mapping on $V \times \ldots \times V \times V^* \times \ldots \times V^*$ with values in $\mathbb{R}$, i.e. a mapping which is linear in each of its arguments separately. Suppose that there are k copies of V and l copies of V^*. The mapping T can be represented by its components $T^{\alpha\ldots\beta}{}_{\gamma\ldots\delta}$ where there are k lower and l upper indices. These components are obtained by evaluating T on k elements of a basis of V and l elements of the dual basis. Considered abstractly this object is a tensor of type (l, k) on the vector space V but in the following the representation with respect to a basis is taken as fundamental.

A symmetric bilinear form g on V has components $g_{\alpha\beta}$ with $g_{\alpha\beta} = g_{\beta\alpha}$. If the matrix of components is invertible then $g_{\alpha\beta}$ is said to be non-degenerate. Correspondingly it is said to be degenerate if the matrix is singular. For reasons which will become clear later the word 'metric' will be used interchangeably with 'non-degenerate bilinear form'. By a suitable choice of basis the matrix of components can be brought into a form where it is diagonal with diagonal elements taken from the set $\{-1, 0, 1\}$ (Sylvester canonical form). The number of these with each value is independent of the basis chosen and is known as the signature of the bilinear form. The bilinear form is degenerate if and only if a zero occurs in this list. It is said to be positive definite if only $+1$ occurs and negative definite if only -1 occurs. Sometimes these two cases are described as signature $(+, +, \ldots, +)$ and $(-, -, \ldots, -)$ respectively. The case of most interest in general relativity is that of signature $(-, +, \ldots, +)$ which is known as Lorentzian signature. An alternative convention leads to the signature $(+, -, \ldots, -)$ for Lorentzian metrics but in this book the convention with only one minus sign is always used. The value of the dimension of most interest in general relativity is four. The three plus signs correspond to the three dimensions of physical space and the minus sign to time. There are two reasons for also considering higher dimensions in general relativity. The first is that certain modern theories of fundamental physics such as string theory naturally lead to the consideration of higher dimensions. The second is that, even if the main goal is to understand the four-dimensional case, considering other dimensions may lead to useful mathematical insights. If v^α is a vector in V then v_α is defined to be equal to $g_{\alpha\beta} v^\beta$. The inverse of $g_{\alpha\beta}$ is denoted by $g^{\alpha\beta}$ and so $g^{\alpha\beta} g_{\beta\gamma} = \delta^\alpha_\gamma$. If η_α is an element of V^* then η^α is defined to be $g^{\alpha\beta} \eta_\beta$. Note the internal consistency of these conventions. This process of 'raising and lowering indices' may also be applied to indices of more complicated tensors, e.g. $T^\alpha{}_{\beta\gamma} = g^{\alpha\delta} T_{\delta\beta\gamma}$. Symmetrization and antisymmetrization are often denoted by brackets around the indices concerned. For instance, for any tensor $T_{\alpha\beta}$,

$$T_{(\alpha\beta)} = \frac{1}{2}(T_{\alpha\beta} + T_{\beta\alpha}), \quad T_{[\alpha\beta]} = \frac{1}{2}(T_{\alpha\beta} - T_{\beta\alpha}). \quad (2.1)$$

The operation of producing new tensors from given ones by summing over certain indices after possibly taking products is called contraction. For instance, contracting the tensor $T_{\alpha\beta}$ with the metric gives the trace $\operatorname{tr} T = g^{\alpha\beta} T_{\alpha\beta}$.

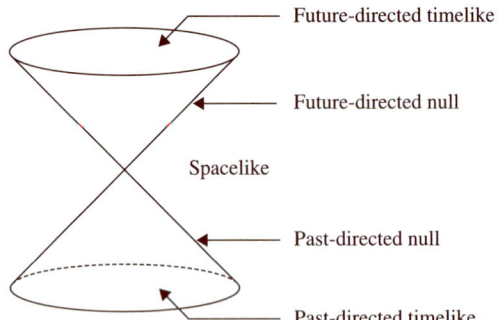

Figure 2.1 *Different types of vectors in a vector space with Lorentzian metric*

Non-zero vectors in a vector space V with Lorentzian metric are called timelike, spacelike or null according to whether $g_{\alpha\beta}v^\alpha v^\beta$ is negative, positive or zero, respectively. This is referred to as the causal character of the vector. The set of null vectors, the null cone, is a double cone with the origin removed and has two connected components. Often the vectors belonging to these two components are called future- and past-directed. Which component receives which name is a matter of convention. The set of timelike vectors is also disconnected. The boundaries of its connected components, with the zero vector removed, are the components of the set of null vectors. The timelike vectors are labelled future- or past-directed according to the labelling of the boundary of the component they lie in (cf. Fig. 2.1). Vectors that are timelike or null are called causal. If v^α and w^β are future-directed causal vectors then $g_{\alpha\beta}v^\alpha w^\beta \leq 0$ with equality only if v^α and w^α are null and proportional. The set of spacelike vectors is connected except in dimension two where it is disconnected. The causal character of a direction is by definition that of any vector spanning it.

In a vector space with a bilinear form any linear subspace has an induced bilinear form obtained by evaluating the original bilinear form on vectors lying in the subspace. In a vector space with a metric subspaces may be classified according to the signature of the induced bilinear form. This will be discussed in detail for the case of Lorentzian signature although corresponding considerations are clearly possible in more general cases. The subspace is said to be timelike, null or spacelike if the induced bilinear form is Lorentzian, degenerate or positive definite, respectively (cf. Fig. 2.2). This classification is exhaustive. A null subspace contains exactly one null direction and all other directions in it are spacelike. The orthogonal complement of a subspace W, defined in the same way as in Euclidean geometry, has dimension $n - \dim W$ and is timelike, null or spacelike when W is spacelike, null or timelike, respectively. When the induced metric of W is

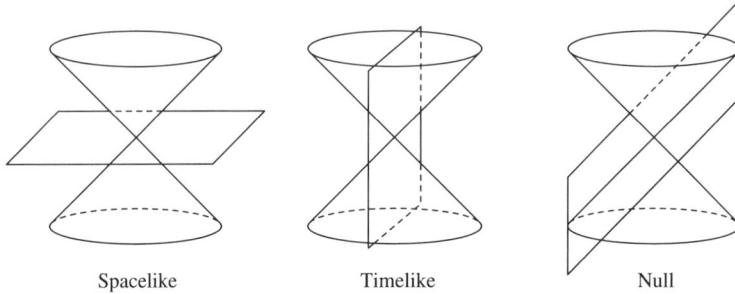

Figure 2.2 *Hyperplanes in a Lorentzian vector space of dimension n*

non-degenerate the whole space V is the direct sum of W and its orthogonal complement. When W is null this is not true since W and its orthogonal complement have non-trivial intersection.

2.1.2 Lorentzian geometry

Consider now a connected smooth manifold M. Any point of M is covered by a local coordinate system $\{x^\alpha\}$. A vector field on M is determined by its components v^α in local coordinates. These are functions of the coordinates. Its components in two overlapping coordinate systems $\{x^\alpha\}$ and $\{x'^\alpha\}$ are related by $v'^\alpha = v^\beta \partial x'^\alpha / \partial x^\beta$. A vector field v can be applied to a smooth function to obtain a new smooth function $v(f) = v^\alpha \partial f / \partial x^\alpha$. Tensor fields can be specified by their components in a way similar to that just done for vector fields. The values of all vector fields at a point x of a manifold of dimension n form a vector space of dimension n, the tangent space at x, which is denoted by $T_x M$. Its dual space is called the cotangent space and is denoted by $T_x^* M$. A tensor field with p lower and no upper indices that is antisymmetric in all its indices is called a p-form. Given a smooth function f a one-form $\eta = df$ called the exterior derivative of f can be defined. It satisfies $\eta_\alpha = \partial f / \partial x^\alpha$ and $df(v) = v(f)$ for any vector field v. If a coordinate system $\{x^\alpha\}$ is given the one-forms dx^α define a basis of the cotangent space at each point. The dual basis is often denoted by $\{\partial/\partial x^\alpha\}$. A smooth *pseudo-Riemannian metric* on the manifold M can be defined by its components $g_{\alpha\beta}$ which define a smooth matrix-valued function. This is supposed to be a non-degenerate bilinear form on each tangent space. It follows from the fact that the metric is non-degenerate and continuous and the connectedness of M that the signature of the matrix $g_{\alpha\beta}(x)$ does not depend on the point x. In general relativity information about lengths and time intervals is encoded in a Lorentzian metric, i.e. a pseudo-Riemannian metric with Lorentzian signature. In that context a manifold with a Lorentzian metric is often referred to as a *spacetime*. A pseudo-Riemannian manifold whose metric is positive definite is a Riemannian manifold, an object more familiar to mathematicians.

The simplest example of a spacetime is Minkowski space. As a manifold it is $\mathbb{R}^4$ and it has a global coordinate system in which the components of the metric are diag$\{-1, 1, 1, 1\}$. Minkowski space is the arena of special relativity. In a notation common in general relativity the Minkowski metric reads

$$-dt^2 + dx^2 + dy^2 + dz^2. \tag{2.2}$$

This notation can be interpreted in the following way. If T and U are two tensors then it is possible to define their tensor product $T \otimes U$. In a coordinate representation the components of the tensor product are the products of the components of the individual tensors. For example, if T_α and U_α are one-forms then $(T \otimes U)_{\alpha\beta} = T_\alpha U_\beta$. In (2.2) the first term dt^2 should be interpreted as $dt \otimes dt$ where dt is treated as a one-form and the other terms should be interpreted similarly. Changing from the Cartesian coordinates (x, y, z) to spherical polar coordinates (r, θ, ϕ) puts the Minkowski metric into the form

$$-dt^2 + dr^2 + r^2(d\theta^2 + \sin^2\theta d\phi^2). \tag{2.3}$$

This trivial transformation serves to demonstrate an important conceptual point. The matrix of components of the metric in the new coordinates becomes degenerate at $r = 0$ and $\theta = 0$ although it is known, due to the form in Cartesian coordinates, that the metric itself is perfectly well-behaved there. This is an example of what is called a coordinate singularity.

Important quantities derived from the metric are the Christoffel symbols which are given by

$$\Gamma^\alpha_{\beta\gamma} = \frac{1}{2}g^{\alpha\delta}(\partial_\beta g_{\delta\gamma} + \partial_\gamma g_{\delta\beta} - \partial_\delta g_{\beta\gamma}), \tag{2.4}$$

where $\partial_\alpha = \frac{\partial}{\partial x^\alpha}$. The corresponding geometrical object is the Levi-Cività connection which is not a tensor. In a neighbourhood of any point $p \in M$ there is a coordinate system (normal coordinates) in which the components of the metric take the same values as in Minkowski space at p and the Christoffel symbols vanish at that point. This fact is related to the physical notion of the equivalence principle which describes how in a certain sense general relativity is approximated locally by special relativity. Associated with the connection is the concept of covariant derivative. If v^α is a vector then the object with components $\partial_\alpha v^\beta$ is not a tensor. A related object which is a tensor is the covariant derivative defined by

$$\nabla_\alpha v^\beta = \partial_\alpha v^\beta + \Gamma^\beta_{\alpha\gamma} v^\gamma. \tag{2.5}$$

More generally the covariant derivatives of tensors can be defined. For instance

$$\nabla_\alpha T^\beta{}_{\gamma\delta} = \partial_\alpha T^\beta{}_{\gamma\delta} + \Gamma^\beta_{\alpha\epsilon} T^\epsilon{}_{\gamma\delta} - \Gamma^\epsilon_{\alpha\gamma} T^\beta{}_{\epsilon\delta} - \Gamma^\epsilon_{\alpha\delta} T^\beta{}_{\gamma\epsilon}. \tag{2.6}$$

The Levi-Città connection is defined in such a way that $\nabla_\alpha g_{\beta\gamma} = 0$. The metric is covariantly constant.

There is another kind of derivative which is defined in an invariant way on a manifold. This is the Lie derivative, denoted by $\mathcal{L}_X$. Let X^α be a vector field. The Lie derivative of the tensor $T^\alpha{}_{\beta\gamma}$ is given in coordinates by

$$(\mathcal{L}_X T)^\alpha{}_{\beta\gamma} = X^\epsilon \partial_\epsilon T^\alpha{}_{\beta\gamma} - \partial_\epsilon X^\alpha T^\epsilon{}_{\beta\gamma} + \partial_\beta X^\epsilon T^\alpha{}_{\epsilon\gamma} + \partial_\gamma X^\epsilon T^\alpha{}_{\beta\epsilon}. \tag{2.7}$$

It is a tensor. Lie derivatives of other tensors can be defined in an analogous way. A geometrical interpretation of the Lie derivative is given in Chapter 4. Because the Lie derivative is a tensor it can be computed in normal coordinates with the consequence that the partial derivatives in (2.7) can be replaced by covariant derivatives with respect to any metric without changing the definition.

A class of curves of key importance for general relativity are the geodesics. Let $\gamma : I \to M$ be a smooth curve in M defined on a parameter interval I. It has tangent vector with components $v^\alpha = \frac{dx^\alpha}{d\lambda}$ where λ is the parameter. A curve is said to have a certain causal character (timelike, null or spacelike) if its tangent vector has this character at each point. A timelike curve has the physical interpretation of the worldline of an observer, i.e. it is defined by the set $\{(t, x^a(t))\}$ of positions of the observer at different times. A curve γ is called a geodesic if its tangent vector satisfies the equation $v^\alpha \nabla_\alpha v^\beta = f v^\beta$ for some function f of λ. In the case that $f = 0$ the parameter is called affine. More generally a vector field w^α defined along a curve γ with tangent vector v^α is said to be parallelly transported along γ if $v^\alpha \nabla_\alpha w^\beta = 0$. Thus an affinely parametrized geodesic is precisely one whose tangent vector is parallelly transported along itself. It is easy to show by considering a change of parameter that each geodesic possesses an affine parameter and that an affine parameter is unique up to affine transformations $\lambda \mapsto a\lambda + b$ for constants a and b. In Minkowksi space the geodesics are the straight lines and the affine parameters are those induced by affine functions of the standard coordinates. It follows from the geodesic equation that in the case of an affine parameter $v^\alpha v_\alpha$ is independent of λ. Thus a geodesic has a well-defined causal character. The arc length of a spacelike curve is defined by integrating the quantity $(v^\alpha v_\alpha)^{1/2}$ with respect to λ. Similarly, the arc length of a timelike curve is defined by integrating $(-v^\alpha v_\alpha)^{1/2}$. The arc length of a timelike curve is also known as proper time and represents physically the time measured by a good clock carried by an observer whose worldline is the given timelike curve. The only reasonable definition of the length of a null curve is zero. For curves that are timelike or spacelike the arc length is a preferred affine parameter. For null curves all affine parameters are equally good. A timelike geodesic is interpreted as the worldline of a freely falling observer, i.e. one who is acted on by no force other than gravity. The concept of observer could be replaced by that of a particle with non-zero rest mass such as an electron. A null geodesic can be interpreted as the worldline of a particle of zero rest mass such as a photon. It can also be interpreted as a light ray. A geodesic is called complete if it is defined globally with respect to an affine parameter. A spacetime in which all (causal) geodesics are complete is called (causally) geodesically complete. If a causal geodesic is incomplete, existing only for a finite amount of affine parameter in some time direction then this has the physical interpretation that a particle suddenly ceases to exist. In general relativity a spacetime which is geodesically incomplete is called singular or is said to contain a *singularity*. The latter terminology does not imply that it is possible to define a mathematical object which is 'the singularity'.

The Christoffel symbols can be used to define the Riemann curvature tensor by

$$R^\alpha{}_{\beta\gamma\delta} = \partial_\gamma \Gamma^\alpha_{\beta\delta} - \partial_\delta \Gamma^\alpha_{\beta\gamma} + \Gamma^\alpha_{\gamma\epsilon}\Gamma^\epsilon_{\beta\delta} - \Gamma^\alpha_{\delta\epsilon}\Gamma^\epsilon_{\beta\gamma} \qquad (2.8)$$

and then the commutator of covariant derivatives is given by

$$\nabla_\gamma \nabla_\delta z^\alpha - \nabla_\delta \nabla_\gamma z^\alpha = R^\alpha{}_{\beta\gamma\delta} z^\beta. \qquad (2.9)$$

Taking traces defines the Ricci tensor $R_{\alpha\beta} = R^\gamma{}_{\alpha\gamma\beta}$ and the Ricci scalar $R = g^{\alpha\beta} R_{\alpha\beta}$. Note that here the same 'kernel' R is used to denote different objects with the difference being made clear by the different indices. It should be pointed out that there are sign conventions involved in defining the curvature quantities and that these are fixed differently by different authors. The conventions just introduced agree with those of [99] and [206]. The Einstein tensor is defined by $G_{\alpha\beta} = R_{\alpha\beta} - \frac{1}{2} R g_{\alpha\beta}$.

The components of the curvature tensor satisfy the following algebraic identities:

$$R_{\alpha\beta\gamma\delta} = -R_{\beta\alpha\gamma\delta}, \qquad (2.10)$$

$$R_{\alpha\beta\gamma\delta} = R_{\gamma\delta\alpha\beta}, \qquad (2.11)$$

$$R_{\alpha\beta\gamma\delta} + R_{\alpha\delta\beta\gamma} + R_{\alpha\gamma\delta\beta} = 0. \qquad (2.12)$$

The last of these is known as the algebraic Bianchi identity. It follows from the second identity that the Ricci tensor is symmetric. In addition there is the (differential) Bianchi identity:

$$\nabla_\alpha R_{\beta\gamma\delta\epsilon} + \nabla_\gamma R_{\alpha\beta\delta\epsilon} + \nabla_\beta R_{\gamma\alpha\delta\epsilon} = 0. \tag{2.13}$$

Taking the trace of this identity twice with the metric shows that $\nabla^\alpha G_{\alpha\beta} = 0$.

There is an algebraic decomposition of the Riemann tensor which is sometimes useful, namely

$$R_{\alpha\beta\gamma\delta} = C_{\alpha\beta\gamma\delta} - \frac{2}{n-2}(R_{\alpha[\delta}g_{\gamma]\beta} + R_{\beta[\gamma}g_{\delta]\alpha}) - \frac{2}{(n-1)(n-2)}Rg_{\alpha[\gamma}g_{\delta]\beta}. \tag{2.14}$$

This has been written down for spacetime dimension $n \geq 3$. It is regarded as defining $C_{\alpha\beta\gamma\delta}$ which is called the Weyl tensor. This tensor has the property that its trace on any pair of indices is zero. It also has an interesting conformal invariance property, namely that the quantity $C^\alpha{}_{\beta\gamma\delta}$ is the same for $g_{\alpha\beta}$ and the conformally related metric $\Omega^2 g_{\alpha\beta}$ for any smooth positive function Ω. If the Weyl tensor is zero in spacetime dimension $n > 3$ then the metric is conformal to a flat metric. In three dimensions the Weyl tensor vanishes automatically and the Riemann tensor is determined algebraically by the Ricci tensor and the metric. In that case conformal flatness must be characterized in a different way. A pseudo-Riemannian metric on a manifold of dimension at least three for which the Riemann tensor can be written as a function times $g_{\alpha[\gamma}g_{\delta]\beta}$ is said to have constant curvature. It follows from the Bianchi identity that the function multiplying the metric is constant. In two dimensions the Riemann tensor of any metric is equal to a function times $g_{\alpha[\gamma}g_{\delta]\beta}$ and the function need not be constant. In that case the metric is said to have constant curvature if that function is constant.

There is a special language used to describe the causal structure of spacetime. The causal future of p, $J^+(p)$, is the set of all points q such that there is a future-directed causal curve from p to q. (The point p itself also belongs to $J^+(p)$.) The chronological future of p, $I^+(p)$, is the set of all points q such that there is a future-directed timelike curve from p to q. The causal (chronological) future of a set S is the set of all points q such that there is a point $p \in S$ and a future-directed causal (timelike) curve from p to q. There are corresponding definitions resulting from the simultaneous replacement of 'future' by 'past' and '+' by '−'. It can be shown that if $q \in I^+(p)$ and $r \in J^+(q)$ then $r \in I^+(p)$. A future set is a set S such that $J^+(S) = S$. If a future set cannot be written as the union of two proper subsets which are themselves future sets it is called an indecomposable future or IF. If an indecomposable future is not the causal future of a point p then it is called a terminal indecomposable future or TIF. There are similar definitions of IP and TIP where 'future' has been replaced by 'past'. These definitions are useful when discussing the concept of a singularity. All of them depend not on the whole metric but only on its conformal class, i.e. its equivalence class under conformal rescalings. The set of all TIPs and TIFs can itself be given a causal structure in a natural way and defines a kind of boundary of the spacetime whose points represent singularities in some cases. It is called the causal boundary or *c*-boundary.

A spacelike hypersurface is one whose tangent space is everywhere spacelike. An important class of spacelike hypersurfaces are those which are asymptotically flat. Physically they represent the geometry of space in a solution modelling an isolated system. It is assumed that the complement of a compact subset of the hypersurface is a disjoint union of sets which are diffeomorphic to the exterior of a closed ball in $\mathbb{R}^3$. These are called the ends of the manifold. After being transported to a subset of Euclidean space by a suitable diffeomorphism the components of the metric of an end should tend to δ_{ab} at infinity. In general some extra assumption is made about the rate of convergence of the metric to the flat metric.

2.1.3 Geodesic deviation and singularity theorems

It is of interest to consider the behaviour of one-parameter families of geodesics in a given spacetime. A family of this kind defined in a region G with the property that each point of G lies on exactly one of the geodesics is known as a *congruence* of geodesics. If the geodesics are causal they can be thought of as families of particles and their behaviour reveals the effect of gravity on a cloud of test particles. To put it another way they show how the behaviour of test particles can be used to probe the geometry of spacetime. In particular they illustrate the effects of curvature as will now be shown. Consider a smooth one-parameter family of timelike geodesics of the form $x^\alpha(s, \tau)$. Here τ is the parameter along the individual geodesics and is chosen to coincide with proper time. The parameter s distinguishes the different geodesics. Let t^α be the tangent vector to the geodesics associated with the given parametrization, i.e. $t^\alpha = \partial x^\alpha / \partial \tau$. Let s^α denote the deviation vector defined by $\partial x^\alpha / \partial s$. Then a computation shows that s^α evolves according to the geodesic deviation equation

$$\frac{d^2 s^\alpha}{d\tau^2} = -R^\alpha{}_{\beta\gamma\delta} t^\beta s^\gamma t^\delta. \tag{2.15}$$

Intuitively this relation shows how the separation of nearby geodesics evolves. It is a system of second order linear ordinary differential equations. The solutions of these equations are called Jacobi fields. Let p be a point and γ a timelike curve passing through p. A point q on γ is called conjugate to p along γ if there is a non-trivial Jacobi field along γ which vanishes at p and q. Intuitively this means that there is a one-parameter family of timelike curves starting at p and ending at the same point q up to an error of linear order in the parameter. A timelike geodesic joining a point p to a point q maximizes proper length among all smooth causal curves close to γ joining p to q, provided q is sufficiently close to p. When q is further away from p this maximization property may be lost but this can only happen when q is further along the geodesic than the first point conjugate to p. It is possible to write $s^\alpha(\tau) = A^\alpha{}_\beta(\tau) s^\beta(\tau_0)$ for some fixed time τ_0 where $A^\alpha{}_\beta$ is a solution of the equation

$$\frac{d^2 A^\alpha{}_\gamma}{d\tau^2} = -R^\alpha{}_{\beta\sigma\delta} t^\beta A^\sigma{}_\gamma t^\delta. \tag{2.16}$$

Let $B^\alpha{}_\beta = A^\gamma{}_\beta \dot{A}^\alpha{}_\gamma$ where here the overdot denotes $d/d\tau$. In fact $B_{\alpha\beta} = \nabla_\alpha t_\beta$ (see [206], p. 217). The tensor $B_{\alpha\beta}$ has the property that its contraction with t^α on either index is zero. Let $h_{\alpha\beta}$ be the projection tensor $g_{\alpha\beta} + t_\alpha t_\beta$. The tensor $B_{\alpha\beta}$ can be decomposed into an antisymmetric part $\omega_{\alpha\beta} = B_{[\alpha\beta]}$, a trace $\theta = B^{\alpha\beta} h_{\alpha\beta}$ and a symmetric tracefree part $\sigma_{\alpha\beta} = B_{(\alpha\beta)} - \frac{1}{3}\theta h_{\alpha\beta}$. The quantities $\omega_{\alpha\beta}$, θ and $\sigma_{\alpha\beta}$ are called the rotation, expansion and shear of the congruence of geodesics respectively. Note that θ can alternatively be expressed as $\theta = (\det A)^{-1} \frac{d}{d\tau}(\det A)$. An equation of key importance in general relativity is the Raychaudhuri equation which follows from the equation of geodesic deviation. It reads

$$d\theta/d\tau = -R_{\alpha\beta} t^\alpha t^\beta + \omega_{\alpha\beta} \omega^{\alpha\beta} - \sigma_{\alpha\beta} \sigma^{\alpha\beta} - \frac{1}{3}\theta^2. \tag{2.17}$$

A spacetime is said to satisfy the *timelike convergence condition* if $R_{\alpha\beta} V^\alpha V^\beta \geq 0$ for any timelike vector V^α. From the Raychaudhuri equation it follows that if the timelike convergence condition is satisfied the expansion of timelike geodesics tends to decrease towards the future as a consequence of the spacetime curvature (hence the name of that condition). The only term on the right hand side of the Raychaudhuri equation which is positive is that containing the rotation. If a congruence of geodesics is initially non-rotating it is always non-rotating. This is in particular true for a family of timelike geodesics which start orthogonal to a spacelike hypersurface. In the non-rotating case the expansion can only decrease provided the timelike convergence condition is satisfied. There is also a Raychaudhuri equation for null geodesics which will not be described in detail here.

The Raychaudhuri equation is well-known for its role in the proofs of the singularity theorems of Penrose and Hawking. These theorems show that under certain conditions which are stable under small perturbations of initial data a spacetime must be geodesically incomplete. There are different inputs required for these theorems but one which is central is the following fact. Consider a non-rotating congruence of timelike geodesics in a spacetime satisfying the timelike convergence condition. Then if θ is ever negative it must diverge to $-\infty$ in finite proper time if the geodesic exists that long. Furthermore if a negative upper bound for the initial value of θ is assumed an explicit upper bound for the time until which θ must diverge is obtained. The proof is to compare a solution of the Raychaudhuri equation, which satisfies the inequality $\dot{\theta} \leq -\frac{1}{3}\theta^2$, with a solution of the equation $\dot{u} = -\frac{1}{3}u^2$. Comparison arguments of this type are explained in Chapter 5.

It is useful to extend the definition of a causal curve in the following way. Let γ be a continuous curve and suppose that given two points p and q on γ there exists a finite sequence of points $r_1, \ldots, r_m$ on γ with $r_1 = p$ and $r_m = q$ such that for each i with $1 \leq i \leq m-1$ there is a smooth future-directed causal curve from r_i to r_{i+1}. Then γ is called future-directed causal. It can be shown that a curve which is causal in this sense automatically satisfies a Lipschitz condition. (The term is defined in Chapter 5.) This follows from the fact that it must always be inside the light cone. It follows from Rademacher's theorem [139] that a curve which satisfies a Lipschitz condition is differentiable except on a set of measure zero. The definition of a future-directed causal continuous curve just made is in fact equivalent to the condition that its tangent vector exists and is future-directed causal almost everywhere. A continuous future-directed causal curve γ is said to be inextendible if it cannot be extended to a continuous future-directed causal curve defined on a longer time interval. It is past or future inextendible if it cannot be extended to the past or future respectively. Very often the causal curves of interest are smooth but it is nevertheless important to consider curves which are merely continuous when discussing inextendibility.

A subset S of a spacetime is called a *Cauchy hypersurface* if it is intersected exactly once by any inextendible causal curve. The fact that it is called a hypersurface is justified by the fact that any subset satisfying this condition is actually a Lipschitz hypersurface. In the following it will normally be assumed that the Cauchy hypersurfaces considered are smooth. A smooth Cauchy hypersurface is the appropriate place to prescribe initial data for a hyperbolic system whose principal symbol is given by the spacetime metric. (For the definition of the principal symbol see Section 7.1.) It is also in a sense the appropriate place to prescribe initial data for the Einstein equations. More information on this can be found in Section 8.1 and Chapter 9. A spacetime is called *globally hyperbolic* if it contains a Cauchy hypersurface. It turns out that the globally hyperbolic spacetimes satisfying the Einstein equations are those which can be uniquely determined by initial data.

In Riemannian geometry an important and useful class of metrics are those which are geodesically complete. The singularity theorems show that very often physically interesting spacetimes do not have this property and so restricting consideration to geodesically complete spacetimes is not appropriate in general relativity. There arises a need for a class of well-behaved spacetimes which can take over the role of the complete metrics in Riemannian geometry. The globally hyperbolic spacetimes do this. A complete Riemannian manifold has the following property of geodesic connectivity. Given any two points p and q there exists a geodesic joining p to q. Globally hyperbolic spacetimes have the following analogous but weaker property. Given two points p and q with p in the chronological past of q there exists a timelike geodesic which joins p to q. They also have a related property involving Cauchy hypersurfaces. Let S be a Cauchy hypersurface and p a point in the chronological future of S. Then there exists a future-directed timelike geodesic γ from a point q of S to p. Moreover the geodesic can be chosen to meet S orthogonally and so that there is no point conjugate to S along the curve before p. This last statement means by definition that there is no Jacobi field along γ which is tangent to S at q and vanishes at a point of γ closer to S than p. In this situation it can be shown that the geodesic maximizes distance to S until p is reached.

With the information which has been collected the proof of one of the singularity theorems, the Hawking singularity theorem, can be sketched. Consider a spacetime which satisfies the timelike convergence condition and contains a compact Cauchy hypersurface S with positive mean curvature tr$k$. (For the definition of the mean curvature see Section 2.3.) The claim is that there is a positive constant C such that no future-directed timelike curve starting on S can be longer than C. This implies in particular that all timelike geodesics are incomplete. Suppose that a future-directed timelike curve γ starting on S is longer than C. Then there is a point p on γ at a distance greater than C from S. As a consequence of the fact that the spacetime is globally hyperbolic there is a timelike geodesic of length greater than C from S to p. Thus it may be assumed w.l.o.g. that γ is a timelike geodesic starting orthogonal to S. The mean curvature has a strictly positive minimum on the compact set S. Consider the congruence of future-directed timelike geodesics starting orthogonal to S. It contains γ. The expansion of this congruence at the point q where γ meets S is related to the mean curvature of S at that point by $\theta = -\text{tr}k/3$. The Raychaudhuri equation then gives an upper bound for the length of γ to the future. Choosing C to be bigger than this bound gives a contradiction.

The first singularity theorem, that of Penrose, is applicable to asymptotically flat spacetimes and uses the concept of *trapped surface* which will now be discussed. Let S be an asymptotically flat hypersurface and Σ a surface in S which is the boundary of a compact set. There are precisely two null directions orthogonal to S at each point. This gives rise to two families of null geodesics starting on S. For these it is possible to define expansions θ and θ' in a way similar to what is done for a congruence of timelike geodesics. The expansions satisfy a Raychaudhuri equation. Due to the fact that Σ has an interior and exterior in S it is possible to distinguish between an ingoing and outgoing family. Let θ be the expansion of the outgoing family. In the case of a sphere in a hyperplane in Minkowski space or a small perturbation of that situation $\theta' < 0$ and $\theta > 0$. The surface is said to be trapped if $\theta \leq 0$. The expansion carries information about the behaviour of the area element of the spheres obtained from Σ by transporting them along the null geodesics. In the case of a trapped surface the area of the spheres initially decreases along the outgoing null geodesics. In that situation the Raychaudhuri equation for null geodesics can be used to show that the area of the surfaces continues decreasing and, together with other arguments, to show the presence of incomplete null geodesics if a suitable curvature condition is satisfied. This is the null convergence condition which says that $R_{\alpha\beta} l^\alpha l^\beta \geq 0$ for any null vector l^α. It is a consequence of the timelike convergence condition.

The singularity theorems have been presented here as theorems in Lorentzian geometry based on certain geometric assumptions. The connection to the Einstein equations is discussed in Chapter 3.

2.1.4 Volume and integration

Associated to a pseudo-Riemannian metric is the volume form which is an n-form in n dimensions. In a basis it can be uniquely defined by requiring that the component $\epsilon_{12...n}$ is equal to the square root of the modulus of the determinant of the metric in that basis. This definition is invariant under orientation-preserving changes of basis, i.e. those which have positive determinant. To put it another way, the sign of the volume form depends on a choice of orientation for the vector space. Given a p-form F it can be contracted with the volume form to get an $(n-p)$-form which, after multiplying by $1/p!$, gives an object $*F$ called the dual of F. The dual of $*F$ satisfies $**F = (-1)^{s+p(n-p)}F$ where s is the number of minus signs in the signature of the metric. Given any tensor with a pair of indices in which it is antisymmetric it is possible to dualize on that pair while leaving the others untouched. For instance, starting with the Riemann tensor it is possible to use the left pair of indices to get an object usually denoted by $*R_{\alpha\beta\gamma\delta}$ and the right pair of indices to get a corresponding object $R^*_{\alpha\beta\gamma\delta}$. The left and right duals of the Weyl tensor are equal, $*C_{\alpha\beta\gamma\delta} = C^*_{\alpha\beta\gamma\delta}$. In the four-dimensional Lorentzian case the product of two copies of the volume form satisfies the following identity

$$\epsilon^{\alpha_1\alpha_2\alpha_3\alpha_4}\epsilon_{\beta_1\beta_2\beta_3\beta_4} = -\det(\delta^{\alpha_i}_{\beta_j}). \tag{2.18}$$

Further identities can be obtained from this one by contraction.

There is an invariant way of integrating a p-form ω over a p-dimensional manifold M. In local coordinates it just corresponds to integrating the component $\omega_{1...p}$ in Euclidean space. The volume of a hypersurface S with non-degenerate induced metric in an n-dimensional pseudo-Riemannian manifold can be defined as follows. First contract the volume form corresponding to the metric with a unit normal vector to S to get an $(n-1)$-form. Then integrate that over the manifold S. A useful result not requiring a metric is Stokes' theorem which says that if ω is an $(n-1)$-form and S a hypersurface bounding a compact region G in M then $\int_G d\omega = \int_S \omega$. When a metric is present this result for an $(n-1)$-form can be dualized to get the statement that $\int_S v_\alpha n^\alpha dV_{n-1} = \int_G \nabla_\alpha v^\alpha dV_n$. Here n^α is the unit normal vector to S and dV_{n-1} and dV_n are the volume forms associated to the metrics on the given spaces.

2.2 The Einstein equations

The Einstein equations for a four-dimensional Lorentzian metric $g_{\alpha\beta}$ can be written as

$$G_{\alpha\beta} + \Lambda g_{\alpha\beta} = 8\pi T_{\alpha\beta}. \tag{2.19}$$

Here $G_{\alpha\beta}$ is the Einstein tensor, Λ is a constant, the cosmological constant, and $T_{\alpha\beta}$ is the energy–momentum tensor. The numerical factor 8π is dependent on the choice of physical units. In general units it would be replaced by $8\pi G/c^4$, where G is the gravitational constant and c is the speed of light. Here so-called geometrical units are used where G and c have the numerical value unity. The energy–momentum tensor describes the matter content of spacetime and depends on the metric and certain matter fields. Since the covariant divergence of the Einstein tensor is zero a necessary consistency condition implied by the Einstein equations is that the divergence of the energy–momentum tensor should be zero. The latter condition has the physical interpretation of conservation of energy–momentum and is satisfied by any reasonable matter model. What the matter fields are depends on the physical situation under consideration. In the case where no matter is present $T_{\alpha\beta} = 0$ and the Einstein equations become $G_{\alpha\beta} + \Lambda g_{\alpha\beta} = 0$. These are known as the vacuum Einstein equations. They are equivalent to $R_{\alpha\beta} = \Lambda g_{\alpha\beta}$. In Minkowski space $R^\alpha{}_{\beta\gamma\delta} = 0$ and the curvature vanishes. For this reason Minkowski space is said to be flat. It is in particular a solution of the vacuum Einstein equations with $\Lambda = 0$.

Since the second and third parts in the decomposition (2.14) of the Riemann tensor are determined algebraically, via the Einstein equations, by the energy–momentum tensor it is natural to think of the Weyl tensor as encoding the degrees of freedom of the gravitational field itself.

The natural generalization of the Einstein equations to other dimensions is to take (2.19) as it stands except that the relevance of the numerical constant 8π is not clear. A simple solution is to replace it by some unspecified positive constant κ. Minkowski space has an obvious analogue in any dimension.

2.3 The 3+1 decomposition

A characteristic feature of special and general relativity is that they are formulated in spacetime and that there is no preferred way of splitting spacetime into space and time. Nevertheless it is often useful to choose a splitting despite the non-uniqueness involved. One reason is that the idea of a process evolving in time can be more easily grasped intuitively than a spacetime picture. In general relativity this process is called a 3+1 decomposition. The resulting formulae have many applications to the subjects treated in this book and they are collected in this section. The 3+1 decomposition is also the standard way of bringing the Einstein equations into a form which is suitable for solving them numerically.

Suppose that a four-dimensional spacetime $(M, g_{\alpha\beta})$ is given and let S be a spacelike hypersurface. The spacetime metric induces a metric on S which is denoted by g_{ab}. The convention is used that Latin indices relate to the hypersurface S and run from 1 to 3 while Greek indices relate to the spacetime manifold M and run from 0 to 3. The index zero corresponds to the time coordinate. Let n^α denote the future-directed unit normal vector to S where the convention is used that the coordinate t increases towards the future. In other words $dt(n) > 0$. Consider now adapted coordinates on M where the first, $t = x^0$, is constant on S and the others x^a coincide with a given coordinate system on S. Then the components of the spacetime metric where the indices take values between 1 and 3 are equal to the components of the induced metric with the same indices. Thus there is no conflict between the coordinate notation and the abstract index notation. There is also another object on S which is invariantly defined by the spacetime metric, namely the second fundamental form k_{ab}. Let x^a and y^a be vector fields tangent to S and extend them in a smooth way to vector fields x^α and y^α on an open neighbourhood of S. Then the contraction $k_{ab}x^a y^b$ of the second fundamental form with x^a and y^a is equal to $x^\alpha \nabla_\alpha y^\beta n_\beta$ where n^α is a unit normal vector to S. It can be shown that this definition is independent of the extension and that the tensor k_{ab} is symmetric. Note that the sign of the second fundamental form is linked to a choice of the normal vector and in the following it is usually chosen to be the future-directed normal. The sign convention used here for the second fundamental form is the opposite of that in [99] and [206] but is often used in the literature, particularly that on numerical relativity. The trace of the second fundamental form tr$k$ is called the mean curvature.

The normal vector n^α is a four-dimensional object invariantly associated to S. Another is the projection tensor $h_{\alpha\beta} = g_{\alpha\beta} + n_\alpha n_\beta$. The mixed form $h^\alpha{}_\beta$ is a linear mapping whose kernel is the normal space and which is the identity on vectors tangent to S. If an index of any tensor is contracted with the projection tensor in its mixed form then this spacetime index can be replaced by a spatial one. The mechanism of this will be explained for vectors and one-forms. If v^α is a vector then $h^\beta{}_\alpha v^\alpha$ is tangent to S. Thus it can be identified with an intrinsic tangent vector on S. On the other hand if θ_α is a one-form then $h^\alpha{}_\beta \theta_\alpha$ is a one-form which agrees with θ_α on vectors tangent to S. Note that for a fully projected tensor raising and lowering indices with $g_{\alpha\beta}$ or $h_{\alpha\beta}$ gives the same result. Furthermore, projecting on a lower index leaves the numerical value of the component in an adapted coordinate system unchanged. One quantity whose projections are of importance is the energy–momentum tensor $T^{\alpha\beta}$. Define

$$\rho = T^{\alpha\beta}n_\alpha n_\beta, \quad j^\alpha = -T^{\beta\gamma}h^\alpha{}_\beta n_\gamma, \quad S^{\alpha\beta} = T^{\gamma\delta}h^\alpha{}_\gamma h^\beta{}_\delta. \qquad (2.20)$$

The minus sign in the definition of j^α ensures that it has the physical interpretation of a matter current. This means that it points in the direction in which matter is flowing. In terms of coordinate expressions $\rho = T^{00}(n_0)^2$, $j_a = -T^0_a n_0$ and $S_{ab} = T_{ab}$. The four-dimensional trace of the energy–momentum tensor is given by $\mathrm{tr}^{(4)}T = -\rho + \mathrm{tr}S$.

Once the induced metric, the second fundamental form and the projection tensor have been introduced they can be used to write down relations between the curvature of spacetime and the curvature of the Riemannian metric on S. These are the Gauss and Codazzi equations. The Gauss equation is

$$^{(4)}R_{abcd} = R_{abcd} + k_{ac}k_{bd} - k_{ad}k_{bc} \qquad (2.21)$$

and the Codazzi equation is

$$^{(4)}R_{\sigma abc}n^\sigma = -\nabla_c k_{ab} + \nabla_b k_{ac}. \qquad (2.22)$$

Further equations can be obtained from these by contraction. From (2.21)

$$^{(4)}R_{ab} = R_{ab} + \mathrm{tr}k k_{ab} - k_{ac}k_b^c - {}^{(4)}R_{\sigma a \tau b}n^\sigma n^\tau \qquad (2.23)$$

and contracting again gives

$$^{(4)}R + 2\,^{(4)}R_{\alpha\beta}n^\alpha n^\beta = R - k_{ab}k^{ab} + (\mathrm{tr}k)^2. \qquad (2.24)$$

Contracting (2.22) gives

$$^{(4)}R_{\sigma a}n^\sigma = -\nabla^a k_{ab} + \nabla_b(\mathrm{tr}k). \qquad (2.25)$$

Combining (2.24) and (2.25) with the Einstein equations $G^{\alpha\beta} + \Lambda g_{\alpha\beta} = 8\pi T^{\alpha\beta}$ gives the constraint equations

$$R - k_{ab}k^{ab} + (\mathrm{tr}k)^2 = 16\pi\rho + 2\Lambda, \qquad (2.26)$$
$$\nabla^a k_{ab} - \nabla_b(\mathrm{tr}k) = 8\pi j_b. \qquad (2.27)$$

These two equations are known as the Hamiltonian and momentum constraints respectively. The names come from the Hamiltonian approach to the Einstein equations which is not discussed in this book.

Choosing a time coordinate in general relativity can be broken down into two steps. The following discussion uses the concept of a foliation which is introduced in Section 2.5. The first step is to choose a family of spacelike hypersurfaces which define a foliation of part of spacetime and the second is to choose a labelling of these hypersurfaces. When this has been done a function t is obtained whose value on a given hypersurface is the real number chosen to label that hypersurface. When a foliation has been chosen there is an induced metric and second fundamental form for each leaf but there is no way of comparing these objects on different leaves without introducing extra structure. If l is the length of dt then the lapse function α is defined to be l^{-1}. In coordinates $\alpha = (-g^{00})^{-1/2}$. If the labelling is changed by replacing t by a smooth monotone function of t then α is multiplied by the derivative of this function. In particular the restriction of α to one hypersurface is multiplied by a constant. This shows to what extent α depends on t and to what extent it only depends on the foliation. The component n_0 is equal to $-\alpha$. Hence $\rho = \alpha^2 T^{00}$ and $j_a = \alpha T_a^0$.

Once a time coordinate (or just the corresponding foliation) has been fixed choosing a spatial coordinate can be broken down into two steps. The first is to specify which points on different hypersurfaces are to have the same spatial coordinates. This is equivalent to specifying a congruence of curves transverse to the foliation. The second is to label these curves. This can be done by specifying a coordinate system on one leaf of the foliation (which in general is only possible locally on the leaf). Given a fixed time coordinate t, a preferred parametrization of the congruence of curves is defined. Let t^α be the field of tangent vectors to the curves when they are parametrized in this way. The shift vector β^a is the image of t^α under the projection $h^\alpha{}_\beta$. It depends genuinely on the time coordinate and not only on the foliation which the time coordinate defines. A quantity which depends only on the foliation is $\alpha^{-1}\beta^a$. The unit normal vector can be written as $\alpha^{-1}(\frac{\partial}{\partial t} - \beta^a \frac{\partial}{\partial x^a})$. The four-dimensional metric can be written concisely as

$$-\alpha^2 dt^2 + g_{ab}(dx^a + \beta^a dt)(dx^b + \beta^b dt). \tag{2.28}$$

An equation for the time derivative of the metric is obtained by rewriting the definition of the second fundamental form. It is

$$\partial_t g_{ab} = -2\alpha k_{ab} + \nabla_a \beta_b + \nabla_b \beta_a. \tag{2.29}$$

It is interesting to note that $\nabla_a \beta_b + \nabla_b \beta_a$ is the Lie derivative of the metric with respect to the vector β^a. If an equation arising in the 3+1 decomposition contains a time derivative in coordinates with zero shift this can be reinterpreted as a Lie derivative with respect to the vector $\partial/\partial t$ which is just α times the unit normal vector to the foliation. Hence the corresponding equation for general α and β^a can be obtained by expressing this Lie derivative in the other coordinates where it is the Lie derivative with respect to $\partial/\partial t$ minus the Lie derivative with respect to the shift vector. This means that the equations for a general shift can be obtained rather easily from those for zero shift. As a consequence of (2.29) the determinant of the metric satisfies

$$\partial_t(\det g) = 2(-\alpha \operatorname{tr} k + \nabla_a \beta^a)(\det g). \tag{2.30}$$

Note also the useful identity

$$\partial_a(\det g) = 2\Gamma^b_{ab}(\det g). \tag{2.31}$$

The Ricci tensor of the spatial metric satisfies the evolution equations

$$\partial_t R_{ab} = \Delta(\alpha k_{ab}) + \nabla_a \nabla_b(\alpha \operatorname{tr} k) - \nabla^d \nabla_a(\alpha k_{db}) - \nabla^d \nabla_b(\alpha k_{da})$$
$$+ \beta^c \nabla_c R_{ab} + R_{ac} \nabla_b \beta^c + R_{cb} \nabla_a \beta^c \tag{2.32}$$

and the scalar curvature satisfies

$$\partial_t R = 2\alpha k^{ab} R_{ab} + 2\Delta(\alpha \operatorname{tr} k) - 2\nabla^a \nabla^b(\alpha k_{ab}) + \beta^a \nabla_a R. \tag{2.33}$$

The equation for the evolution of the second fundamental form is derived using the Einstein equations and reads

$$\partial_t k_{ab} = -\nabla_a \nabla_b \alpha$$
$$+ \alpha \left[R_{ab} + (\text{tr}k) k_{ab} - 2k_{ac} k^c_b - 8\pi \left(S_{ab} - \frac{1}{2} g_{ab} \text{tr}S \right) - 4\pi \rho g_{ab} - \Lambda g_{ab} \right]$$
$$+ \beta^c \nabla_c k_{ab} + k_{ac} \nabla_b \beta^c + k_{cb} \nabla_a \beta^c. \tag{2.34}$$

Note that the combination of matter terms which occurs in this equation is just that which, via the field equations, is equal to $-{}^{(4)}R_{ab}$. Using this fact (2.34) could be converted to a purely geometrical equation. Combining that equation with (2.23) gives a 3+1 expression for the curvature tensor components ${}^{(4)}R_{0a0b}$. Alternative evolution equations for k_{ab} may be obtained by combining (2.34) with the Hamiltonian constraint. The evolution equation for the mixed version of the second fundamental form is

$$\partial_t k^a{}_b = -\nabla^a \nabla_b \alpha + \alpha \left[R^a{}_b + (\text{tr}k) k^a{}_b - 8\pi \left(S^a{}_b - \frac{1}{2} \delta^a_b \text{tr}S \right) - 4\pi \rho \delta^a_b - \Lambda \delta^a_b \right]$$
$$+ \beta^c \nabla_c k^a{}_b + k^a{}_c \nabla_b \beta^c - k^c{}_b \nabla^a \beta_c. \tag{2.35}$$

The trace of (2.35) gives

$$\partial_t (\text{tr}k) = -\Delta \alpha + \alpha [R + (\text{tr}k)^2 + 4\pi \text{tr}S - 12\pi\rho - 3\Lambda] + \beta^a \nabla_a (\text{tr}k). \tag{2.36}$$

Here Δ is the Laplacian $\nabla_\alpha \nabla^\alpha$. Combining the last equation with the Hamiltonian constraint gives

$$\partial_t (\text{tr}k) = -\Delta \alpha + \alpha [k_{ab} k^{ab} + 4\pi(\rho + \text{tr}S) - \Lambda] + \beta^a \nabla_a (\text{tr}k). \tag{2.37}$$

The components of the spacetime metric are given by

$${}^{(4)}g_{00} = -\alpha^2 + \beta_a \beta^a, \tag{2.38}$$
$${}^{(4)}g_{0a} = \beta_a, \tag{2.39}$$
$${}^{(4)}g_{ab} = g_{ab} \tag{2.40}$$

and the components of its inverse are

$${}^{(4)}g^{00} = -\alpha^{-2}, \tag{2.41}$$
$${}^{(4)}g^{0a} = \alpha^{-2} \beta^a, \tag{2.42}$$
$${}^{(4)}g^{ab} = g^{ab} - \alpha^{-2} \beta^a \beta^b. \tag{2.43}$$

2.3 The 3+1 decomposition

The determinants of the three- and four-dimensional metrics are related by $\det{}^{(4)}g = -\alpha^2 \det g$. The four-dimensional Christoffel symbols are

$$^{(4)}\Gamma^0_{00} = \alpha^{-1}(\partial_t \alpha + \beta^a \partial_a \alpha - k_{ab}\beta^a \beta^b), \tag{2.44}$$

$$^{(4)}\Gamma^0_{0a} = \alpha^{-1}(\partial_a \alpha - k_{ab}\beta^b), \tag{2.45}$$

$$^{(4)}\Gamma^0_{ab} = -\alpha^{-1} k_{ab}, \tag{2.46}$$

$$^{(4)}\Gamma^a_{00} = \partial_t \beta^a + \beta^b \nabla_b \beta^a - 2\alpha k^a_c \beta^c + \alpha \nabla^a \alpha$$
$$- \alpha^{-1}(\partial_t \alpha + \beta^b \partial_b \alpha - k_{bc}\beta^b \beta^c)\beta^a, \tag{2.47}$$

$$^{(4)}\Gamma^a_{0b} = -\alpha k^a_b + \nabla_b \beta^a - \alpha^{-1}\beta^a \nabla_b \alpha + \alpha^{-1} k_{bc}\beta^c \beta^a, \tag{2.48}$$

$$^{(4)}\Gamma^a_{bc} = \Gamma^a_{bc} + \alpha^{-1} k_{bc}\beta^a. \tag{2.49}$$

The contracted Christoffel symbols, defined by $^{(4)}\Gamma^\alpha = (^{(4)}g^{\beta\gamma})^{(4)}\Gamma^\alpha_{\beta\gamma}$, are

$$^{(4)}\Gamma^0 = \alpha^{-3}(\partial_t \alpha - \beta^a \partial_a \alpha + \alpha^2 \mathrm{tr}\,k), \tag{2.50}$$

$$^{(4)}\Gamma^a = \Gamma^a - \alpha^{-1}\nabla^a \alpha - \alpha^{-2}(\partial_t \beta^a - \beta^b \partial_b \beta^a) - {}^{(4)}\Gamma^0 \beta^a. \tag{2.51}$$

It is sometimes useful when studying the Einstein–matter equations to use components in an orthonormal frame, i.e. a smooth choice of orthonormal basis of the tangent space, rather than coordinate components. Now some equations relevant to calculations in frame components will be written out. Let e^a_i be an orthonormal frame on the spatial manifold. It can be completed to an orthonormal frame in spacetime by adjoining the unit normal vector n^α. The components of the resulting frame are

$$^{(4)}e^0_0 = \alpha^{-1}, \tag{2.52}$$

$$^{(4)}e^a_0 = -\alpha^{-1}\beta^a, \tag{2.53}$$

$$^{(4)}e^0_i = 0, \tag{2.54}$$

$$^{(4)}e^a_i = e^a_i \tag{2.55}$$

and the components of the corresponding dual coframe are

$$^{(4)}\theta^0_0 = \alpha, \tag{2.56}$$

$$^{(4)}\theta^0_a = 0, \tag{2.57}$$

$$^{(4)}\theta^i_0 = \beta^a \theta^i_a, \tag{2.58}$$

$$^{(4)}\theta^i_a = \theta^i_a. \tag{2.59}$$

The rotation coefficients are defined by

$$^{(4)}\gamma^\kappa_{\lambda\mu} = \eta^{\kappa\sigma} g(\nabla_{e_\lambda} e_\mu, e_\sigma) \tag{2.60}$$

$$= \eta^{\kappa\sigma} g_{\alpha\beta} e^\alpha_\sigma e^\delta_\lambda (\partial_\delta e^\beta_\mu + \Gamma^\beta_{\epsilon\delta} e^\epsilon_\mu). \tag{2.61}$$

In the last equation the superscript (4) has been omitted for typographical reasons. In 3+1 form the rotation coefficients are

$$^{(4)}\gamma^0_{0i} = {}^{(4)}\gamma^i_{00} = \alpha^{-1} e_i(\alpha), \tag{2.62}$$

$$^{(4)}\gamma^0_{ij} = {}^{(4)}\gamma^j_{i0} = -k_{ab} e^a_i e^b_j, \tag{2.63}$$

$$^{(4)}\gamma^i_{0j} = -\alpha^{-1} \gamma^i_{kj} \theta^k_a \beta^a + \frac{1}{2}\alpha^{-1}(e^a_j \nabla_a \beta^b \theta^i_b - \delta^{is} e^a_s \nabla_a \beta^b \theta^t_b \delta_{jt}$$
$$- c^i_j + \delta^{is} c^t_s \delta_{jt}), \tag{2.64}$$

$$^{(4)}\gamma^i_{jk} = \gamma^i_{jk}. \tag{2.65}$$

Here $c^i_j = e^a_j \partial_t \theta^i_a$. The evolution equation for the second fundamental form looks identical in coordinate and frame components.

The 3+1 decomposition of the equation $\nabla_\alpha T^{\alpha\beta} = 0$ is

$$\partial_t \rho - \beta^a \nabla_a \rho - \alpha (\mathrm{tr} k) \rho + \alpha^{-1} \nabla_a (\alpha^2 j^a) - \alpha k_{ab} T^{ab} = 0, \tag{2.66}$$

$$\partial_t j^a - \beta^b \nabla_b j^a - \nabla^a \beta_b j^b - \alpha (\mathrm{tr} k) j^a + \nabla_b (\alpha T^{ab})$$
$$- 2\alpha k^a{}_b j^b + \rho \nabla^a \alpha = 0. \tag{2.67}$$

If the Einstein evolution equation (2.35) and the equation $\nabla_\alpha T^{\alpha\beta} = 0$ are satisfied then the constraint quantities

$$C = R - k_{ab} k^{ab} + (\mathrm{tr} k)^2 - 16\pi \rho - 2\Lambda, \tag{2.68}$$

$$C_b = \nabla^a k_{ab} - \nabla_b (\mathrm{tr} k) - 8\pi j_b \tag{2.69}$$

satisfy

$$\partial_t C = 2\alpha (\mathrm{tr} k) C - 2\alpha^{-1} \nabla^a (\alpha^2 C_a), \tag{2.70}$$

$$\partial_t C_a = \alpha (\mathrm{tr} k) C_a - \frac{1}{2} \alpha^{-1} \nabla_a (\alpha^2 C) \tag{2.71}$$

in coordinates with zero shift.

2.4 Conformal rescalings

A simple way of obtaining a new metric from a given one which has important applications is to do a conformal rescaling. Here some formulae for conformal rescalings will be collected. Let $g_{\alpha\beta}$ be a Lorentz metric on a manifold M of dimension n. Let $\tilde{g}_{\alpha\beta} = \Omega^2 g_{\alpha\beta}$ for some function Ω. Then

$$\tilde{\Gamma}^{\alpha}_{\beta\gamma} = \Gamma^{\alpha}_{\beta\gamma} + \Omega^{-1}(\delta^{\alpha}_{\beta}\nabla_{\gamma}\Omega + \delta^{\alpha}_{\gamma}\nabla_{\beta}\Omega - g_{\beta\gamma}g^{\alpha\epsilon}\nabla_{\epsilon}\Omega). \tag{2.72}$$

The Ricci tensor of $g_{\alpha\beta}$ is given by

$$\tilde{R}_{\alpha\beta} = R_{\alpha\beta} + (n-2)[-\Omega^{-1}\nabla_{\alpha}\nabla_{\beta}\Omega + 2\Omega^{-2}\nabla_{\alpha}\Omega\nabla_{\beta}\Omega]$$
$$- [\Omega^{-1}\nabla_{\gamma}\nabla_{\epsilon}\Omega + (n-3)\Omega^{-2}\nabla_{\gamma}\Omega\nabla_{\epsilon}\Omega]g^{\gamma\epsilon}g_{\alpha\beta}. \tag{2.73}$$

2.5 Covering spaces and foliations

Let M and N be connected manifolds of the same dimension. A smooth mapping ϕ from M onto N is called a *covering map* if each point $x \in N$ has an open neighbourhood U with the property that $\phi^{-1}(U)$ is the disjoint union of sets U_α so that the restriction of ϕ to U_α is a mapping onto U which has a smooth inverse. In this case M is said to be a covering of N. Any manifold has a universal covering manifold, a covering which itself has no non-trivial covering. For example the mapping $\theta \mapsto e^{i\theta}$ is a covering map from $\mathbb{R}$ to the unit circle in the complex plane. In fact it defines the universal covering.

A *foliation* of a manifold M of dimension m is a set of submanifolds N_α of dimension n, the leaves, depending on a parameter $\alpha \in I$, with I an index set, such that $M = \cup_{\alpha \in I} N_\alpha$ and each point $p \in M$ has a neighbourhood admitting a coordinate system with the following property. In these coordinates $x^1, \ldots, x^m$ the intersections of the manifolds N_α with the domain of definition of the coordinates coincide with the sets where the first $m - n$ coordinates are constant.

2.6 Further reading

Standard references for the material in this section are [99] and [206]. A very thorough mathematical treatment of the structures needed for general relativity is [96]. Another mathematical text covering some of the same ground which can be recommended is [207]. Spacetime singularities are discussed in a general way in [177].

3 Matter models

General relativity models the interaction of material bodies due to the gravitational field. In addition to the Einstein equations which describe the field a description of the matter is necessary. There are many different matter models of relevance to different physical situations and in this chapter a number of the most important of these are surveyed. To specify a matter model the following three components must be fixed:

- the nature of the mathematical objects (matter fields) which describe the particular type of matter;
- the form of the energy–momentum tensor as a function of the basic matter fields and the metric;
- the equations of motion of the matter – these are partial differential equations for the matter fields which also depend on the metric.

These components together define the Einstein–matter equations for the given choice of matter model. This is a system of partial differential equations for the metric and the matter fields consisting of the Einstein equations with the given energy–momentum tensor coupled to the equations of motion of the matter. It is this coupled system which must be solved.

In physics it is common to define a matter model in terms of a Lagrangian L which is a function of the matter fields and the metric. In this case the energy–momentum tensor is defined in terms of the variational derivative of L with respect to the metric

$$T_{\alpha\beta} = -2\delta L/\delta g^{\alpha\beta} + L g_{\alpha\beta}, \tag{3.1}$$

and the equations of motion are defined by the variational derivatives with respect to the matter field ϕ

$$\nabla^\alpha(\delta L/\delta(\nabla^\alpha \phi)) = \delta L/\delta \phi. \tag{3.2}$$

Up to now the formulation in terms of a Lagrangian has not played a central role in the study of the Einstein equations as a system of PDE and it is the equations of motion which have mainly been used. At the same time it is known that a Lagrangian formulation of a PDE and the associated energy–momentum tensor can be useful for discovering conserved quantities – see the discussion in Section 10.3. Note that when the matter model is defined in terms of a Lagrangian L_m the entire Einstein–matter system consists of the Euler–Lagrange equations corresponding to the Lagrangian $(16\pi)^{-1}(R - 2\Lambda) + L_m$. In the vacuum case $L_m = 0$ and assuming $\Lambda = 0$ the action defined in this way is called the Einstein–Hilbert action.

There are some general inequalities which are satisfied by a wide variety of different matter models and these will now be discussed. The first is the *weak energy condition* (WEC) which says that $T_{\alpha\beta}v^\alpha v^\beta$ is non-negative for all future-directed causal vectors v^α. An equivalent definition is to require the same inequality for all timelike vectors since then the inequality for causal vectors follows by continuity. A timelike vector can be thought of as the tangent vector to the worldline of some observer and the quantity $T_{\alpha\beta}v^\alpha v^\beta$ has the physical interpretation of the energy density measured in the frame of reference of that observer. Thus the interpretation of the weak energy condition is that every observer measures a non-negative energy density. The *dominant energy condition* (DEC) says that $T_{\alpha\beta}v^\alpha w^\beta \geq 0$ for all future-directed causal vectors v^α and w^α. Again causal vectors can be replaced by timelike vectors without changing the definition. An equivalent formulation is that $T_{\alpha\beta}v^\beta$ is past-directed causal or zero whenever v^α is future-directed causal. Connected with this is the interpretation of the DEC that energy cannot flow with more than the speed of light. The physical necessity of this condition is perhaps less obvious than that of the WEC but it is satisfied by most matter models. The reason for the name 'dominant energy condition' is that under this condition, in terms of components in an orthonormal frame, T^{00} dominates all other components of the energy–momentum tensor. In 3+1 notation this says that $(j^i j_i)^{1/2}$ and $(S^{ik}S_{ik}/3)^{1/2}$ can be bounded by ρ. If a matter field satisfies the dominant energy condition then it can be shown that the vanishing of its energy–momentum tensor on a Cauchy hypersurface implies that the energy–momentum tensor vanishes on the whole spacetime, so that the vacuum Einstein equations hold (see [99], p. 94). The *strong energy condition* is the condition that

$$\left(T_{\alpha\beta} - \frac{1}{2}g^{\gamma\delta}T_{\gamma\delta}g_{\alpha\beta}\right)v^\alpha v^\beta \geq 0, \tag{3.3}$$

for any causal vector v^α. In the case that the cosmological constant Λ is zero this is equivalent, via the Einstein equations, to the condition $R_{\alpha\beta}v^\alpha v^\beta \geq 0$, the timelike convergence condition. The latter condition essentially means that gravity is attractive. This can be seen by looking at the role it plays in the Raychaudhuri equation. The fact that the cosmological expansion is accelerated indicates that this condition does not hold in the real world, so that either $\Lambda > 0$ or the strong energy condition is violated. Nevertheless familiar matter models satisfy the strong energy condition and a matter model which violates it deserves to be called exotic. It is clear that the DEC implies the WEC. On the other hand the SEC neither implies nor is implied by the DEC or WEC. The *null energy condition* says that $T_{\alpha\beta}l^\alpha l^\beta \geq 0$ for any null vector l^α. It is equivalent to the null convergence condition, whether or not the cosmological constant vanishes, and follows from the WEC. It is important because of its role in the Raychaudhuri equation for null geodesics.

In the Hawking singularity theorem as discussed in Subsection 2.1.3 the timelike convergence condition plays a key role. For a solution of the Einstein equations with vanishing cosmological constant this can be replaced by the strong energy condition, thus replacing a geometrical hypothesis by a physical one. The singularity theorems were a central topic in the development of mathematical relativity in the 1960s and 1970s. Note that the analytic input for these theorems comes from ODE theory. The Einstein–matter system itself is only used to obtain an inequality, the timelike convergence condition. PDE theory plays no role. Physically it would seem reasonable to expect that the disappearance of particles associated to geodesic incompleteness should be accompanied by some extreme physical conditions like diverging energy density or curvature. The singularity theorems give no information about this. Getting such information is likely to require applying PDE theory to the Einstein–matter system and can almost certainly not be based on an inequality alone. The most important known physical phenomena involving spacetime singularities are the big bang and black holes. To get a thorough mathematical understanding of these things it is necessary to go beyond the framework of the singularity theorems.

There is an interesting analogue of the Hawking singularity theorem for globally hyperbolic spacetimes with negative cosmological constant and matter content satisfying the strong energy condition. The cosmological constant contributes a negative constant to the Raychaudhuri equation and this implies, using a comparison argument as before, that the expansion can only remain positive in the future or negative in the past for a predictable finite time. Thus every timelike geodesic is incomplete in both time directions and there are singularities both in the past and in the future. The upper bound on the time of existence of the geodesics depends on the value of Λ but is otherwise universal.

Consider the situation where the entire past of a compact Cauchy hypersurface can be covered by a foliation with compact leaves whose mean curvature tends uniformly to minus infinity in the past direction. (A similar definition can be made with 'past' replaced by 'future'.) Then the spacetime is said to have a *crushing singularity* in the past. It follows from the arguments used in the proof of the Hawking singularity theorem that as a timelike curve approaches a crushing singularity its remaining arc length tends to zero. This means that a globally hyperbolic spacetime can never be extended beyond a crushing singularity. It can be shown that in a spacetime with a compact Cauchy hypersurface and a crushing singularity which satisfies the strong energy condition the foliation in the definition of the crushing singularity can always be replaced by a foliation each of whose leaves is compact and has constant mean curvature.

A crude but convenient distinction is that between field theoretic and phenomenological matter models. It is not sharply defined. Intuitively, the field theoretic models are regarded as being fundamental while the phenomenological models are thought of as effective theories, derived by approximation from more fundamental ones, which are useful for applications to some particular class of physical situations. In the context of classical general relativity this distinction should not be taken too seriously since a truly fundamental theory would have to include quantum effects.

3.1 Scalar fields

The first type of matter model to be discussed is the scalar field. This does not correspond to any type of matter encountered in everyday life and indeed no such field has yet been observed directly in nature. On the other hand there are indirect indications that scalar fields have an important place in physics. The reason for starting with this model here is that it is the simplest example from a mathematical point of view.

A scalar field ϕ is a real-valued function on spacetime. There are different types of scalar field depending on the choice of equations of motion. An important class is that of minimally coupled scalar fields. The definition depends on the choice of a function $V : \mathbb{R} \to \mathbb{R}$, the potential, with some suitable degree of smoothness. The energy–momentum tensor is

$$T_{\alpha\beta} = \nabla_\alpha \phi \nabla_\beta \phi - \left(\frac{1}{2} \nabla_\gamma \phi \nabla^\gamma \phi + V(\phi)\right) g_{\alpha\beta}, \qquad (3.4)$$

and the equation of motion is $\nabla^\alpha \nabla_\alpha \phi = V'(\phi)$. This matter model can be derived from the Lagrangian $L = -\frac{1}{2} \nabla^\alpha \phi \nabla_\alpha \phi - V$. It is usual to assume that the potential V is non-negative which is a kind of positive energy assumption but in recent years models where this is violated have been considered in cosmology. Important special cases are $V = 0$, a massless linear scalar field, and $V(\phi) = \frac{1}{2} m^2 \phi^2$, a massive linear scalar field of mass m. Exponential potentials are also common in the cosmology literature. The minimally coupled scalar field satisfies the dominant and weak energy conditions for any choice of non-negative potential. In the case $V = 0$ the equation of motion written on Minkowski space is the ordinary wave equation which is the model example of a hyperbolic equation. Hyperbolic equations occur frequently among the equations of motion of matter in general relativity and the Einstein equations themselves can be reduced by suitable procedures to a system of hyperbolic equations. This is explained in Chapter 9.

A simple generalization of the above model is a multiscalar field. In this case the basic quantity is a mapping with values in $\mathbb{R}^k$ or, equivalently, a set of k real-valued functions ϕ^I. There is a potential which is a real-valued function on $\mathbb{R}^k$, usually assumed non-negative. The energy–momentum tensor is

$$T_{\alpha\beta} = \nabla_\alpha \phi^I \nabla_\beta \phi^J \delta_{IJ} - \left(\frac{1}{2} \nabla_\gamma \phi^I \nabla^\gamma \phi^J \delta_{IJ} + V(\phi) \right) g_{\alpha\beta} \tag{3.5}$$

and the equations of motion are $\nabla^\alpha \nabla_\alpha \phi^I = \partial V / \partial \phi^I$. The matter model can be derived from the Lagrangian $L = -\frac{1}{2} \nabla^\alpha \phi^I \nabla_\alpha \phi^J \delta_{IJ} - V$. For a non-negative potential it satisfies the DEC and WEC. Among the commonest potentials found in the literature are those which are sums of products of exponentials of the individual fields ϕ^I.

A further generalization is obtained if $\mathbb{R}^k$ thought of as a Euclidean space is replaced by a manifold N with a metric h_{IJ}. The energy–momentum tensor is obtained from that of the multiscalar field by replacing δ_{IJ} by h_{IJ}. The same is true of the Lagrangian. The equations of motion involve the Christoffel symbols Γ^I_{JK} of the metric h_{IJ}. Their explicit form is:

$$\nabla^\alpha \nabla_\alpha \phi^I + \Gamma^I_{JK}(\phi) \nabla^\alpha \phi^J \nabla_\alpha \phi^K = h^{IJ} \frac{\partial V}{\partial \phi^J}. \tag{3.6}$$

The special case where the potential is zero is particularly well-known. In the physics literature it is called a nonlinear σ-model and in the mathematics literature it is called a hyperbolic harmonic map or *wave map*. In this book the last of these names is used. A harmonic map, a concept more widespread in mathematics, is a solution of the analogous equation with a Riemannian manifold as domain instead of a Lorentzian manifold. The more general case of (3.6) is referred to as a wave map with potential. A choice of the manifold N (the target manifold) which is particularly important is the hyperbolic plane H^2. This can be defined by the following metric on $\mathbb{R}^2$:

$$dP^2 + e^{2P} dQ^2. \tag{3.7}$$

It is a metric of constant negative curvature. In this case the wave map equations on Minkowski space have the explicit form:

$$\partial_t^2 P = \Delta P + e^{2P}((\partial_t Q)^2 - \delta_{ab}\nabla^a Q \nabla^b Q), \tag{3.8}$$

$$\partial_t^2 Q = \Delta Q - 2(\partial_t P \partial_t Q - \delta_{ab}\nabla^a P \nabla^a Q). \tag{3.9}$$

Another direction of generalization is to consider a Lagrangian with a more general dependence on $X = -\frac{1}{2}\nabla^\alpha \phi \nabla_\alpha \phi$ as well as ϕ. This kind of model is known in cosmology as k-essence. The k in the name comes from 'kinetic' due to the possibility of non-standard dependence on the kinetic energy. A general Lagrangian $L = L(\phi, X)$ leads to the energy–momentum tensor

$$T^{\alpha\beta} = \frac{\partial L}{\partial X}\nabla^\alpha \phi \nabla^\beta \phi + L g^{\alpha\beta} \tag{3.10}$$

and the equations of motion

$$\left(\frac{\partial L}{\partial X}g^{\alpha\beta} - \frac{\partial^2 L}{\partial X^2}\nabla^\alpha \phi \nabla^\beta \phi\right)\nabla_\alpha \nabla_\beta \phi + \frac{\partial^2 L}{\partial \phi \partial X}g^{\alpha\beta}\nabla_\alpha \phi \nabla_\beta \phi = -\frac{\partial L}{\partial \phi}. \tag{3.11}$$

This energy–momentum tensor can violate any of the energy conditions. It is possible to derive explicit conditions which characterize when the different energy conditions are satisfied. Of course further generalizations with several fields or dependence of the Lagrangian on higher derivatives are possible. Here only one further example will be mentioned. This is the (generalized) Skyrme model which is derived from the Lagrangian

$$L = -\frac{1}{2}S_\alpha^\alpha + \frac{1}{4}(S_{\alpha\beta}S^{\alpha\beta} - S_\alpha^\alpha S_\beta^\beta), \tag{3.12}$$

where $S_{\alpha\beta} = h_{IJ}\partial_\alpha \phi^I \partial_\beta \phi^J$ and h_{IJ} is a Riemannian metric. In the original Skyrme model this was the metric of a round sphere. For information about interesting properties of this model see [32].

Next curvature-coupled scalar fields will be considered. These are modifications of the minimally coupled scalar field which are encountered when quantum corrections are taken into account. Often complex scalar fields are considered in this context but here the equations are only written down for the case of a real scalar field. This matter model can be derived from the Lagrangian

$$L = -\frac{1}{2}\nabla_\alpha \phi \nabla^\alpha \phi - \frac{1}{2}\xi R \phi^2, \tag{3.13}$$

where ξ is a constant. The equation of motion of the scalar field is

$$\nabla^\alpha \nabla_\alpha \phi = \xi R \phi. \tag{3.14}$$

Note that it contains the scalar curvature and hence second derivatives of the metric. The energy–momentum tensor is given by

$$T_{\alpha\beta} = (1 - 2\xi)\nabla_\alpha\phi\nabla_\beta\phi + \left(2\xi - \frac{1}{2}\right)\nabla_\gamma\phi\nabla^\gamma\phi g_{\alpha\beta}$$
$$- 2\xi\phi\nabla_\alpha\nabla_\beta\phi + 2\xi\phi\nabla_\gamma\nabla^\gamma\phi g_{\alpha\beta} + \xi\phi^2 G_{\alpha\beta}. \quad (3.15)$$

This also contains second derivatives of the metric. In systems of partial differential equations the highest order derivatives occurring usually play a dominant role. In all the matter models introduced previously second derivatives of the metric occur only on the left hand side of the Einstein equations. For the curvature-coupled scalar field this is not the case and this fact leads to difficulties in the analysis of the equations. For this reason it is useful to rearrange the equations as follows. The dependence of the energy–momentum tensor on the curvature is through the Einstein tensor and for $1 - 8\pi\xi\phi^2 > 0$ the Einstein equations can be solved for $G_{\alpha\beta}$. This leads to an equation of the form $G_{\alpha\beta} = 8\pi \hat{T}_{\alpha\beta}$ with a modified energy–momentum tensor $\hat{T}_{\alpha\beta}$ defined by

$$\hat{T}_{\alpha\beta} = \frac{T_{\alpha\beta} - \xi\phi^2 G_{\alpha\beta}}{1 - 8\pi\xi\phi^2}. \quad (3.16)$$

The curvature-coupled scalar field has an interesting behaviour under conformal rescalings. Assuming that $1 - 8\pi\xi\phi^2 > 0$, define a conformal factor by $\Omega = \sqrt{1 - 8\pi\xi\phi^2}$. Let P be the function defined by

$$P(x) = \int_0^x \frac{\sqrt{1 - 8\pi(1 - 6\xi)\xi y^2}}{1 - 8\pi\xi y^2} dy. \quad (3.17)$$

Let $\tilde{g}_{\alpha\beta} = \Omega^2 g_{\alpha\beta}$ and $\tilde{\phi} = P(\phi)$. Then it turns out that if $g_{\alpha\beta}$ and ϕ satisfy the Einstein equations with a curvature coupled scalar field as source then $\tilde{g}_{\alpha\beta}$ and $\tilde{\phi}$ satisfy the Einstein equations minimally coupled to a linear scalar field, i.e. the system for $\xi = 0$. Thus mathematically problems for the curvature coupled scalar field can be transformed to corresponding problems for a minimally coupled scalar field. Note, however, that this correspondence is in general destroyed by any other matter field present in spacetime. If the matter field satisfies the usual equations of motion before conformal rescaling it satisfies modified equations of motion involving the conformal factor after rescaling. Even without extra matter the correspondence breaks down for solutions with $1 - 8\pi\xi\phi^2 < 0$. Note that in the case without extra matter it is possible to compute the scalar curvature by using the field equation for the geometry and substituting the result back into the equation of motion to get the equation

$$[1 - 8\pi\xi(1 - 6\xi)\phi^2]\nabla_\alpha\nabla^\alpha\phi = 8\pi\xi(1 - 6\xi)(\nabla_\alpha\phi\nabla^\alpha\phi)\phi. \quad (3.18)$$

In particular in the case $\xi = \frac{1}{6}$ this is just the linear wave equation.

There is a class of alternative theories to general relativity which can be reduced (in the absence of matter) by a conformal transformation to the Einstein equations coupled to a nonlinear scalar field. In this class of theories the Einstein–Hilbert Lagrangian, which is just the scalar curvature R, is replaced by $f(R)$ for some function f. In contrast to the Einstein equations the equations of motion corresponding to this Lagrangian generally contain derivatives of order four. For this reason these theories are often called higher order gravity theories. Suppose that a smooth function f is given and that f' and f'' are non-zero. Let $F = f'$. The potential for the nonlinear scalar field is defined as follows. Let $\phi = \sqrt{\frac{3}{16\pi}} \log F$ and express the function $\frac{f - Rdf/dR}{16\pi(df/dR)^2}$ of R as a function of ϕ. Since f'' is non-zero this is possible and V is smooth due to the inverse function theorem. Define $\tilde{g}_{\alpha\beta} = F(R)g_{\alpha\beta}$. Then if $g_{\alpha\beta}$ satisfies the field equations of the $f(R)$-theory it follows that $\tilde{g}_{\alpha\beta}$ satisfies the Einstein equations with the energy–momentum tensor of a nonlinear scalar field with the potential just discussed. Conversely if $\tilde{g}_{\alpha\beta}$ satisfies the Einstein equations with this energy–momentum tensor then $g_{\alpha\beta}$ satisfies the field equation of the $f(R)$ theory. Note that other alternative theories of gravity such as that with the Lagrangian $R_{\alpha\beta}R^{\alpha\beta}$ cannot be reduced to an Einstein–matter system in this simple way and are more problematic from a PDE point of view.

Perhaps the most famous alternative theory to general relativity is the Brans–Dicke theory. (The names Jordan and Fierz are sometimes also included.) This is known as a scalar-tensor theory since the basic quantities describing the gravitational field are a scalar field ϕ and a Lorentzian metric $g_{\alpha\beta}$. One reason for the fame of the Brans–Dicke theory is that it played the role of an alternative to general relativity in observational tests. The action is

$$L = \frac{1}{16\pi}\int \phi R - \frac{\omega}{\phi}g^{\alpha\beta}\nabla_\alpha\phi\nabla_\beta\phi + L_m. \quad (3.19)$$

Here ω is a constant and L_m is the Lagrangian of any matter present. Since the scalar field ϕ is part of the description of the gravitational field in this theory it is not counted as a matter field. The scalar field ϕ is supposed positive. The field equations for the quantities describing the gravitational

3: Matter models

field are

$$\nabla_\alpha \nabla^\alpha \phi = \frac{8\pi}{2\omega + 3} \mathrm{tr}\, T, \tag{3.20}$$

$$G_{\alpha\beta} = \frac{8\pi}{\phi} T_{\alpha\beta} + \frac{\omega}{\phi^2}\left(\nabla_\alpha \phi \nabla_\beta \phi - \frac{1}{2}\nabla_\gamma \phi \nabla^\gamma \phi g_{\alpha\beta}\right)$$
$$+ \frac{1}{\phi}(\nabla_\alpha \nabla_\beta \phi - g_{\alpha\beta}\nabla_\gamma \nabla^\gamma \phi). \tag{3.21}$$

The energy–momentum tensor and the equations of motion of the matter have exactly the same form as in the Einstein theory. In some sense general relativity is the limit of Brans–Dicke theory for $\omega \to \infty$ although there seems to be some confusion in the literature about which sense this is. The experimental tests give lower bounds on ω which have grown over the years with the increasing precision of the experiments to reach a value of several thousand. As a consequence the theory became less and less plausible to many physicists and interest in it declined. More recently interest in this theory has increased again due to relations to string theory.

By a conformal transformation the equations (3.20) and (3.21) can be brought into a form where they look like the Einstein equations coupled to certain matter fields. The Lagrangian is

$$\frac{\tilde{R}}{16\pi} - \frac{1}{2}\tilde{g}^{\alpha\beta}\tilde{\nabla}_\alpha \tilde{\phi}\tilde{\nabla}_\beta \tilde{\phi} + \exp\left(-8\sqrt{\frac{\pi}{2\omega + 3}}\tilde{\phi}\right) L_m(\tilde{g}). \tag{3.22}$$

The evolution equations for $\tilde{g}_{\alpha\beta}$ and $\tilde{\phi}$ are

$$\tilde{\nabla}_\alpha \tilde{\nabla}^\alpha \tilde{\phi} = \sqrt{\frac{8\pi}{2\omega + 3}}\, \mathrm{tr}\,\tilde{T}, \tag{3.23}$$

$$G_{\alpha\beta} = 8\pi \tilde{T}_{\alpha\beta} + \left(\tilde{\nabla}_\alpha \tilde{\phi} \tilde{\nabla}_\beta \tilde{\phi} - \frac{1}{2}\tilde{\nabla}_\gamma \tilde{\phi} \tilde{\nabla}^\gamma \tilde{\phi} \tilde{g}_{\alpha\beta}\right). \tag{3.24}$$

Here $\tilde{g}_{\alpha\beta} = \phi g_{\alpha\beta}$ and $\tilde{\phi} = \sqrt{\frac{2\omega+3}{16\pi}}\log \phi$. The notation $L_m(\tilde{g})$ means that in this expression the metric $\tilde{g}_{\alpha\beta}$ has been used in the matter Lagrangian. Similarly $\mathrm{tr}\,\tilde{T}$ means that $\tilde{g}_{\alpha\beta}$ has been used in the expression for the energy–momentum tensor and in forming the trace. The equations of motion of the matter are thus those of matter moving in the geometry $g_{\alpha\beta}$ and as a consequence depend on the $\tilde{\phi}$ when expressed in the new variables. The formulation using the equations (3.20) and (3.21) is referred to as the Jordan conformal frame while that using (3.23) and (3.24) is called the Einstein frame. There are more general scalar-tensor theories whose Lagrangians are obtained from that of Brans–Dicke theory by replacing the constant ω

by a function $\omega(\phi)$ and multiplying the scalar curvature by a function of ϕ. In an even more general class of theories an arbitrary function of R and ϕ is introduced into the action. This then includes the $f(R)$ theories as a special case.

3.2 The Maxwell and Yang–Mills equations

The Maxwell equations describe the electromagnetic field. The basic matter quantity is a tensor $F_{\alpha\beta}$ which is antisymmetric. The relation to the formulation of electromagnetism in terms of electric and magnetic fields is as follows. In Minkowski space define $E_i = F_{0i}$, $B_1 = F_{32}$, $B_2 = F_{13}$ and $B_3 = F_{21}$. The vector fields E and B defined in this way which are purely spatial are the electric and magnetic fields respectively. The equations of motion of the Maxwell field, the Maxwell equations, are $\nabla^\alpha F_{\alpha\beta} = 0$ and $\nabla_\alpha F_{\beta\gamma} + \nabla_\gamma F_{\alpha\beta} + \nabla_\beta F_{\gamma\alpha} = 0$. These are the source-free Maxwell equations, i.e. the equations which apply in the absence of charged matter. When charge is present the second Maxwell equation is unchanged but the first has a non-zero right hand side given by $4\pi J^\alpha$ where J^α is the four-current. In Minkowski space J^0 corresponds to the electric charge density while the spatial vector defined by the remaining components J^a is the electric current. The second Maxwell equation can be solved by writing $F_{\alpha\beta} = \nabla_\alpha A_\beta - \nabla_\beta A_\alpha$ for an arbitrary one-form A_α, the four-potential. Different choices of A_α define the same field $F_{\alpha\beta}$ if they are related by a gauge transformation $\tilde{A}_\alpha = A_\alpha + \nabla_\alpha \phi$ for some function ϕ. The energy–momentum tensor is

$$T_{\alpha\beta} = \frac{1}{4\pi}\left[F_\alpha{}^\gamma F_{\beta\gamma} - \frac{1}{4}F^{\gamma\delta}F_{\gamma\delta}g_{\alpha\beta}\right]. \tag{3.25}$$

The equations can be obtained from the Lagrangian $L = -\frac{1}{16\pi}F^{\alpha\beta}F_{\alpha\beta}$. In the Lagrangian formulation of the Maxwell equations the four-potential must be regarded as the basic matter field. The energy–momentum tensor of the Maxwell field satisfies the weak, strong and dominant energy conditions. The proof of the DEC is not simple. The source-free Maxwell equations are conformally invariant. If $F_{\alpha\beta}$ is a solution of the Maxwell equations with the metric $g_{\alpha\beta}$ it is also a solution of the Maxwell equations with the metric $\Omega^2 g_{\alpha\beta}$ for an arbitrary smooth positive function Ω. This can be seen most easily on the level of the Lagrangian, using the fact that the dimension of spacetime is four. The energy–momentum tensor of the Maxwell field is trace-free and in fact this is a general property of energy–momentum tensors arising from conformally invariant Lagrangians.

The $3+1$ decomposition of the Maxwell equations is as follows. Let $E^\alpha = F^{\alpha\beta}n_\beta$ and $B^\alpha = {}^*F^{\alpha\beta}n_\beta$ where n^α is the unit normal vector to the spatial foliation. These are purely spatial vectors which may be denoted by E^a and B^a and are the electric and magnetic field respectively. They satisfy the constraint equations $\nabla_a E^a = 0$ and $\nabla_a B^a = 0$ as a consequence of the source-free Maxwell equations. The latter also imply the evolution equations

$$\partial_t E^a - (\alpha \operatorname{tr} k) E^a - \beta^b \nabla_b E^a + \nabla_b \beta^a E^b = \epsilon^{abc} \nabla_b (\alpha B_c), \tag{3.26}$$

$$\partial_t B^a - (\alpha \operatorname{tr} k) B^a - \beta^b \nabla_b B^a + \nabla_b \beta^a B^b = -\epsilon^{abc} \nabla_b (\alpha E_c), \tag{3.27}$$

where ϵ_{abc} is the volume element of the spatial metric.

The Yang–Mills equations generalize the Maxwell equations. Their direct physical applications lie in the domain of quantum theory but the classical equations are of mathematical interest. The data needed to define the Yang–Mills equations include a finite-dimensional Lie algebra and a left-invariant metric on the algebra. (Some elementary remarks on Lie algebras can be found in Chapter 4.) The basic matter field is the gauge potential, a Lie-algebra-valued one-form A_α^I. From this it is possible to define a field strength, a Lie-algebra-valued two-form, by

$$F_{\alpha\beta}^I = \nabla_\alpha A_\beta^I - \nabla_\beta A_\alpha^I + C_{JK}^I A_\alpha^J A_\beta^K, \tag{3.28}$$

where C_{JK}^I are the structure constants of the Lie algebra. This is the analogue of the Maxwell field. In fact the Yang–Mills equations reduce to the Maxwell equations in the case that the Lie algebra is one-dimensional. However the Yang–Mills equations, in contrast to the Maxwell equations, cannot in general be expressed in terms of the field strength alone. The equations of motion are

$$\nabla^\alpha F_{\alpha\beta}^I + C_{JK}^I A^{J\alpha} F_{\alpha\beta}^K = 0 \tag{3.29}$$

and these generalize part of the Maxwell equations. The field strength satisfies the Bianchi identity which corresponds to the rest of the Maxwell equations. It reads

$$\nabla_\alpha F_{\beta\gamma}^I + \nabla_\gamma F_{\alpha\beta}^I + \nabla_\beta F_{\gamma\alpha}^I + C_{JK}^I (A_\alpha^J F_{\beta\gamma}^K + A_\gamma^J F_{\alpha\beta}^K + A_\beta^J F_{\gamma\alpha}^K) = 0. \tag{3.30}$$

The energy–momentum tensor is

$$T_{\alpha\beta} = \frac{1}{4\pi} \left[F_\alpha{}^{\gamma I} F_{\beta\gamma}^J - \frac{1}{4} F^{\gamma\delta I} F_{\gamma\delta}^J g_{\alpha\beta} \right] k_{IJ}. \tag{3.31}$$

Here k_{IJ} is the given metric on the Lie algebra. The energy conditions are satisfied provided the metric k_{IJ} is positive definite. This restricts the Lie algebra to be the product of a semisimple and an Abelian algebra. The Yang–Mills field can be obtained from the Lagrangian $L = -\frac{1}{16\pi} F^{\alpha\beta I} F_{\alpha\beta}^J k_{IJ}$.

The Yang–Mills equations are conformally invariant. They are also invariant under the gauge transformations defined by the formula

$$\tilde{A}_\alpha^I = (g^{-1}A_\alpha g)^I + (g^{-1}\nabla_\alpha g)^I, \qquad (3.32)$$

where g is a function with values in a Lie group G with the given Lie algebra. The terms on the right hand side of (3.32) require some explanation. In the case that G is a group of matrices and as a consequence the elements of the Lie algebra can also be interpreted as matrices it can be interpreted directly in terms of matrix multiplication. From a more sophisticated point of view this gauge invariance arises from the geometrical interpretation of the Yang–Mills potential as a local representation of a connection on a principal G-bundle with G a Lie group. The Lie algebra occurring in the above form of the Yang–Mills equations is the Lie algebra of G.

3.3 Continuum mechanics

One of the most frequently used matter models in general relativity, both in cosmology and in the study of isolated systems, is the perfect fluid. The matter fields are the energy density μ, the four-velocity u^α of the fluid, which is a unit timelike vector, and the entropy density s. They are related by an equation of state $p = f(\mu, s)$. Often the isentropic case is considered where s is a constant and can be omitted from the equations. The energy–momentum tensor is

$$T_{\alpha\beta} = (\mu + p)u_\alpha u_\beta + p g_{\alpha\beta}. \qquad (3.33)$$

The equations of motion, the relativistic Euler equations, are equivalent to the equation $\nabla^\alpha T_{\alpha\beta} = 0$ in the isentropic case. In the general case these must be supplemented by the equation $u^\alpha \nabla_\alpha s = 0$. Note that while the equations of motion of a matter model always imply that the energy–momentum tensor is divergence-free it is in general not the case that the divergence-free property of the energy–momentum tensor implies the equations of motion. The isentropic perfect fluid is an exception in this respect. It is part of the definition of the matter model that μ should be non-negative. This implies that the WEC is satisfied. The dominant energy condition follows from the additional assumption that $|p| \leq \mu$. For an isentropic fluid this is a consequence of the conditions that $p(0) = 0$ and $|dp/d\mu| \leq 1$ which are physically motivated. The second of these says that the speed of sound in the fluid is less than the speed of light. The speed of sound is given by $(dp/d\mu)^{1/2}$. This only gives a sensible answer provided $dp/d\mu$ is non-negative. It will be seen in Section 8.2 that when the formally computed speed of sound is imaginary the Euler equations have bad mathematical properties.

A perfect fluid with zero pressure, i.e. one for which the function f is identically zero, is called dust. There are very many papers on dust in the literature of general relativity. The reason is not that dust has properties which are specially desirable from a physical point of view. It is rather the fact that dust allows a lot more explicit calculations to be carried out than most matter models. Analysing models which have more realistic physical properties without making approximations requires serious use of the theory of partial differential equations as presented in this book. The problem with dust is that, due to the absence of pressure even for very high densities, it can easily happen that the density becomes infinite in the course of the evolution. In the simplest case with high symmetry this is known as shell-crossing. Another frequently encountered equation of state is that of a radiation fluid $p = \mu/3$. It is important in models of the early universe.

An important special class of solutions of the Euler equations are those which are irrotational. The defining condition is that $u_{[\alpha}\nabla_\beta u_{\gamma]} = 0$. Geometrically it is equivalent to the condition that the hyperplanes orthogonal to the four-velocity are tangent to hypersurfaces. If a smooth solution of the isentropic Euler equations satisfies this condition on a Cauchy hypersurface then it satisfies it everywhere. In that case there is a relationship between the fluid and a scalar field which will now be explained. Consider first a solution ϕ of a k-essence model with Lagrangian L depending only on X. If the gradient of ϕ is timelike in some region define u^α to be the corresponding unit timelike vector. Then the energy–momentum tensor of the k-essence field takes the form

$$T^{\alpha\beta} = 2X\frac{\partial L}{\partial X}u^\alpha u^\beta + Lg^{\alpha\beta}. \tag{3.34}$$

This is the energy–momentum tensor of a perfect fluid with energy density $2X\frac{\partial L}{\partial X} - L$ and pressure L. The four-velocity of this fluid, being proportional to the gradient of a scalar, is irrotational. Provided L can be written as a function of $2X\frac{\partial L}{\partial X} - L$ so as to define an equation of state a solution describing an irrotational fluid is obtained. If it is assumed that $\frac{\partial L}{\partial X} > 0$ and $2X\frac{\partial^2 L}{\partial X^2} + \frac{\partial L}{\partial X} > 0$ then this is possible. Conversely, if an isentropic fluid is irrotational its four-velocity is proportional to the gradient of a scalar function so that $u^\alpha = \nabla^\alpha\phi/\sqrt{-\nabla_\beta\phi\nabla^\beta\phi}$. Comparing the energy–momentum tensors shows that if the two matter models are to coincide the relation

$$2Xf'(\rho)d\rho/dX = \rho + f(\rho) \tag{3.35}$$

must be satisfied. Integrating this gives

$$X(\rho) = X_0 \exp \int_{\rho_0}^{\rho} \frac{2f'(\sigma)}{\sigma + f(\sigma)} d\sigma. \tag{3.36}$$

Supposing that this relation can be inverted a Lagrangian can be defined by $L(X) = f(\rho(X))$. In this way an irrotational fluid gives rise to a corresponding k-essence model. As a simple example, if the equation of state of the fluid is $p = w\rho$ then the general solution of (3.35) is $\rho = (X/X_0)^{\frac{1+w}{2w}}$ so that $L = w(X/X_0)^{\frac{1+w}{2w}}$.

A straightforward generalization of the perfect fluid which is important in cosmology is a mixture of several non-interacting perfect fluids. Each fluid satisfies the Euler equations in the given spacetime with its own equation of state. The energy–momentum tensor is the sum of the energy–momentum tensors for the individual fluids.

In general relativity an elastic solid can be described in the following way. Let B be a three-dimensional manifold called the material manifold. Its points represent the particles of the elastic body. The basic unknown describing the solid is a mapping f from spacetime to B. It can be expressed in terms of coordinates on B so that it is represented by three functions f^A. The mapping f is supposed to have the maximal rank three so that its kernel is one-dimensional. Furthermore the kernel is assumed timelike. The future-directed unit vector spanning the kernel is the four-velocity of the particles of the solid. The equations of motion and the energy–momentum tensor are determined by a Lagrangian $\mu(f^A, \partial_\alpha f^A, g_{\alpha\beta})$. This definition is required to be covariant and it can be shown that this implies that in fact μ is a function of f^A and H^{AB} where $H^{AB} = g^{\alpha\beta} \partial_\alpha f^A \partial_\beta f^B$. The reason for calling the Lagrangian μ is that it turns out to be the energy density of the matter in the rest frame of the particles of the solid. If a volume form is chosen on B then it is possible to define a particle number density n on spacetime. It can be expressed in terms of H^{AB} and the volume form on B. A special choice of the Lagrangian is that where it only depends on n. The resulting matter model is equivalent to a perfect fluid. This provides one way of writing a perfect fluid as a Lagrangian field theory.

3.4 Kinetic theory

In kinetic theory matter is treated as a collection of particles which are described statistically. The basic quantity is a function $f(x^\alpha, p^\alpha)$, the distribution function, which is the density of particles at a given spacetime point with given four-momentum. The four-momentum is required to be future-directed causal. Because of its physical interpretation this function should be non-negative. Frequently particles with a fixed rest mass m are considered. This means that the function f is defined on the set determined by the equation $g_{\alpha\beta}p^\alpha p^\beta = -m^2$. This set is called the mass shell. Often physical units are chosen so as to set $m = 1$ and the treatment in this book is restricted to that case. The energy–momentum tensor is

$$T^{\alpha\beta} = -\int f(x^\alpha, p^a) p^\alpha p^\beta |\det g|^{1/2}/p^0 \, dp^a. \tag{3.37}$$

Here $\det g$ is the determinant of the spacetime metric and the idea is to regard the mass shell relation as defining p^0 in terms of p^a and the metric so as to eliminate the dependence of f on p^0. It is also possible to consider particles of zero rest mass by setting $m = 0$ in the mass shell relation.

The equation of motion of particles is as follows:

$$\partial f/\partial t + (p^a/p^0)\partial f/\partial x^a - (\Gamma^a_{\alpha\beta} p^\alpha p^\beta / p^0)\partial f/\partial p^a = Q(f). \tag{3.38}$$

The expression $Q(f)$ describes collisions between the particles. The simplest case is that of collisionless matter where $Q(f) = 0$. In this case (3.38) is called the *Vlasov equation* and the equation says that f is constant along geodesics. Later other examples of collision terms will be given. An explicit formula can be given for the solution of the Vlasov equation in terms of the solution of the geodesic equation. Let f_0 be the restriction of the solution f to the hypersurface $t = 0$. Define $X^a(s, t, x^b, p^b)$ and $P^a(s, t, x^b, p^b)$ to be the solution of the geodesic equation

$$dX^a/ds = P^a/P^0, \tag{3.39}$$

$$dP^a/ds = -\Gamma^a_{\beta\gamma} P^\beta P^\gamma / P^0 \tag{3.40}$$

with initial data $X^a(t, t, x^b, p^b) = x^a$ and $P^a(t, t, x^a, p^b) = p^a$. That a unique solution of this type exists follows from the local existence and uniqueness theorem for systems of ODE discussed in Chapter 5. Then

$$f(t, x^a, p^a) = f_0(X^a(0, t, x^a, p^a), P^a(0, t, x^a, p^a)). \tag{3.41}$$

It turns out that the Vlasov equation defines a matter model which has very nice properties when coupled to the Einstein equations. It lacks the tendency to form singularities unrelated to gravity which is seen in the case of dust. In fact solutions of the Einstein–dust system can be thought of as very singular solutions of the Einstein–Vlasov system. They are obtained formally by choosing f to be of the form

$$f(t, x^a, p^a) = \mu(t, x^a)\delta(p^a - \bar{p}^a(t, x^a)). \tag{3.42}$$

Thus this choice of f has a Dirac δ dependence on the momentum variables. The statement about the better behaviour of solutions of the

Einstein–Vlasov system holds for regular solutions f which are continuously differentiable.

The most common type of collision terms are those which lead to the Boltzmann equation. Only binary collisions are taken into account. Here three versions of the Boltzmann equation will be considered. The first concerns particles which are treated as classical in the sense that quantum effects are neglected. If quantum effects are taken into account it is possible under certain circumstances to get an effective equation which is independent of any notions of quantum theory but is different from the classical equation. In fact there are two different equations of this type which apply to fermions and bosons respectively. These equations are important in the study of the early universe in cosmology.

Consider an elastic collision of particles with initial four-momenta p^α and q^α. Let the four-momenta after the collision be p'^α and q'^α. These are related by the conservation of the total four-momentum, $p^\alpha + q^\alpha = p'^\alpha + q'^\alpha$. This corresponds to the conservation of energy and momentum in the non-relativistic case. Let C be the collision manifold which is defined by the conservation of four-momentum and the conditions that the four-momentum of each particle is a future-directed unit timelike vector. This is a five-dimensional manifold. The metric endows it with a natural volume element. In the case of classical particles the collision term in the Boltzmann equation takes the form

$$Q(f)(p^\alpha) = \int_C k(p^\alpha, q^\alpha, p'^\alpha, q'^\alpha)[f(p'^\alpha)f(q'^\alpha) - f(p^\alpha)f(q^\alpha)], \quad (3.43)$$

where k is the collision cross-section and the integral is taken with respect to the natural volume element on C. The dependence of k on the momentum variables is not arbitrary. It should be Lorentz invariant which means concretely that it should depend only on the invariants $s = (p^\alpha + q^\alpha)(p_\alpha + q_\alpha)$ and $t = (p^\alpha - p'^\alpha)(p_\alpha - p'_\alpha)$. These are two of the Mandelstam variables. The quantity s is the total energy of the system in the frame where the total three-momentum is zero. Another quantity which has a direct physical interpretation in that frame is Θ, the scattering angle, which satisfies the relation

$$\cos\Theta = 1 + \frac{2t}{4-s}. \quad (3.44)$$

The collision term can be expressed directly in terms of these variables. The result is

$$Q(f)(q^\alpha) = \int [f(p'^\alpha)f(q'^\alpha) - f(p^\alpha)f(q^\alpha)] v_M \sigma(s, \Theta) d\Sigma dp. \quad (3.45)$$

In this formula $v_M = \frac{\sqrt{s(s-4)}}{2p_0q_0}$ is called the Møller velocity and

$$p'^i = p^i + a\Omega^i, \tag{3.46}$$

$$q'^i = q^i - a\Omega^i, \tag{3.47}$$

where Ω^i is a point on the unit sphere and a is function depending on p^i, q^i and ω^i. The expression $d\Sigma$ denotes the volume form on the standard unit sphere. As the notation indicates, these formulae are to be interpreted as expressed in terms of the momentum in an orthonormal frame. This has the advantage that the formulae can be taken over directly from those in special relativity. Explicitly

$$a(p^i, q^i, \Omega^i) = \frac{2(p^0 + q^0)(\delta_{ij}\Omega^i(q^j p^0 - p^j q^0))}{(p^0 + q^0)^2 - [\delta_{ij}\Omega^i(p^j + q^j)]^2}. \tag{3.48}$$

More details can be found in [91].

Two mathematically simple choices for σ are the case where it is constant (corresponding to hard spheres in the non-relativistic case) and where it is equal to s, called Israel molecules (corresponding to Maxwell molecules in the non-relativistic case). In the quantum Boltzmann equation the expression $f(p'^\alpha)f(q'^\alpha) - f(p^\alpha)f(q^\alpha)$ is replaced by

$$[f(p'^\alpha)f(q'^\alpha)(1 + \tau f(p^\alpha))(1 + \tau f(q^\alpha)) - f(p^\alpha)f(q^\alpha)$$
$$\times (1 + \tau f(p'^\alpha))(1 + \tau f(q'^\alpha))], \tag{3.49}$$

where $\tau = 1$ for bosons and $\tau = -1$ for fermions. The correction in the case of fermions is related to the Pauli exclusion principle. During a collision a fermion is prevented from entering a state which is already occupied. The correction in the case of bosons is related to stimulated emission resulting in the production of additional particles during collisions.

3.5 Other matter models

An easy way to create new matter models from old is to combine two models in the following way. The equations of motion for each matter field remain unchanged while the new energy–momentum tensor is the sum of those for the individual matter models. If the individual matter models are defined by Lagrangians L_1 and L_2 then the new matter model is defined by the Lagrangian $L_1 + L_2$. Physically the combined matter model describes a non-interacting mixture of the two individual types of matter. For instance a scalar field can be combined in this way with any of the other matter models already presented. If a Maxwell field is combined in this way with another type of matter this means physically that the other matter is uncharged. Different perfect fluids combined in this way, for instance dust and radiation, are common in cosmology. In kinetic theory it is possible to introduce several species of particles, each satisfying its own Vlasov (or Boltzmann) equation. The possibilities are endless.

Different matter models can also be combined in such a way that they describe interacting mixtures. For instance there is the Yang–Mills–Higgs system. In addition to a Yang–Mills field the unknowns in this system include a multi-scalar field ϕ which, from a geometrical point of view, is a section of a vector bundle associated to the principal G-bundle where the gauge potential lives. Concretely this means that there is a representation of the Lie algebra on a vector space E in which the Higgs field ϕ takes its values. The representation allows an element of the Lie algebra to be multiplied by an element of E to get another element of E. This is a bit messy to write in components. The connection defined by the gauge potential defines a related connection on the associated bundle. This allows the definition of a covariant derivative $D_\alpha \phi$. The Lagrangian for the Yang–Mills–Higgs system is the sum of the Yang–Mills Lagrangian, minus the squared length of $D_\alpha \phi$ and, possibly, a potential which is a function on E which is invariant under the representation of the Lie algebra. A special case of all this where it is easy to write explicit equations is the charged scalar field. This is the case where the group G is $U(1)$, corresponding to a Maxwell field, and the potential is zero. The matter fields are a four-potential A_α and a complex-valued function ϕ. Strictly speaking, to comply with the terminology used up to now, ϕ should be considered as a pair of real scalar fields. In complex notation the energy–momentum tensor of the charged scalar field is

$$T_{\alpha\beta} = \frac{1}{2}[D_\alpha \phi \overline{D_\beta \phi} + \overline{D_\alpha \phi} D_\beta \phi - (D_\gamma \phi \overline{D^\gamma \phi})g_{\alpha\beta}], \qquad (3.50)$$

where $D_\alpha \phi = \nabla_\alpha \phi + iA_\alpha \phi$. The complex notation is convenient but its importance should not be overestimated. The equation of motion of ϕ is

$$D^\alpha D_\alpha \phi = 0 \qquad (3.51)$$

while the Maxwell field satisfies

$$\frac{1}{4\pi}\nabla^\alpha F_{\alpha\beta} = \frac{1}{2i}(\phi \overline{D_\beta \phi} - \bar{\phi} D_\beta \phi). \qquad (3.52)$$

The energy–momentum tensor for the whole system is the sum of that of a charged scalar field given above and a Maxwell field. The equations of motion of a charged scalar field coupled to a Maxwell field are known in the PDE literature as the Maxwell–Klein–Gordon equations.

3: Matter models

A matter model arising in string theory is the Maxwell–dilaton theory. The matter fields are a two-form $F_{\alpha\beta}$ and a scalar field ϕ. The equations of motion are

$$\nabla_\alpha \nabla^\alpha \phi = \frac{\lambda}{16\pi} F_{\alpha\beta} F^{\alpha\beta} e^{\lambda\phi}, \tag{3.53}$$

$$\nabla_\alpha (e^{\lambda\phi} F^{\alpha\beta}) = 0, \tag{3.54}$$

$$\nabla_{[\alpha} F_{\beta\gamma]} = 0, \tag{3.55}$$

where λ is a constant. The energy–momentum tensor is

$$T_{\alpha\beta} = \nabla_\alpha \phi \nabla_\beta \phi - \frac{1}{2} \nabla_\gamma \phi \nabla^\gamma \phi g_{\alpha\beta} + \frac{1}{4\pi} \left[F_\alpha{}^\gamma F_{\beta\gamma} - \frac{1}{4} F^{\gamma\delta} F_{\gamma\delta} g_{\alpha\beta} \right] e^{\lambda\phi}. \tag{3.56}$$

This matter model is defined by the Lagrangian $L = -\frac{1}{2} \nabla_\alpha \phi \nabla^\alpha \phi - \frac{1}{16\pi} e^{\lambda\phi} F_{\alpha\beta} F^{\alpha\beta}$. There is an interesting relation between the Einstein–Maxwell–dilaton system and the Einstein vacuum equations (with $\Lambda = 0$) in higher dimensions. This is related to the notion of Kaluza–Klein reduction. Consider a solution of the Einstein vacuum equations in five dimensions on a manifold of the form $M \times S^1$. Suppose that the action of the circle S^1 on itself by translations induces an action by isometries on $M \times S^1$. Let adapted coordinates be chosen such that $x^0, \ldots, x^3$ project to coordinates on M while x^4 is a standard coordinate on the circle. Denote the metric in five dimensions by $\bar{g}$. Introduce the notation $\bar{g}_{44} = e^{2\phi}$, $\bar{g}_{\alpha 4} = A_\alpha$ and $\bar{g}_{\alpha\beta} = g_{\alpha\beta}$. Define a conformally rescaled metric by $\tilde{g}_{ab} = e^{\phi} g_{ab}$. The Einstein vacuum equations for $\bar{g}$ are equivalent to the Einstein–Maxwell–dilaton system for $\tilde{g}_{\alpha\beta}, A_\alpha, \phi$ with a suitably chosen value of λ. Here A_α is interpreted as (a constant multiple of) the four-potential of the Maxwell field. If the metric $\bar{g}$ is also invariant under reflection in the coordinate x^4 then $A_\alpha = 0$ and the system reduces to the Einstein equations coupled to a massless scalar field. The equations before conformal transformation are related to the equations of Brans–Dicke theory.

A physically important model involving a charged fluid is that of relativistic magnetohydrodynamics. There are several astrophysical situations where it has been conjectured to play a role. One case is that of γ-ray bursts which are the most energetic explosions known. The energy–momentum tensor in this case is the sum of the familiar energy–momentum tensors for a perfect fluid and a Maxwell field but the two parts are not divergence free. Magnetohydrodynamics corresponds to a fluid of infinite conductivity and this is expressed by the condition $F_{\alpha\beta} u^\beta = 0$. This is equivalent to the statement that the electric field vanishes in the rest frame of the fluid.

More general charged fluids lead to severe mathematical complications and are not considered further here. The Navier–Stokes equations, which model dissipative fluids in Newtonian physics, have infinite propagation speed and it is not evident how to generalize them to the relativistic case. There are relativistic models for dissipative fluids with hyperbolic field equations. The problem seems to be that there are too many of them and no good criterion for deciding between them. In kinetic theory interactions can be considered between different species of matter. In the early universe, for instance, it is important to consider the interaction of electrons and photons via Compton scattering. In this case there are two distribution functions f_1 and f_2 and a collision term $Q(f_1, f_2)$ depending on both. There are models for fermions such as the Dirac equation which can be coupled to the Einstein equations if desired.

3.6 Further reading

The theorem on the existence of constant mean curvature foliations near a crushing singularity is proved in [89]. Mathematical background on k-essence models can be found in [179]. The treatment of the curvature coupled scalar field in the text follows [30]. For information on scalar-tensor theories see [86] and [79]. The 3+1 decomposition for the Maxwell equations is discussed in [191]. More information on the elastic solid can be found in the paper [22] of Beig and Schmidt. For an expository account of the quantum Boltzmann equation see [78]. Einstein–matter systems coming from string theory including the Einstein–Maxwell–dilaton system are discussed in [74]. A good source of information about magnetohydrodynamics containing physical background and the full equations in $3+1$ form is [21]. References for the models briefly mentioned in the last paragraph above are given in section 5.4 of [85].

4 Symmetry classes

Let ψ be a diffeomorphism of the manifold M, i.e. a smooth mapping from M to itself which has a smooth inverse. The mapping ψ can be used to transport vector fields v in such a way that the transported vector field $\psi_* v$ satisfies the relation $\psi_* v(f) = v(f \circ \psi)$ for any smooth function f. Other tensors can be transported in a compatible way. For instance the metric is transported to the pull-back metric $\psi^* g$ which satisfies $\psi^* g(v, w) = g(\psi_* v, \psi_* w)$ for all vector fields v and w. If $\psi^* g = g$ then ψ is called an isometry. The intuitive interpretation is that ψ is a symmetry of the metric g. More generally, if $\psi^* g$ is related to g by a conformal rescaling then ψ is called a conformal transformation. Suppose now that $\psi(\lambda)$ is a smooth one-parameter family of diffeomorphisms with $\psi(0)$ equal to the identity. Then the vector field v^α defined by

$$v^\alpha(x^\beta) = \frac{d}{d\lambda}(\psi^\alpha(\lambda, x^\beta))|_{\lambda=0} \tag{4.1}$$

is called the generator of the family. Given a smooth vector field v^α any solution of the system of equations

$$\frac{d}{d\lambda}(\psi^\alpha(\lambda, x^\beta)) = v^\alpha(\psi^\gamma(\lambda, x^\beta)) \tag{4.2}$$

with $\psi^\alpha(0, x^\beta) = x^\alpha$ defines a local one-parameter family of diffeomorphisms which agrees with the identity for $\lambda = 0$. There are two senses in which it may only be local. It may only exist for a finite range of λ and the diffeomorphisms may only be defined on some subset of the manifold. The solution is only defined on a subset of the space coordinatized by (λ, x^β) containing the points where $\lambda = 0$. Equation (4.2) is a system of ODE depending on x^β as parameters and the existence of a one-parameter family of local diffeomorphisms with the given vector field as generator follows from the local existence theory discussed in Chapter 5. If the diffeomorphisms are isometries the generator satisfies the Killing equation

$$\nabla_\alpha v_\beta + \nabla_\beta v_\alpha = 0 \tag{4.3}$$

and is called a Killing vector field. It can be thought of as an infinitesimal symmetry. If the diffeomorphisms are conformal transformations it satisfies the conformal Killing equation

$$\nabla_\alpha v_\beta + \nabla_\beta v_\alpha = f g_{\alpha\beta}, \tag{4.4}$$

for some function f. If the function f is constant the vector field is called homothetic. Using a one-parameter family of diffeomorphisms $\psi(\lambda)$ with $\psi(0)$ equal to the identity to transport a tensor field gives rise to a one-parameter family of tensor fields. If the family of diffeomorphisms is generated by a vector field X then the derivative of this family with respect to the parameter at $\lambda = 0$ is the Lie derivative of the tensor field with respect to X. Given two vector fields v and w it is possible to form their commutator or Lie bracket, defined by $[v, w](f) = v(w(f)) - w(v(f))$. It is equal to the Lie derivative $\mathcal{L}_v w$.

The set of all isometries of a pseudoriemannian metric forms a group under composition, the isometry group. In fact it is a Lie group. This means that it admits the structure of a manifold in such a way that the group operations (composition and taking the inverse) are smooth. The generators of one-parameter subgroups of a Lie group G form a vector space of the same dimension as the group. Taking the Lie brackets of two of these vector fields gives another vector field of the same type. In this way the vector space of generators of one-parameter subgroups gets an algebraic structure which makes it into a Lie algebra. In general a Lie algebra is a vector space V with an antisymmetric bilinear mapping $[\,,\,]$ which satisfies the Jacobi identity

$$[[X, Y], Z] + [[Z, X], Y] + [[Y, Z], X] = 0. \tag{4.5}$$

It is a fundamental result of the theory of Lie groups that there exists a unique connected and simply connected Lie group corresponding to any finite-dimensional Lie algebra. If X_I is a basis of the Lie algebra then the structure constants C^I_{JK} are defined by

$$[X_J, X_K] = C^I_{JK} X_I. \tag{4.6}$$

If x is a point of M and h an element of a group G of isometries of a metric g on M then let $\psi(h, x)$ be the image of x under the isometry h. This defines a mapping ψ from $G \times M$ to M which is called the action of the group G on M. The orbit of a point x under a group action is the set of all points of the form $\psi(h, x)$ for some $h \in G$. If the whole of M is a single orbit the action is said to be transitive. The isotropy group of x is the set of all $h \in G$ such that $\psi(h, x) = x$. The orbit is a submanifold and the sum of the dimensions of orbit and isotropy group is always equal to the dimension of the group. The tangent space at x inherits a Lorentzian metric from the spacetime. The derivative of each element of the isotropy group at x is a linear mapping of the tangent space onto itself. These mappings together define an action of the isotropy group on the tangent space. The derivative can only be the identity if the isometry inducing it is the identity. Hence the isotropy group is realized as a subgroup of the Lorentz group acting on the tangent space at x.

4.1 Static and stationary models

A spacetime is called *stationary* if there is a timelike Killing vector field v^α. If in addition the equation $v_{[\alpha} \nabla_\beta v_{\gamma]} = 0$ holds then the spacetime is said to be *static*. This condition is equivalent (locally in spacetime) to the existence of a foliation whose leaves are orthogonal to v^α. This is a consequence of Frobenius' theorem (see [96], p. 93).

The simplest example of a static spacetime is Minkowski space where the Killing vector v can be chosen to have the components $(1, 0, 0, 0)$ in standard coordinates. A less trivial example of a static spacetime is given by the Schwarzschild metric

$$-(1 - 2m/r)dt^2 + (1 - 2m/r)^{-1}dr^2 + r^2(d\theta^2 + \sin^2\theta d\phi^2), \quad (4.7)$$

which is static in the region $r > 2m$ where $\partial/\partial t$ is timelike. This spacetime represents a black hole whose mass is equal to the parameter $m > 0$. The case $m < 0$ is unphysical and for $m = 0$ the metric reduces to that of Minkowski space in polar coordinates. The coordinate singularities of polar coordinates are present in this form of the metric but are as easy to understand as in Minkowski space. More interesting is what happens when $r = 2m$ where the matrix of components of the metric in these coordinates is also irregular. In the early days of general relativity it was not understood what happens there. According to our modern understanding $r = 2m$ represents the event horizon which is the surface of the black hole. This can be seen by introducing other coordinates. For instance the coordinate t can be replaced by $v = t + r + 2m \log(r - 2m)$ in the region $r > 2m$. The coordinates (v, r, θ, ϕ) are known as ingoing Eddington–Finkelstein coordinates. In these coordinates the metric is given by

$$-(1 - 2m/r)dv^2 + 2dvdr + r^2(d\theta^2 + \sin^2\theta d\phi^2). \quad (4.8)$$

This form of the metric makes it clear that it can be extended smoothly through $r = 2m$ and that this coordinate hypersurface is null. The region $r > 2m$ represents the exterior of the black hole while the region $r < 2m$ is the interior. A future-directed causal geodesic which starts in the interior can never reach the exterior. This means that no observer or particle can leave

4.1 Static and stationary models

the black hole and this is in fact the characteristic property of black holes. It is also possible to introduce the coordinate $u = t - r - 2m\log(r-2m)$ together with v. Then the metric takes the form

$$-\left(1 - \frac{2m}{r}\right) du\,dv + r^2(d\theta^2 + \sin^2\theta\,d\phi^2). \tag{4.9}$$

Here the function r is defined implicitly by the relation

$$\frac{1}{2}(v - u) = r + 2m\log(r - 2m). \tag{4.10}$$

All the coordinate systems introduced so far cover only part of the full black hole spacetime. A coordinate system which covers the whole spacetime, Kruskal coordinates, can be obtained by introducing $u' = -\exp(-u/4m)$ and $v' = \exp(v/4m)$. The metric becomes

$$-\frac{16m^2}{r}\exp(-r/2m) du'\,dv' + r^2(u',v')(d\theta^2 + \sin^2\theta\,d\phi^2). \tag{4.11}$$

The function $r(u',v')$ is determined by the relation

$$u'v' = -(r - 2m)\exp(r/2m). \tag{4.12}$$

The apparent singularity at $r = 2m$ in the Schwarzschild coordinates (t, r) is known to be a coordinate singularity but if this was not known how could it be tested for? One way of doing so is to use scalar invariants of the metric. Consider for instance the Kretschmann scalar $R_{\alpha\beta\gamma\delta}R^{\alpha\beta\gamma\delta}$. For any smooth metric $g_{\alpha\beta}$ this is a smooth function. Whether or not it is bounded in some region is independent of the coordinates used to compute it. Thus in the approach to a coordinate singularity it must remain bounded. In the Schwarzschild solution the Kretschmann scalar is given by $48m^2/r^6$. It remains bounded as $r = 2m$ is approached as it must due to the fact that there is only a coordinate singularity there. This formula also shows that for $m \neq 0$ the Kretschmann scalar blows up at $r = 0$ and this reveals that there is a geometrical singularity there.

The Schwarzschild solution can be characterized as the only static solution of the vacuum Einstein equations with $\Lambda = 0$ other than Minkowski space satisfying certain physically reasonable boundary conditions. This is sometimes called the no hair theorem due to John Wheeler's formulation of this fact, 'A black hole has no hair'. It is also true that any spherically symmetric solution of the vacuum Einstein equations is locally isometric to the Schwarzschild solution, a fact known as Birkhoff's theorem. A more general discussion of spherically symmetric spacetimes can be found in Section 4.3.

A generalization of the Schwarzschild solution to the Einstein–Maxwell equations is the Reissner–Nordström solution:

$$-(1 - 2m/r + e^2/r^2)dt^2 + (1 - 2m/r + e^2/r^2)^{-1}dr^2 + r^2(d\theta^2 + \sin^2\theta d\phi^2). \tag{4.13}$$

It represents a charged black hole with mass m and charge e provided $|e| < m$. With one exception any spherically symmetric solution of the source-free Einstein–Maxwell equations must be locally isometric to the Reissner–Nordström solution. The exception is the Bertotti–Robinson solution whose metric can be written in the form:

$$-dt^2 + \sin^2 t\, dx^2 + (d\theta^2 + \sin^2\theta d\phi^2). \tag{4.14}$$

A black hole which is rotating is stationary but not static. There is an explicit solution of the vacuum Einstein equations, the Kerr solution, which is a rotating generalization of the Schwarzschild solution. In one particular coordinate system (Boyer–Lindquist coordinates) it takes the form

$$-\left(1 - \frac{2mr}{\Sigma}\right)dt^2 - \frac{4amr\sin^2\theta}{\Sigma}dt d\phi + \frac{\Sigma}{\Delta}dr^2$$
$$+ \Sigma d\theta^2 + \left(r^2 + a^2 + \frac{2a^2mr\sin^2\theta}{\Sigma}\right)\sin^2\theta d\phi^2, \tag{4.15}$$

where $\Sigma = r^2 + a^2\cos^2\theta$ and $\Delta = r^2 - 2mr + a^2$. It represents a black hole provided $|a| < m$. There is a no hair theorem stating that the Kerr solution is the only stationary solution of the vacuum Einstein equations satisfying certain physical boundary conditions. There is an expectation (but no theorem) that when matter collapses completely to a black hole the result should be well approximated at late times by a Kerr solution. An important difference between the Kerr solution and the Schwarzschild solution is that in the former case the stationary Killing vector field which tends to one generating time translations at infinity changes its causal character and becomes spacelike in a region outside the black hole. This region is called the *ergosphere*.

These explicit solutions have not been listed for their own sake but because they are believed, and in restricted circumstances known, to play an important role in the dynamics of more general solutions of the Einstein equations. Twenty years ago it was probably the case that most astrophysicists were sceptical that black holes are common in our universe, or even that they exist at all. Now this has changed completely. For instance it is generally believed on the basis of solid observational evidence that there is a black hole in the centre of most galaxies. The evidence for the presence of a black hole is particularly strong in the case of our own galaxy. The black hole itself cannot be seen but it is possible to observe stars orbiting around a central point. The amount of mass concentrated in a very small region revealed by this motion cannot be explained by anything other than the presence of a black hole. The black holes at the centres of galaxies have large masses of the order a million times that of the sun. At the same time it is believed that many smaller black holes exist with masses a few times that of the sun.

Static solutions of the Einstein equations can also be used to model objects less exotic than black holes. Ordinary stars, which consist of concentrations of gas held together by their own gravity, can be accurately described in terms of Newtonian gravity. However in the late stages of their evolution many stars approach states where gravity is much stronger such as white dwarves, neutron stars and black holes. In an ordinary star the tendency of gravity to pull the matter together is compensated by the pressure of the gas which is at high temperature. This high temperature is maintained by fusion reactions in the interior. Once the nuclear fuel of a star is largely exhausted it must, provided it does not explode, either support itself by other mechanisms (which is what happens in a white dwarf or neutron star) or it must collapse completely, leading to a black hole. An important consequence of general relativity is that there is an upper limit to the mass of an object made of the kind of matter found in a white dwarf or neutron star known as the Chandrasekhar mass. It is not easy to translate this idea from astrophysics into a clear mathematical statement. A discussion of the topic of mass limits for stars can be found in section 6.2 of [206].

A simple description of the matter in a star is given by a perfect fluid which is taken to be isentropic. Provided the star is not rotating it is reasonable to suppose that it is symmetric under all rotations about its centre. Thus mathematically this is a spherically symmetric static solution of the Einstein–Euler system. Everything can be expressed in terms of functions of a radial variable. The metric can be written in the form

$$-e^{2\mu}dt^2 + e^{2\lambda}dr^2 + r^2(d\theta^2 + \sin^2\theta d\phi^2), \tag{4.16}$$

where μ and λ depend only on r. It is useful to introduce a function $m(r)$ by the relation $1 - 2m/r = e^{-2\lambda}$. The key field equation, known as the Tolman–Oppenheimer–Volkov equation (TOV equation) is

$$\frac{dp}{dr} = -\frac{1}{r^2}(\mu + p)(m + 4\pi r^3 p)(1 - 2m/r)^{-1}. \tag{4.17}$$

The problem of constructing solutions of this kind comes down to solving an ODE. It is complicated by the fact that this equation becomes singular at $r = 0$.

Related mathematical problems can be posed for matter described by collisionless matter or elasticity. In the former case there are astrophysical objects such as globular clusters which are well described by these models. However they do not usually require the use of general relativity. A physical situation where general relativity is relevant is the central bulge of a spiral galaxy where the population of stars could be modelled by a cloud of collisionless matter surrounding a massive black hole. The description of objects which can be modelled as elastic bodies is typically possible without the use of general relativity although neutron stars have a solid crust for which elasticity theory is the appropriate description.

4.2 Spatially homogeneous models

The simplest models in cosmology are those which are spatially homogeneous. What this means is the following. There is a group of isometries G whose orbits are three-dimensional and spacelike. They form a foliation of spacetime and if this foliation is labelled by a parameter t then this defines a time coordinate. It is always possible to choose this time coordinate such that the metric is of the form

$$-dt^2 + g_{ab}(t,x^c)dx^a dx^b. \qquad (4.18)$$

Coordinates with this property (in any spacetime) are called Gaussian. In the homogeneous case if an orbit is thought of as space at a given time then there is an isometry which maps any point of space to any other. In this sense all points of space are equivalent and this is the intuitive meaning of homogeneity. In a four-dimensional spacetime which is spatially homogeneous the dimension of the orbits is three and so the dimension of the isometry group is at least three. It turns out that the only dimensions which can occur for the isometry group are three, four and six. If the isometry group is six-dimensional then the isotropy group at a point x must be three-dimensional and must therefore have the full rotation group $SO(3)$ as a subgroup. There is an element of the isotropy group which maps any vector at x tangent to the orbit to any other such vector. In this case the spacetime is said to be isotropic. All directions at a point are equivalent and this is the intuitive meaning of isotropy. The homogeneous and isotropic spacetimes are those most frequently applied in cosmology. This is because on the one hand they are the simplest models and on the other hand they seem to agree well with astronomical observations after averaging on a suitably large scale. These models are sometimes called FLRW models after Friedmann, Lemaitre, Robertson and Walker. Sometimes only a subset of these authors are named. The FLRW models can be classified according to the sign of the scalar curvature R of the orbits. A model with a given sign of the curvature can be transformed by multiplying the metric with a constant factor to one where the curvature has an arbitrary modulus and the same sign. This rescaling can be thought of as a change of units and is thus not physically significant. Thus the spacetimes can be labelled by a curvature parameter K taking the values $(-1, 0, 1)$. The case $K = 0$ is referred to as spatially flat. Sometimes in the cosmology literature it is confusingly referred to as flat although the spacetime curvature is not zero.

A well-known FLRW model is the Einstein–de Sitter model with metric

$$-dt^2 + t^{4/3}(dx^2 + dy^2 + dz^2). \tag{4.19}$$

It is spatially flat and satisfies the Einstein equations with zero cosmological constant coupled to dust. The four-velocity has components $(1, 0, 0, 0)$ and the energy density is proportional to t^{-2}. This solution should not be confused with the de Sitter spacetime which can be written in the form

$$-dt^2 + e^{2Ht}(dx^2 + dy^2 + dz^2). \tag{4.20}$$

It is a solution of the vacuum Einstein equations with positive cosmological constant $\Lambda = 3H^2$. It is the analogue of Minkowski space for $\Lambda > 0$. It plays an important role as a model for the late-time dynamics of rather general solutions of the Einstein equations with positive cosmological constant and matter. By analogy with the case of black holes this is sometimes called the 'cosmic no hair theorem'.

The class of FLRW models is defined purely by their symmetry properties and not by a specific choice of matter model. The symmetry implies that the energy–momentum tensor automatically has the form of that of a perfect fluid. The metric takes the form

$$-dt^2 + a^2(t)d\Sigma^2, \tag{4.21}$$

where $d\Sigma^2$ is a Riemannian metric of constant curvature. The function a is called the scale factor. Suppose that the matter satisfies the strong and dominant energy conditions and that the cosmological constant is zero. Suppose further that the mean curvature of at least one of the homogeneous hypersurfaces is negative. Then the sign of the curvature of the spatial metric determines the qualitative dynamics for any reasonable matter model. If $K \leq 0$ then a is always increasing, tends to zero at some finite time in the past (corresponding to the big bang) and tends to infinity as $t \to \infty$. The spacetime is geodesically complete in the future. If $K > 0$ then for a wide class of matter models a increases to a maximum after which it decreases.

It tends to zero in finite time both in the past and in the future (corresponding to big bang and big crunch singularities). In an FLRW model $T_{ij} = pg_{ij}$ for some function p of t. A sufficient condition implying that a solution with $K > 0$ cannot have $\dot{a} > 0$ for an infinite time is that $p \geq 0$. For matter models with a well-behaved global Cauchy problem this implies that there is a time at which $\dot{a} = 0$.

Physically the timelike geodesics normal to the homogeneous hypersurfaces are interpreted as the worldlines of typical galaxies. Correspondingly the homogeneous hypersurfaces themselves are interpreted as the rest frame of these galaxies. If $\dot{a} > 0$ this means that the galaxies are receding from each other – the universe is expanding. Due to astronomical observations it can be concluded that $\dot{a} > 0$ now. With the conventions used here this means that $\text{tr} k < 0$ and this is the reason for considering models with $\text{tr} k < 0$ at some time in the above. Mathematically, the behaviour of models where $\text{tr} k$ is always positive can be obtained from what has already been said by replacing t by $-t$.

The spacetimes that are spatially homogeneous but not isotropic can be classified. This follows from the classification of homogeneous three-dimensional Riemannian manifolds carried out by Bianchi at the end of the nineteenth century. In the majority of cases the isometry group has a three-dimensional subgroup which acts transitively. The one exception is the case of metrics on $S^2 \times S^1$ which are invariant under the obvious action of the product of the rotation group $SO(3)$ with the circle considered as a group under addition modulo 2π. The latter can also be thought of as the group of complex numbers of unit modulus under multiplication and denoted by $U(1)$. Spacetimes with this type of symmetry are known in general relativity as Kantowski–Sachs spacetimes. The cases where the isometry group acts transitively on the orbits can be classified in terms of the Lie algebra of the isometry group. Any three-dimensional Lie algebra can occur. In general relativity it is standard to use the terminology of Bianchi who distinguished types I–IX. In fact types VI and VII are one-parameter families of non-isomorphic Lie algebras and are often denoted by VI$_h$ and VII$_h$ to make this parameter explicit. The Bianchi types fall into two broad classes A and B. Class A consists of types I, II, VI$_0$, VII$_0$, VIII and IX while all the rest belong to class B. By definition the Lie algebras belonging to Class A are those which are unimodular, which means that the trace of the structure constants C_{IJ}^I is zero. The property of belonging to Class A is closely connected with the possibility of the spatial manifold being a covering of a compact manifold. It may be remarked in passing that there is no known analogue of the Bianchi classification which applies to general dimensions. A good source for the available information concerning five-dimensional spatially homogeneous cosmological models (the first step) is [104]. Some more information about homogeneous cosmological models in higher dimensions is given in Section 5.9.

Some readers may wonder how the Bianchi classification is related to the geometrization conjecture of Thurston and the topological classification of three-dimensional manifolds. This will be discussed briefly. In Thurston's work three-dimensional manifolds are cut into pieces that admit locally homogeneous Riemannian structures. The Thurston classification of these uses the isometry group of maximal dimension which a metric admits while the Bianchi classification uses the minimal group which acts transitively. Bianchi types I and VII$_0$ both correspond to the Euclidean geometry in Thurston's classification. Both these Lie groups are subgroups of the Euclidean group. Type II corresponds to Thurston's type Nil. Type III (and type VI$_{-1}$ which is the same thing) corresponds to $H^2 \times \mathbb{R}$. Geometries of this Bianchi type, which is Class B, can only be compactified if they admit a four-dimensional family of local isometries. Bianchi types V and VII$_h$ for $h \neq 0$ correspond to hyperbolic geometry. The geometries of this type can only be compactified if they admit a six-dimensional family of isometries. Type VI$_0$ corresponds to Sol in Thurston's classification. Type VIII corresponds to $SL(2, \mathbb{R})$ and type IX to the spherical geometry. Kantowski–Sachs corresponds to $S^2 \times \mathbb{R}$. Since the other Bianchi types cannot be compactified they do not occur in Thurston's classification. It has been conjectured that there might be a connection between the geometrization conjecture and the long-time behaviour of forever expanding cosmological models but the evidence for this is at present scant.

Now the Einstein equations will be written down in $3 + 1$ form for a general Bianchi model. The spatial manifold is the Lie group G itself. The Lie group acts on itself by left translations, $\psi(g, h) = gh$. There is a left-invariant basis. It is assumed that the metric on each time slice is left-invariant so that its components g_{ij} in the left-invariant basis depend only on t. Tensors such as the second fundamental form and the spatial Ricci tensor are expressed in terms of their components k_{ij} and R_{ij} in the left-invariant basis. The Einstein equations become $\partial_t g_{ij} = -2k_{ij}$ and

$$\partial_t k^i{}_j = R^i{}_j + (\mathrm{tr}k)k^i{}_j - 8\pi \left(S^i{}_j - \frac{1}{2}\delta^i_j \mathrm{tr} S\right) - 4\pi \rho \delta^i_j. \qquad (4.22)$$

For fixed values of the matter variables this system only contains derivatives with respect to t – it is a system of ordinary differential equations. The components of the spatial Ricci tensor depend algebraically on the components g_{ij} and the structure constants C^i_{jk}. The only dependence of the equations on the Bianchi type comes through these structure constants. The rotation coefficients of the left invariant frame can be expressed as follows:

$$\gamma^i_{jk} = \frac{1}{2}g^{il}(C_{jkl} + C_{lkj} + C_{ljk}), \tag{4.23}$$

$$C_{ijk} = g_{kl}C^l_{ij}. \tag{4.24}$$

The equations for the matter quantities are in general also ordinary differential equations in the spatially homogeneous case. For instance for a scalar field the equation of motion is

$$\ddot{\phi} + 3H\dot{\phi} = -V'(\phi), \tag{4.25}$$

where $H = -\mathrm{tr}k/3$ and a dot denotes a time derivative. In cosmology H is called the Hubble parameter. An exception to the statement that the field equations become ODE is the case of kinetic matter models where the distribution function retains its dependence on the momentum variables in the spatially homogeneous case. Thus for instance the Vlasov equation is still a PDE in that situation.

In thinking about cosmology from a mathematical point of view it is often useful to start with spacetimes which admit a compact Cauchy hypersurface. Unfortunately if the two conditions of spatial compactness and homogeneity are applied at the same time not very much is left. The only Bianchi types which are possible are I and IX. To get a more general class of spacetimes it is possible to consider locally homogeneous spacetimes. This means that the symmetry group corresponding to the given Bianchi type does not act on the spacetime itself but only on the metric obtained by pulling back the spacetime metric to the universal covering manifold. This allows many more Bianchi types to be included, in particular all those of Class A.

One general property of spatially homogeneous solutions of the Einstein equations with $\Lambda = 0$ coupled to matter satisfying the dominant energy condition is the following. By homogeneity the scalar curvature is constant on any one of the hypersurfaces of constant time. If there is a time when $\mathrm{tr}k = 0$ then the Hamiltonian constraint implies that this constant is non-negative and that it is positive unless both $\rho = 0$ and $k_{ab} = 0$. By the dominant energy condition the first of these implies $T_{\alpha\beta} = 0$. It can be shown that the only Bianchi type for which the scalar curvature R can be positive is type IX. If the volume of space reaches a maximum at some time then $\mathrm{tr}k = 0$ at that time. Thus a Bianchi model which recollapses must be type IX unless at some time the model is vacuum and has vanishing second fundamental form. It turns out that except in type IX the condition $R = 0$ implies that the metric is flat. What this means is that the initial data at the given time coincides with data for flat space. If it is assumed that for the given matter model the Einstein–matter equations have a well-posed initial value problem then it follows that the spacetime is flat and static. This is not a useful cosmological model. Thus it may be said that the only Bianchi type compatible with recollapse under the given assumptions is type IX. All other Bianchi models expand for ever if they are expanding at some time. Kantowski–Sachs models can also recollapse. In fact they must do so under reasonable conditions. It has been proved that a solution of the Einstein–matter equations with $\Lambda = 0$ satisfying the dominant and strong energy conditions and the inequality $\mathrm{tr} S \geq 0$ (non-negative average pressure) must recollapse if it is of Bianchi type IX or Kantowski–Sachs. More precisely what has been proved is that it cannot expand for an infinite proper time. For matter models which have a globally well-behaved Cauchy problem in the homogeneous case this implies that it must recollapse. Matter models of this type are plentiful and include, for instance, vacuum, perfect fluids with rather general equations of state, and collisionless matter.

4.3 Surface symmetry

This section is concerned with spacetimes admitting the action of a group of isometries which is three-dimensional and has two-dimensional spacelike orbits. Under these circumstances each orbit is a homogeneous and isotropic Riemannian manifold. As in the corresponding three-dimensional case these can be classified by the sign of their curvature which is distinguished by a parameter κ taking the values 1, 0 and -1. The three symmetry types are called spherical, plane and hyperbolic symmetry, respectively. Collectively they are denoted by the term surface symmetry. The orbits themselves are referred to as surfaces of symmetry. The universal cover of any orbit is isometric up to a constant rescaling to the standard sphere, the Euclidean plane or the hyperbolic plane, according to the sign of the curvature. The case of interest in the following is that in which the surfaces of symmetry are compact. This is automatic in the case $\kappa = 1$ and in the other two cases the situation of interest is that of cosmological solutions where compactness is a desirable property mathematically. In fact, by analogy with the case of locally homogeneous Bianchi models, it is too restrictive to demand the symmetry group to act on the spacetime itself and this condition must be relaxed to the requirement that the symmetry group acts on the universal cover.

To write down the explicit form of the metric the following convention will be used. Define the function $\sin_\kappa$ to be sin for $\kappa = 1$, 1 for $\kappa = 0$ and sinh for $\kappa = -1$. Then the following metric has surface symmetry:

$$-e^{2\mu}dt^2 + e^{2\lambda}dr^2 + r^2(d\theta^2 + \sin^2_\kappa \theta d\phi^2). \tag{4.26}$$

Here μ and λ are functions of t and r. Note that the Schwarzschild and Reissner–Nordström metrics belong to this class. The metric (4.26) is not the most general with surface symmetry as will now be explained. In (4.26) the area of a surface of symmetry is proportional to r^2. For this reason the coordinate r is called the area radius. This quantity is defined in any surface symmetric spacetime and is called the area function in general. If the gradient of r is spacelike at some point then coordinates as in (4.26) can be introduced in a neighbourhood of that point. If the gradient of r is timelike at some point, which is common in cosmological solutions, then similar coordinates can be introduced where the factor r^2 in the metric of the orbits is replaced by t^2. In that case t is called an areal time coordinate. Then the metric takes the form

$$-e^{2\mu}dt^2 + e^{2\lambda}dr^2 + t^2(d\theta^2 + \sin^2_\kappa \theta d\phi^2). \tag{4.27}$$

In general the causal character of the gradient of r changes from point to point of the manifold and neither of the possibilities mentioned so far gives a global coordinate system. A more versatile coordinate system is given by double null coordinates. These are discussed further in Chapter 11. The Schwarzschild solution was expressed in coordinates of this type in (4.9).

It is of great interest to understand how a black hole which can be described by the Schwarzschild solution is formed by the collapse of matter. The simplest situation in which this can be considered is the spherically symmetric case. There is an explicit solution of the Einstein equations coupled to dust which models this process and which has formed ideas about what the dynamics is like. This is the Oppenheimer–Snyder solution which was found in 1939 [156]. At that time the concept of black hole did not exist and the meaning of the solution was poorly understood. A disadvantage of this solution is the fact that its matter content is described by dust. The solution itself avoids unphysical effects because the energy density is homogeneous but arbitrarily small inhomogeneous perturbations can lead to pathological behaviour (shell-crossing singularities). On the other hand, numerical work indicates that the Oppenheimer–Snyder solution is an excellent model for the collapse of more realistic types of matter under rather general circumstances. For this reason it is well worth having a thorough analytical understanding of the Oppenheimer–Snyder solution.

The coordinates used to obtain an exact solution are different from those in (4.26). The spacetime is the union of an interior region where the matter density is non-zero and an exterior region where it is zero. By Birkhoff's theorem the exterior region is isometric to part of the Schwarzschild solution.

The coordinates in the interior region include a comoving coordinate which follows the dust particles. This means that the four-velocity of the dust is tangent to the hypersurfaces of constant radial coordinate R. The time coordinate is chosen so that its level hypersurfaces are orthogonal to this four-velocity. It coincides with proper time along the worldline of any dust particle. These coordinates are tied to the fact that the matter is dust and there is no obvious unique analogue for any other matter model. The energy density at any fixed time is constant within a ball about the origin and vanishes outside. In the region where the density is non-zero the solution is isometric to part of the Einstein–de Sitter solution with the time direction reversed. It has the explicit form

$$-dT^2 + T^{4/3}(dR^2 + R^2(d\theta^2 + \sin^2\theta d\phi^2)), \qquad (4.28)$$

and is defined on the region determined by the inequalities $T < 0$ and $R \leq R_0$ for some constant $R_0 > 0$. It must be matched to the exterior region in a suitable way. The idea is to do the matching so that the metric solves the Einstein equations everywhere without introducing a distributional matter source on the joining hypersurface. There is a well-known method of treating this question which comes down to ensuring that there exist coordinates making components of the matched metric C^1. It turns out that this determines the hypersurface which bounds the part of the Schwarzschild solution defining the exterior region. It is a spherically symmetric region whose boundary is composed of radial timelike geodesics. The Oppenheimer–Snyder solution can in principle be transformed to Schwarzschild coordinates but it does not seem to be possible to do this explicitly.

In higher dimensions spherical symmetry can be defined in an analogous way. There is an analogue of the Schwarzschild solution and the analogue of Birkhoff's theorem holds. There are, however, interesting generalizations of spherical symmetry of a kind which does not exist in four dimensions and these give rise to a class of Cauchy problems for the vacuum Einstein equations which have essentially one space dimension. These examples result from the fact that a group smaller than the full rotation group $SO(n)$ can act transitively on the sphere S^n when $n > 2$. For instance there are transitive actions of $SU(2)$ on S^3 and $SO(5) \times SU(2)$ on S^7. More information can be found in [33].

4.4 T^2 symmetry

Consider now the case of a spacetime where a two-dimensional Abelian group of isometries acts. In the case of cosmological solutions it is natural to choose the group to be the compact group T^2, the two-dimensional torus, acting in the standard way on T^3. This situation is referred to as T^2 symmetry. Then the orbits are compact and are two-dimensional tori. There is also a more general case which may be called local T^2 symmetry where a global action of T^2 only exists on the universal covering manifold. The topology is that of a torus bundle over the circle S^1. The manifold can be obtained by taking the product of a closed interval I with T^2 and identifying the ends by means of an element of $SL(2,\mathbb{Z})$, i.e. a 2×2 matrix with integer entries and unit determinant. The topology obtained depends on the matrix chosen. If it is diagonalizable and not equal to plus or minus the identity (which up to a finite covering would give a torus) the spacetimes obtained generalize the models of Bianchi type VI_0 while if it is not diagonalizable they generalize the Bianchi models of type II.

Denote the area of the orbit on which a point lies by t. The metric can be written in the following form with an areal time coordinate

$$e^{2(\eta-U)}(-\alpha dt^2 + d\theta^2) + e^{2U}[dx + Ady + (G + AH)d\theta]^2$$
$$+ e^{-2U}t^2[dy + Hd\theta]^2. \quad (4.29)$$

Here U, A, α, η, G and H are functions of t and θ. In the case of the torus topology they are periodic in θ. In the case of the other topologies they change in a way defined by the $SL(2,\mathbb{Z})$ matrix after a certain interval of θ. Let k^α and l^α be the Killing vectors in a spacetime with T^2 symmetry. The *twists* are the functions defined by

$$\epsilon^{\alpha\beta\gamma\delta}\nabla_\alpha k_\beta k_\gamma l_\delta \quad (4.30)$$

and the corresponding quantity with the roles of k^α and l^α interchanged. In a vacuum spacetime these functions are constant. There are important subclasses with extra symmetry. If in addition to the translations in the coordinates y and z the rotations in the (y, z)-plane are isometries then the result is plane symmetry which was already discussed in the last section. In that case the functions A, G and H vanish and $U = \frac{1}{2}\log t$. Note that the coordinate t here corresponds to t^2 in the usual notation for surface symmetry. This notational conflict reflects common usage in the literature. Other classes can be defined by discrete symmetry groups. Suppose there is an isometry which maps any vector v^α tangent to the orbits to $-v^\alpha$ and any vector orthogonal to the orbits to itself. Then there are coordinates in which this symmetry is realized by the mapping $y \mapsto -y$ and $z \mapsto -z$. With a suitable coordinate choice the functions G and H vanish, at least after passing to the universal cover. A metric of this type is said to have Gowdy symmetry. In spacetimes with Gowdy symmetry the twists vanish. The name of this class of spacetimes is derived from the Gowdy spacetimes which are the solutions of the vacuum Einstein equations with vanishing cosmological constant having this symmetry. In the case of the Gowdy spacetimes it is possible to choose $\alpha = 1$ and the metric can be simplified to

$$t^{-1/2}e^{\lambda/2}(-dt^2 + d\theta^2) + t[e^P(dx + Qdy)^2 + e^{-P}dy^2], \qquad (4.31)$$

where the names of some quantities have been changed so as to agree with those common in the literature on Gowdy spacetimes. Specifically, $Q = A$, $P = 2U - \log t$ and $\lambda = 4(\eta - U) + \log t$. Here the metric in the (t, θ)-plane is proportional to the flat metric – it is conformally flat. In this case the coordinates are said to be conformal. It is important to realize that for almost all types of matter an areal time coordinate is not compatible with conformal coordinates. The compatibility condition can be stated as follows in the case $\Lambda = 0$. Let $\hat{T}$ be the projection of the energy–momentum tensor orthogonal to the symmetry orbits. Then the condition is that tr$\hat{T} = 0$. The compatibility condition also fails for the vacuum equations with non-zero cosmological constant as can be seen by considering the cosmological constant as defining an effective matter field with energy–momentum tensor $-\frac{\Lambda}{8\pi}g_{\alpha\beta}$. There is a large class of solutions of the Einstein–Maxwell equations with Gowdy symmetry for which conformal coordinates are compatible with having an areal time coordinate. In that case the electromagnetic field can be derived from a potential whose only non-vanishing components are $A_2 = \omega$ and $A_3 = \chi$ corresponding to the coordinates x and y. This class can be related to symmetry under the transformation $y \mapsto -y$ and $z \mapsto -z$ but a small trick is necessary. It is based on the fact that if $(g_{\alpha\beta}, F_{\alpha\beta})$ is a solution of the Einstein–Maxwell equations $(g_{\alpha\beta}, -F_{\alpha\beta})$ is too. It is the combination of this transformation with the diffeomorphism above which should be used. The solutions which are invariant under the composition are precisely those for which F_{01} and F_{23} vanish and these are analogues of Gowdy solutions in the electromagnetic case.

If there is a discrete symmetry in a T^2-symmetric spacetime which is a reflection and reverses exactly one direction in the orbit then the spacetime is called polarized. This is the case where $A = H = 0$. A Gowdy spacetime where the reflections in y and z are both symmetries is a polarized Gowdy spacetime. Then $Q = 0$. The name comes from the fact that the Gowdy spacetimes can be interpreted physically as describing gravitational waves moving in the θ-direction. These have two polarizations and the restriction $Q = 0$ kills one of the polarizations. The Gowdy spacetimes have the pleasant property that there are equations for P and Q which decouple from the other Einstein equations. These equations are

$$\partial_t^2 P + t^{-1}\partial_t P = \partial_\theta^2 P + e^{2P}((\partial_t Q)^2 - (\partial_\theta Q)^2), \tag{4.32}$$

$$\partial_t^2 Q + t^{-1}\partial_t Q = \partial_\theta^2 Q - 2(\partial_t P \partial_t Q - \partial_\theta P \partial_\theta Q) \tag{4.33}$$

and are very similar to the equations for a wave map on Minkowski space with values in the hyperbolic plane which only depends on one of the spatial variables. In fact these equations can be realized as a wave map on a simple non-flat background with metric

$$-dt^2 + d\theta^2 + t^2 d\phi^2. \tag{4.34}$$

Wave maps from this metric to the hyperbolic plane which do not depend on the coordinate ϕ correspond to solutions of the Gowdy equations. Once the above equations have been solved for P and Q the whole spacetime can be constructed by integration of the equations

$$\partial_t \lambda = t\{(\partial_t P)^2 + (\partial_\theta P)^2 + e^{2P}[(\partial_t Q)^2 + (\partial_\theta Q)^2]\}, \tag{4.35}$$

$$\partial_\theta \lambda = 2t(\partial_t P \partial_\theta P + e^{2P}\partial_t Q \partial_\theta Q). \tag{4.36}$$

This formulation can be generalized to the case of the solutions of the Einstein–Maxwell equations with Gowdy symmetry. The evolution equations are

$$\partial_t^2 P + t^{-1}\partial_t P = \partial_\theta^2 P + e^{2P}((\partial_t Q)^2 - (\partial_\theta Q)^2)$$
$$- \frac{2}{t}\left\{-e^P[(Q\partial_t\omega - \partial_t\chi)^2 - (Q\partial_\theta\omega - \partial_\theta\chi)^2]\right.$$
$$\left.+ e^{-P}((\partial_t\omega)^2 - (\partial_\theta\omega)^2)\right\}, \tag{4.37}$$

$$\partial_t^2 Q + t^{-1}\partial_t Q = \partial_\theta^2 Q - 2(\partial_t P \partial_t Q - \partial_\theta P \partial_\theta Q)$$
$$+ \frac{4}{t}e^{-P}\left[((\partial_t\omega)^2 - (\partial_\theta\omega)^2)Q - (\partial_t\omega\partial_t\chi - \partial_\theta\omega\partial_\theta\chi)\right], \tag{4.38}$$

$$\partial_t^2 \chi = \partial_\theta^2 \chi - \partial_t P \partial_t \chi + \partial_\theta P \partial_\theta \chi$$
$$+ (2Q\partial_t P + \partial_t Q)\partial_t\omega - (2Q\partial_\theta P + \partial_\theta Q)\partial_\theta\omega$$
$$- e^{2P}Q[\partial_t Q(Q\partial_t\omega - \partial_t\chi) - \partial_\theta Q(Q\partial_\theta\omega - \partial_\theta\chi)], \tag{4.39}$$

$$\partial_t^2 \omega = \partial_\theta^2 \omega + \partial_t P \partial_t \omega - \partial_\theta P \partial_\theta \omega$$
$$- e^{2P}[\partial_t Q(Q\partial_t\omega - \partial_t\chi) - \partial_\theta Q(Q\partial_\theta\omega - \partial_\theta\chi)]. \tag{4.40}$$

A polarized special case can be defined by setting $Q = 0$, $\chi = 0$. The remaining equations for P and ω are

$$\partial_t^2 P + t^{-1} \partial_t P = \partial_\theta^2 P - 2t^{-1} e^{-P}((\partial_t \omega)^2 - (\partial_\theta \omega)^2), \qquad (4.41)$$

$$\partial_t^2 \omega = \partial_\theta^2 \omega + \partial_t P \partial_t \omega - \partial_\theta P \partial_\theta \omega. \qquad (4.42)$$

Introducing $\psi = \frac{1}{2}(P + \log t)$ shows that they are equivalent to the equations in [42] with $\chi = 0$. They are also also equivalent to the Gowdy equations as is made clear by introducing the variable $\tilde{P} = -\frac{1}{2}(P + \log t) = -\psi$ in the equations of motion for the case with electromagnetism. For this reason any result on the global behaviour of general Gowdy spacetimes gives rise to a corresponding result for the class of polarized solutions of the Einstein–Maxwell equations. It should however be noted that the quantities which correspond to each other in the sense that they appear in the same way in the equations of motion have different physical interpretations in the two cases.

The Gowdy spacetimes can be seen in a wider context by considering an analogue of T^2 symmetry in higher dimensions. Consider a solution of the vacuum Einstein equations with $\Lambda = 0$ in $n + 2$ dimensions with an action of T^n by isometries. The symmetries can be represented as translations of coordinates y_1 to y_n. Suppose that simultaneous reflection in all these coordinates is an isometry. Then the essential field equations take the form

$$\partial_t^2 X^I + t^{-1} \partial_t X^I = \partial_\theta^2 X^I + \Gamma^I_{JK}(\partial_t X^J \partial_t X^K - \partial_\theta X^J \partial_\theta X^K) = 0, \qquad (4.43)$$

where Γ^I_{JK} are the Christoffel symbols of a left-invariant metric on the quotient space $SL(n, \mathbb{R})/SO(n)$. In the case $n = 2$ this reduces to the Gowdy case since $H^2 = SL(2, \mathbb{R})/SO(2)$. In the case $n = 3$ the target space metric can be written in the explicit form

$$dP_1^2 + dP_2^2 + e^{2P_1}(dQ_1 - Q_3 dQ_2)^2 + e^{P_1 - \sqrt{3} P_2} dQ_2^2 + e^{P_1 + \sqrt{3} P_2} dQ_3^2. \qquad (4.44)$$

It should be mentioned that there is another type of Gowdy symmetry. In that case the solution is invariant under an action of T^2 on S^3 or $S^2 \times S^1$. Each of the Killing vectors vanishes on a one-dimensional submanifold which may be called an axis. The equations written on the quotient of the original space by the symmetry become singular on the axis. For this reason the 'spherical' Gowdy spacetimes are analytically much more difficult to handle than those on a torus. As a consequence most mathematical work on Gowdy spacetimes has concentrated on the case of the torus.

A type of symmetry related to T^2 symmetry is cylindrical symmetry, where there is one translational symmetry and one rotational symmetry which commutes with it. The rotational Killing vector vanishes on an axis. This actually forces the twist constants to vanish in vacuum. Although the orbits have infinite volume it is still possible to define a function which is analogous to the area of the orbits in the Gowdy case. The gradient of the function is spacelike in cylindrical symmetry in contrast to the case of Gowdy solutions on T^2.

The field equations for vacuum spacetimes with cylindrical symmetry can be reduced to the equations for a wave map with values in the hyperbolic plane. The domain of this wave map is three-dimensional Minkowski space and it is invariant under rotations about an axis. There is a polarized case where the wave map reduces to a solution of the linear wave equation. The corresponding solutions are known as Einstein–Rosen waves. This formulation is related to the reduction for $U(1)$ symmetry discussed in the next section. A similar procedure can be carried out for solutions of the Einstein–Maxwell equations with cylindrical symmetry. The equations can be reduced to those for a wave map with values in a four-dimensional manifold, the complex hyperbolic plane.

4.5 $U(1)$ symmetry

Consider a spacetime with a one-dimensional isometry group which has compact spacelike orbits. This is often called $U(1)$ symmetry since the group can be assumed to be the one-dimensional unitary group. The metric can be written in the form

$$e^{-2\gamma}[-N^2 dt^2 + g_{ab}(dx^a + v^a dt)(dx^b + v^b dt)] + e^{2\gamma}(dx^3 + A_\alpha dx^\alpha)^2. \tag{4.45}$$

In this section Latin indices take the values $1, 2$ and Greek indices the values $0, 1, 2$. All functions occurring in the parametrization of the metric are functions of t and x^a only. The Killing vector is $\partial/\partial x^3$. The form of the metric used emphasizes that the spacetime is thought of in terms of a quotient space which is a three-dimensional Lorentzian manifold. The three-dimensional geometry is written in terms of a $2 + 1$ decomposition.

Everything can be expressed in terms of three-dimensional quantities. There is the metric $g_{\alpha\beta}$, the one-form A_α and the scalar function γ. It is possible to think of A_α as representing a connection which defines a Maxwell field (Abelian Yang–Mills field) on the three-dimensional spacetime. It has a curvature $F_{\alpha\beta}$. In fact the one-form is in general only defined locally and it is only the connection which makes sense globally. The spacetime is of the form $\mathbb{R} \times N$ where N is the total space of a circle bundle over a compact surface Σ. By a suitable choice of bundle models of Bianchi types I (trivial bundle over T^2), II (non-trivial bundle over T^2), III (trivial bundle over a higher genus surface), VII$_0$ (trivial bundle over T^2), VIII (non-trivial bundle over a higher genus surface), IX (non-trivial bundle over S^2) and Kantowski–Sachs (trivial bundle over S^2) can be included. The Maxwell field $F_{\alpha\beta}$ can be decomposed into electric and magnetic parts. In a three-dimensional spacetime the electric field is a vector while the magnetic field is a scalar.

What has been said up to now is independent of the choice of matter model. In the vacuum case the equations can be put into a very special form involving a wave map. One of the vacuum Einstein equations shows that the electric field is divergence-free. Hence it can be written as the sum of the exterior derivative of a scalar and a harmonic form h_a, i.e. one whose Laplacian vanishes. Consider now the case that the harmonic part vanishes which is an assumption compatible with the Einstein equations. This is analogous to assuming that the twist vanishes in a vacuum spacetime with T^2 symmetry. Then $E_a = \nabla_a \omega$ for a scalar function ω. It now turns out that the vacuum Einstein equations imply that γ and ω together satisfy a wave map equation on the background of the metric $g_{\alpha\beta}$ with values in the hyperbolic plane. The relevant form of the hyperbolic metric is $2d\gamma^2 + \frac{1}{2}e^{4\gamma}d\omega^2$. Meanwhile the metric itself satisfies the Einstein equations (in 2+1 dimensions) with the energy–momentum tensor of the wave map just introduced. Thus in total an Einstein–wave map system in three dimensions is obtained. There is a polarized case where $\omega = 0$ and the wave map reduces to the linear wave equation. The polarized case looks deceptively simple – it is 'just' the Einstein equations in three dimensions coupled to a linear wave equation and in three dimensions the gravitational field has no local degrees of freedom of its own. (The Weyl tensor vanishes.) In reality it turns out to be a very tough analytical problem to analyse the solutions of these equations.

4.6 Further reading

The recollapse theorems for Kantowski–Sachs and Bianchi IX solutions are proved in [38] and [135] respectively. For the existence question in homogeneous spacetimes see [169]. The possible connection between the geometrization theorem and the Einstein equations is discussed in [80]. Local T^2 symmetry was introduced in [170]. For the Einstein–Maxwell equations with Gowdy symmetry see [150] which actually discusses the more general Einstein–Maxwell–dilaton–axion system. For the metric on

$SL(3, \mathbb{R})/SO(3)$ see [151]. This metric is also of relevance for some problems arising in string cosmology. The wave map formulation of the Einstein–Maxwell system in the case of cylindrical symmetry is described in [27]. The equations for $U(1)$ symmetry and their special properties in the vacuum case are discussed in [50].

5 Ordinary differential equations

A system of first order ordinary differential equations (ODE) takes the form

$$dX_i/dt = f_i(t, X_1, \ldots, X_n), \qquad (5.1)$$

where $f_i, i = 1, 2, \ldots n$, are functions defined on an open subset U of $(n+1)$-dimensional Euclidean space which are denoted collectively by f. If f does not depend on t the equations are called autonomous. In the following f is always assumed continuous but it is often helpful to assume some additional regularity, as will be discussed in more detail later. The aim is to describe solutions of the equation (5.1). These are functions $X_1(t), \ldots, X_n(t)$ on an interval I of real numbers with values in $\mathbb{R}$ which satisfy (5.1). To make sense of the equation the solution should be differentiable. Then it is automatically continuously differentiable. If the interval I is the whole real line the solution is said to be global while if it is a smaller interval the solution is said to be local. In a system of ODE of order k the left hand side of (5.1) is replaced by $d^k X_i/dt^k$ while the right hand side is allowed to depend on all derivatives of order up to $k - 1$. A system of this kind can be reduced to the form (5.1) by introducing the derivatives of order up to $k - 1$ as additional variables and so is not essentially more general.

A fundamental task in understanding the equation (5.1) is to parametrize the local solutions. This can be done by means of initial data. If the function f is suitably regular then for any set of numbers X_i^0 such that $(0, X_i^0) \in U$ there is a solution $X_i(t)$ on an interval $I = (t_-, t_+)$ containing $t = 0$ with $X_i(0) = X_i^0$. Any other fixed value $t = t_0$ could have been chosen instead of $t = 0$. If the interval I is fixed and f is sufficiently regular then there is only one solution with the prescribed initial data. These are the basic existence and uniqueness statements. They are so important that, in contrast to the usual practice in this book, a complete theorem will be stated and proved. The proof is of interest not only in itself but as a model of a kind of proof which is of great importance in the theory of partial differential equations. It is given in the next section.

If the function f is only assumed to be continuous then an existence theorem can be proved but uniqueness generally fails. The latter fact can be made clear by the simple example with $n = 1$ and $f(X_1) = |X_1|^\alpha$ for $0 < \alpha < 1$ where there are different solutions with the same initial data. To see this consider the function $X_1(t)$ which is identically zero for $t \leq 0$ and equal to $[(1-\alpha)t]^{\frac{1}{1-\alpha}}$ for $t \geq 0$. It is a C^1 solution of the equation and has the same initial datum at $t = 0$ as the solution which is identically zero. This kind of situation can be avoided by imposing a Lipschitz condition. The function f is said to be locally Lipschitz if there is a neighbourhood of any X and a constant K such that $|f(X) - f(Y)| \leq K|X - Y|$ for all Y in the neighbourhood. Here $X = (X_1, \ldots, X_n)$ and $|X| = (X_1^2 + \cdots + X_n^2)^{1/2}$. By the mean value theorem this holds in particular if f is continuously differentiable.

The local solutions cannot always be extended to global solutions. Consider the equation $dx/dt = x^2$. All solutions can be written down explicitly. Apart from the solution which is identically zero they are given by $x(t) = (x_0^{-1} - t)^{-1}$ where $x_0 = x(0)$. If $x_0 > 0$ the solution corresponding to this initial value at $t = 0$ exists globally in the past but only up to $t = x_0^{-1}$ in the future. If $x_0 < 0$ then the corresponding statement holds with past and future interchanged. Only the solution with $x_0 = 0$ exists globally. When an initial value for a system of ODE is prescribed at $t = t_0$ there is a maximal interval of existence (t_-, t_+) on which a corresponding solution exists. Suppose that $t_+ < \infty$. Then it can be shown that either $|X|$ is unbounded as $t \to t_+$ or that $(t, X(t))$ approaches the boundary of U in this limit. This can be formulated equivalently by saying that $(t, X(t))$ leaves every compact subset of U as $t \to t_+$. This statement provides a criterion for global existence. For if a solution on the maximal interval of existence (t_-, t_+) remains in a compact subset of U as t approaches t_+ then it follows that $t_+ = \infty$. Of course analogous statements hold for the past direction.

Comparison principles for ODE are a very useful tool. A principle of this type will now be formulated for the scalar equation $dX/dt = F(t, X)$ where F is a continuous function defined on some open subset U of $\mathbb{R}^2$. Let $X(t)$ be the solution of this equation with $X(t_0) = X_0$ for some $(t_0, X_0) \in U$. Let Y be a C^1 function defined on the same interval as $X(t)$ with $dY/dt \leq F(t, Y)$ and $Y(t_0) \leq X_0$. Then $Y(t) \leq X(t)$ for all t in the interval with $t \geq t_0$. Replacing X and Y by $-X$ and $-Y$ gives a corresponding statement with the inequalities reversed. Replacing t by $-t$ gives analogous statements for the past time direction. These statements are a standard way of obtaining information on functions satisfying differential inequalities. If the equation obtained from the differential inequality by replacing $\leq$ by $=$ can be solved then an estimate is obtained. Differential inequalities to which this procedure can be applied come up regularly in existence and uniqueness proofs for ODE and PDE. There are also comparison principles for integral inequalities. Suppose that a function F is given as before but that now it is assumed that F is non-decreasing as a function of its second argument. If Y is continuous function which satisfies

$$Y(t) \le Y_0 + \int_{t_0}^{t} F(t, Y(s)) ds \qquad (5.2)$$

for some $Y_0 \le X_0$ and $X(t)$ is as before then $Y(t) \le X(t)$ for $t \ge t_0$.

A related result is Gronwall's inequality. Let u, a and b be non-negative continuous real-valued functions on an interval of real numbers. Assume that for $t > t_0$

$$u(t) \le a(t) + \int_{t_0}^{t} b(s) u(s) ds. \qquad (5.3)$$

Then

$$u(t) \le a(t) + \int_{t_0}^{t} a(s) b(s) \exp\left(\int_{s}^{t} b(\sigma) d\sigma\right) ds. \qquad (5.4)$$

These results say that under suitable circumstances a bound on a function in terms of itself implies an absolute bound on that function.

5.1 Existence and uniqueness

In this section the basic existence and uniqueness theorem for ordinary differential equations is stated and proved. The proof uses the method of successive approximations which is widely applicable in PDE theory as well.

Theorem 5.1 *Let f be a Lipschitz mapping from $[t_0, t_0 + a] \times B$ to B where B is the closed ball of radius b about $X_0 \in \mathbb{R}^n$. Let M be a bound for $|f|$ and $\alpha = \min(a, b/M)$. Then the equation (5.1) has a unique solution $X(t)$ on $[t_0, t_0 + \alpha]$ with $X(t_0) = X_0$.*

Proof Let an iteration be defined as follows. Let $X_0(t) = X_0$. Suppose that $X_k(t)$ has been defined on the interval $[t_0, t_0 + \alpha]$ and satisfies $|X_k(t) - X_0| \le b$ for $k \le m$. Define

$$X_{m+1}(t) = X_0 + \int_{t_0}^{t} f(s, X_m(s)) ds. \qquad (5.5)$$

It is clear that X_{m+1} is defined and continuous on $[t_0, t_0 + \alpha]$. Moreover

$$|X_{m+1}(t) - X_0| \le \int_{t_0}^{t} |f(s, X_m(s))| ds \le M\alpha \le b. \qquad (5.6)$$

Hence X_m can be defined iteratively for all m as continuous functions on $[t_0, t_0 + \alpha]$ and $|X_m(t) - X_0| \leq b$. Let K be a Lipschitz constant for f. It will be shown by induction that

$$|X_{m+1}(t) - X_m(t)| \leq \frac{MK^m(t-t_0)^{m+1}}{(m+1)!} \tag{5.7}$$

for all m and $t \in [t_0, t_0 + \alpha]$. That this inequality holds for $m = 0$ is obvious. Assuming that it holds up to $m - 1$ it follows from the relation

$$X_{m+1}(t) - X_m(t) = \int_{t_0}^{t} [f(s, X_m(s)) - f(s, X_{m-1}(s))], \tag{5.8}$$

which holds for $m \geq 1$ that

$$|X_{m+1}(t) - X_m(t)| \leq K \int_{t_0}^{t} |X_m(s) - X_{m-1}(s)| ds$$

$$\leq \frac{MK^m}{m!} \int_{t_0}^{t} (s - t_0)^m ds$$

$$= \frac{MK^m(t-t_0)^{m+1}}{(m+1)!}. \tag{5.9}$$

This estimate shows that $\{X_m\}$ is a Cauchy sequence in the Banach space $C([t_0, t_0 + \alpha], \mathbb{R}^n)$. Hence by completeness it converges uniformly to some $X(t)$. (Banach spaces and completeness are introduced in Section 6.1.) It is then possible to use the uniform convergence to pass to the limit in (5.5) to obtain

$$X(t) = X_0 + \int_{t_0}^{t} f(s, X(s)) ds. \tag{5.10}$$

It follows that $X(t)$ is a solution of the differential equation with the correct initial data. To see that the solution is unique note that any solution $Y(t)$ with the given initial condition satisfies

$$Y(t) = X_0 + \int_{t_0}^{t} f(s, Y(s)) ds. \tag{5.11}$$

By induction it can be shown that

$$|X_m(t) - Y(t)| \leq \frac{MK^m(t-t_0)^{m+1}}{(m+1)!}. \tag{5.12}$$

This proves that $Y(t) = X(t)$ and hence uniqueness. □

This theorem can be refined in various ways. If the function f is C^k then the solution is also C^k. The solution depends continuously on the initial data and the time at which data is prescribed and does so in a C^k manner if f does so. If the function f depends continuously on parameters, i.e. it is a continuous function $F(x, \lambda)$ where λ belongs to $\mathbb{R}^m$ then the solutions depend continuously on λ. If the function f depends smoothly on all its arguments including the parameters then the same is true of the solution.

5.2 Dynamical systems

Dynamical systems are autonomous systems of ordinary differential equations seen from a geometrical point of view. The quantities X_i in (5.1) are thought of as coordinates on a subset U of $\mathbb{R}^n$ and the functions f_i as the components of a vector field on U. The image of a solution of the system of ODE is a curve in space which is an integral curve of the given vector field. Prescribing initial data corresponds to requiring that the curve passes through a particular point of U at a certain time. If $X(t)$ is a global solution then a point X_* is said to be an ω-limit point of the solution if there is a sequence $\{t_n\}$ with $t_n \to \infty$ such that $X(t_n) \to X_*$. The notion of an α-limit point is obtained by replacing ∞ by $-\infty$ in this definition. The α- and ω-limit sets of the solution are the sets of α- and ω-limit points respectively. The dynamical systems picture helps to visualize the relation between different solutions. For example, if $X(t)$ is a solution with ω-limit point X_* then there is a solution $Y(t)$ passing through X_* which lies entirely in the ω-limit set of $X(t)$. This fact is often useful in determining the qualitative behaviour of solutions of a dynamical system. A special role is played by the zeroes of the vector field defining the dynamical system. They correspond to time-independent or stationary solutions. A key step in determining the qualitative behaviour of the solutions of a dynamical system is to find the stationary solutions.

In some cases the α- or ω-limit set of a solution is a stationary solution but there are also various other possibilities. It can, for instance, be a periodic solution. It can also be a network of connecting orbits between stationary solutions. A connecting orbit is one whose α- and ω-limit sets each consist of a single point which is then necessarily a stationary solution. If the limit sets are different the orbit is called heteroclinic. If they are the same it is called homoclinic. Much more complicated limit sets are also possible and often they can only be described to a limited extent. This kind of situation is associated with the terms 'chaos' and 'strange attractors'. The most famous strange attractor is the Lorenz attractor which occurs in a three-dimensional dynamical system. In two dimensions chaos is not possible. This is ensured by the Poincaré–Bendixson theory which implies that in two dimensions the only possible limit sets are given by stationary solutions, periodic solutions and networks of connecting orbits. The proof is based on the Jordan curve theorem. Another consequence of Poincaré–Bendixson theory is that for a two-dimensional dynamical system on a simply connected region each periodic solution must contain a stationary solution in its interior.

A powerful tool in the qualitative analysis of dynamical systems are monotone functions or, as they are also called, Lyapunov functions. A function F on U is called a Lyapunov function if for any solution $X(t)$ the function $F(X(t))$ is decreasing. In a region where a Lyapunov function exists there can be no stationary or periodic solutions and strange attractors are also excluded. Another useful statement is the monotonicity principle which will now be stated. Consider a C^1 dynamical system defined on an open subset of $\mathbb{R}^n$ and let S be a set which is invariant under the dynamics. In other words any solution which starts in S stays in S. Let F be a C^1 function defined on S which is a Lyapunov function for the given system with range contained in the interval (a, b), $a < b$. Here a and b are allowed to be infinite. Then the α-limit set of any solution is disjoint from the set of points in $X_0 \in \bar{S}$ such that $\lim_{X \to X_0} F$ is equal to a. The corresponding statement holds if the α-limit set is replaced by the ω-limit set and at the same time a is replaced by b.

5.3 Formal power series solutions and asymptotic expansions

In some ODE and PDE problems where it has not yet proved possible to obtain existence theorems it can be useful to consider formal power series satisfying the equation. These are not genuine solutions but they can contribute to the understanding of a particular problem. Formal power series solutions of ordinary differential equations are discussed in this section. It is important to realize that the notion of a formal power series solution can be defined in a way which is completely mathematically rigorous. In this section formal power series solutions are explained in the simple case of the scalar equation $dX/dt = f(X)$, where the function f is smooth.

Before coming to the formal power series it is useful to collect some ideas concerning asymptotic expansions. Let f be a function on an interval I of real numbers and let x_* be an endpoint of the interval, finite or infinite. If g is another function the statement that $f(x) = O(g(x))$ as $x \to x_*$ means that there is an open neighbourhood U of x_* and a constant C such that $|f(x)| \le C|g(x)|$ for all $x \in I \cap U$. The statement that $f(x) = o(g(x))$ as $x \to x_*$ means that given $\epsilon > 0$ there exists an open neighbourhood U of

5.3 Formal power series solutions and asymptotic expansions

x_* such that $|f(x)| \leq \epsilon |g(x)|$ for all $x \in I \cap U$. It follows from Taylor's theorem that if f is a C^k function which is defined in a neighbourhood of the origin then

$$f(x) = \sum_{j=0}^{k} f^{(j)}(0) x^j / j! + o(x^k), \tag{5.13}$$

where $f^{(j)}(x) = d^j f / dx^j$. This short statement sums up a lot of the content of Taylor's theorem in this case and shows how efficient the Landau notation using the order symbols O and o can be in encoding information. It is a simple example of an asymptotic expansion. It makes precise in what sense a function is approximated by its Taylor polynomial. If f is C^∞ it is convenient to write

$$f(x) \sim \sum_{j=0}^{\infty} \frac{f^{(j)}(0) x^j}{j!}. \tag{5.14}$$

This is to be interpreted as the statement that (5.13) holds for all values of k. Asymptotic relations may in general be integrated term by term, taking account of the constant of integration which may arise. In contrast, asymptotic relations may not in general be differentiated term by term.

The statement (5.14) can be expressed in words by saying that the function f is asymptotic to the series $\sum_{j=0}^{\infty} f^{(j)}(0) x^j / j!$. This carries no implication that the series converges. In fact it does not converge for a general smooth function. By definition the functions for which the series converges are those which are *analytic* in a neighbourhood of zero. The series here is a *formal series*. Strictly speaking it is just a sequence of coefficients $f^{(j)}(0)/j!$. Any sequence of coefficients a_j can be thought of as a formal series $\sum_{j=0}^{\infty} a_j x^j$. At this point the question arises whether any formal series is the sequence of Taylor coefficients of a smooth function. Borel's theorem is the statement that this is the case. It is possible to define various algebraic operations on formal power series. If $\{a_j\}$ and $\{b_j\}$ are formal series then their sum is the formal series $\{a_j + b_j\}$. This definition satisfies the natural condition that the jth coefficient of the formal Taylor series of the sum of two functions is equal to the sum of the formal Taylor series of the individual functions. The product of the formal series can be defined by the rule that the jth coefficient of the product is equal to $a_0 b_j + \cdots + a_j b_0$. Then the formal Taylor series of the product of two functions is equal to the product of the formal Taylor series of the individual functions. It is also possible to compose the formal series $\{b_j\}$ with the formal series $\{a_j\}$ provided that $a_0 = 0$. The composition $\{c_j\}$ is defined by giving the coefficient c_j as a suitable function of the coefficients a_k and b_k with $k \leq j$. What exactly this function is is uniquely determined by the compatibility condition that if $\{a_j\}$ is the formal Taylor series of f and $\{b_j\}$ is the formal Taylor series of g then $\{c_j\}$ is the formal Taylor series of $g \circ f$. It can thus be seen that there are natural algebraic operations on formal series which are compatible with operations on smooth functions.

Formal power series can also be differentiated. The derivative of $\sum_{j=0}^{\infty} a_j x^j$ is $\sum_{j=0}^{\infty} (j+1) a_{j+1} x^j$. Again the result is uniquely determined by the compatibility with Taylor series. When all these operations have been defined it is straightforward to say what it means to substitute a formal power series into an ordinary differential equation. It is just necessary to interpret all the relevant operations (addition, multiplication, composition and differentiation) in the way indicated above. The end result is a formal power series and if this last series is identically zero (i.e. all its coefficients are zero) then the original formal series is said to satisfy the ODE. For this kind of formal series it is also possible to prescribe initial data at the origin. This just means prescribing the coefficient with index zero. This discussion explains what it means to say that an ordinary differential equation has a formal power series solution with prescribed initial data. If there is a genuine solution with these initial data its Taylor series at the origin is a formal series solution. However the existence of a formal power series solution is no guarantee for the existence of a genuine solution – in general it is a weaker concept. Those formal power series $\{a_n\}$ which are the Taylor series of analytic functions defined on a neighbourhood of the origin can be characterized by the property that

$$|a_n| \leq C^{n+1} n! \tag{5.15}$$

for some positive constant C.

Various generalizations of the above are possible and some of them are of relevance for general relativity. The real-valued functions can be replaced by vector-valued ones. The interval of real numbers can be replaced by an open subset of $\mathbb{R}^n$ and ODE replaced by PDE. It is also possible to treat formal series more general than power series, for example series of the form

$$\sum_{j=0, l=0}^{\infty} a_{j,l} x^j (\log x)^l. \tag{5.16}$$

The theory of formal series has been presented here in a rather heuristic way. It really belongs to the realm of abstract algebra. The reader who wishes to know more is referred to [29].

5.4 Linearization and the Hartman–Grobman theorem

Let X_0 be a stationary solution of a dynamical system. The qualitative behaviour of solutions close to the stationary solution can be investigated by linearizing the system at X_0. What this means is the following. Let $X(t, \lambda)$ be a family of solutions depending on a parameter λ with $X(t, 0) = X_0$. Let $\tilde{X}$ be the derivative of this family with respect to λ, evaluated at $\lambda = 0$. By the chain rule it satisfies the linear equation $d\tilde{X}/dt = Df(X_0)\tilde{X}$. The linearization could have been defined directly by this last equation but the picture with the one-parameter family is very useful when describing linearizations in more complicated contexts for PDE and it is helpful to understand it in the simple ODE context.

A key question is when the qualitative behaviour of solutions of the nonlinear system close to the stationary solution is the same as that of solutions of the linearization at the stationary solution. An alternative way of considering the linearization is to say that it is obtained from the full system by discarding terms which are small, in fact those which are $O(|X - X_0|^2)$. This means passing to an approximation of the equations and the question is whether the solutions of the original system are also approximated by those of the simplified system. Criteria for this can be obtained by examining the eigenvalues of the matrix $Df(X_0)$ which also control the qualitative behaviour of solutions of the linearized system.

What is meant by saying that the qualitative behaviour is the same? The idea is that there should be a continuous mapping ϕ from a neighbourhood of X_0 to a neighbourhood of the origin in the tangent space at X_0 with a continuous inverse which transforms the nonlinear system into the linearized system in a sense which will now be explained. The mapping sending a time t and an initial datum X_0 to the value $\Phi(t, X_0)$ at time t of the solution which starts at the point X_0 when $t = 0$ is called the flow of the dynamical system. The mapping ϕ is required to satisfy the relation $\tilde{\Phi}(t, \phi(X)) = \phi(\Phi(t, X))$ where $\tilde{\Phi}$ is the flow corresponding to the linearized system. In this case the flows Φ and $\tilde{\Phi}$ are said to be topologically equivalent. If the mapping ϕ can be chosen so that it and its inverse are C^k or smooth then the flows are said to be C^k or smoothly equivalent respectively.

A fundamental theorem on the qualitative behaviour of solutions of a dynamical system is the Hartman–Grobman theorem. It says that if no eigenvalue of the linearization at X_0 is purely imaginary then in a neighbourhood of X_0 the flow of the system is topologically equivalent to that of the linearized system at X_0. In this case the stationary solution is said to be hyperbolic. The theorem does not state that an equivalence of class C^1 exists and the question of C^k equivalence is a complicated one. It is connected with the phenomenon of resonances. If X_0 is a hyperbolic stationary solution of a smooth dynamical system then $\mathbb{R}^n$ can be written as a direct sum of two subspaces E_+ and E_-, the unstable and stable subspaces, in such a way that these subspaces are invariant under $A = Df(X_0)$ and that the eigenvalues of the restrictions of A to E_+ and E_- have positive and negative real parts respectively. There are invariant submanifolds M_+ and M_- passing through X_0, the unstable and stable manifolds of X_0, whose tangent spaces at X_0 are E_+ and E_- respectively. These submanifolds are locally unique.

5.5 Examples (Bianchi models)

5.5.1 The Wainwright–Hsu system

In this section some examples of dynamical systems arising from Bianchi models are analysed. There is a very useful formulation of the Bianchi models of Class A due to Wainwright and Hsu which will now be presented. It applies to vacuum models and to the case of a perfect fluid which has a linear equation of state and which is untilted. The latter term means by definition that the four-velocity is orthogonal to the homogeneous hypersurfaces. In this case the energy density ρ of the fluid in the frame of the homogeneous hypersurfaces is equal to the energy density μ of the fluid in its rest frame. There are many other formulations of the same problem but this one is singled out by the fact that it has been used so successfully in proving rigorous results. An important feature of this formulation is the use of dimensionless variables. If a solution has a singularity at some time then in general many variables describing the solution go to infinity in the limit as the singularity is approached. On the other hand in a spacetime describing a phase of continuing cosmological expansion many variables go to zero in the limit of infinite expansion. It is much more useful when determining the asymptotic behaviour of solutions to use natural ratios of the basic variables which can be expected to remain bounded and bounded away from zero in the regimes just mentioned.

Consider a Bianchi model defined on $\mathbb{R} \times G$ where G is a simply connected three-dimensional Lie group. Let $\tilde{e}_i$ be a left-invariant frame on the group. The spatial metric can be described by its components in this basis which are functions of t. Alternatively linear combinations e_i of the $\tilde{e}_i$ with time-dependent coefficients can be chosen which form an orthonormal basis. Adjoining a vector field e_0 defined by $\partial/\partial t$ completes the spatial frame to an orthonormal frame in spacetime. Suppose that the Lie algebra of G is of Class A which means by definition that the trace C^i_{ij} of the structure constants is zero. A Lie group of any dimension with this property

5.5 Examples (Bianchi models)

is called unimodular. The rotation coefficients of the frame are defined by (2.61). Let σ_j^i be the tracefree part of γ_{0j}^i. In a Class A Bianchi model it is possible to write γ_{jk}^i in the form $\epsilon_{jkl}n^{il}$ for a symmetric matrix n^{ij} where ϵ_{ijk} is a three-form with $\epsilon_{123} = 1$. In vacuum Class A Bianchi models the components n^{ij} can be assumed diagonal. The different Bianchi types are distinguished by how many diagonal elements are positive, negative or zero. The components σ_j^i can also be assumed diagonal. The same statements hold for Bianchi models with a perfect fluid provided the fluid is non-tilted. Define

$$\sigma_+ = \frac{1}{2}(\sigma_{22} + \sigma_{33}), \quad \sigma_- = \frac{1}{2\sqrt{3}}(\sigma_{22} - \sigma_{33}) \tag{5.17}$$

and $n_i = n_{ii}$ (no sum) for $i = 1, 2, 3$. The field equations can be expressed entirely in terms of the variables $(\sigma_+, \sigma_-, n_1, n_2, n_3)$, the Hubble parameter $H = -\frac{1}{3}\mathrm{tr}k$ and matter variables, if any. Dimensionless variables are introduced by

$$\Sigma_\pm = \sigma_\pm/H, \quad N_i = n_i/H. \tag{5.18}$$

A new time variable τ is defined in terms of the Gaussian time t by $dt/d\tau = H^{-1}$. Then, with $d/d\tau$ denoted by a prime, the following system is obtained:

$$N_1' = (q - 4\Sigma_+)N_1, \tag{5.19}$$

$$N_2' = (q + 2\Sigma_+ + 2\sqrt{3}\Sigma_-)N_2, \tag{5.20}$$

$$N_3' = (q + 2\Sigma_+ - 2\sqrt{3}\Sigma_-)N_3, \tag{5.21}$$

$$\Sigma_+' = -(2-q)\Sigma_+ - S_+, \tag{5.22}$$

$$\Sigma_-' = -(2-q)\Sigma_- - S_-, \tag{5.23}$$

where

$$q = 2(\Sigma_+^2 + \Sigma_-^2) + \frac{1}{2}(3w+1)\Omega, \tag{5.24}$$

$$S_+ = \frac{1}{6}[(N_2 - N_3)^2 - N_1(2N_1 - N_2 - N_3)], \tag{5.25}$$

$$S_- = \frac{1}{2\sqrt{3}}(N_3 - N_2)(N_1 - N_2 - N_3). \tag{5.26}$$

By definition $\Omega = \rho/3H^2$ and the equation of state of the fluid is $p = w\mu$. The vacuum equations are defined by the constraint

$$\Sigma_+^2 + \Sigma_-^2 + K = 1, \tag{5.27}$$

where

$$K = \frac{1}{12}[N_1^2 + N_2^2 + N_3^2 - 2(N_1 N_2 + N_2 N_3 + N_3 N_1)]. \quad (5.28)$$

In the case with matter $\Omega = 1 - \Sigma_+^2 - \Sigma_-^2 - K$. The Hubble parameter has dropped out of the system of evolution equations completely. If desired it can be computed at the end of the day by solving the equation

$$dH/d\tau = -(1+q)H. \quad (5.29)$$

for H and then reconstructing t. This approach has the attractive feature that all the Bianchi types of Class A are contained in a single dynamical system, being distinguished by the signs of the N_i.

Bianchi type I corresponds to the vanishing of all N_i. The vacuum subset is the circle defined by the equation $\Sigma_+^2 + \Sigma_-^2 = 1$. It is known as the Kasner circle since the vacuum solutions of Bianchi type I are well-known explicit solutions called the Kasner solutions. In the dynamical systems picture the whole Kasner circle consists of stationary solutions. This corresponds to the fact that the Kasner solutions are self-similar. This means that these solutions admit a homothetic vector field. The explicit form of the Kasner solutions is

$$-dt^2 + t^{2p_1}dx^2 + t^{2p_2}dy^2 + t^{2p_3}dz^2. \quad (5.30)$$

The constants p_i are called Kasner exponents and they satisfy the two Kasner relations given by $p_1+p_2+p_3 = 1$ and $p_1^2+p_2^2+p_3^2 = 1$. There are two special solutions with a four-dimensional symmetry group which because of their one-dimensional isotropy group are called locally rotationally symmetric (LRS). Their Kasner exponents are $(-\frac{1}{3}, \frac{2}{3}, \frac{2}{3})$ and $(1, 0, 0)$ or permutations of these. The second solution is actually flat. A useful concept for more general spacetimes is that of *generalized Kasner exponents*. Given a spacetime with a foliation by spacelike surfaces let λ_i be the eigenvalues of the second fundamental form of one of these surfaces with respect to the induced metric. (Cf. Section 8.3 for this terminology.) These are functions on spacetime. Suppose that the mean curvature $\mathrm{tr} k = \lambda_1 + \lambda_2 + \lambda_3$ is everywhere nonzero and define $p_i = \lambda_i/\mathrm{tr} k$. These quantities automatically satisfy the first Kasner relation but in general not the second. In the case of the Kasner spacetime itself foliated by the homogeneous hypersurfaces they are constant and coincide with the Kasner exponents, hence their name.

Non-vacuum perfect fluid solutions are contained in the interior of the Kasner circle due to the positivity of the energy density. The radial lines in the (Σ_+, Σ_-)-plane are invariant and Ω is monotone increasing. Using these facts it is elementary to read off the main qualitative properties of the dynamics. The origin is a stationary point and corresponds to a spatially flat FLRW model. All other solutions move in a monotone fashion from the Kasner circle to the origin as τ goes from $-\infty$ to ∞. In terms of a spacetime picture $\tau \to -\infty$ corresponds to the big bang and $\tau \to \infty$ to unlimited expansion. Roughly speaking, the solution looks like a Kasner solution near the singularity and becomes isotropic in the distant future.

5.5.2 Models of Bianchi types II and VI$_0$

Next the dynamics of Bianchi type II vacuum solutions will be analysed. In the dynamical systems picture the Bianchi type II solutions are characterized by the fact that two of the N_i are zero and one is non-zero. For definiteness the choice $(+, 0, 0)$ will be made. Other choices are related to this one by symmetry transformations of the system and so are not essentially different. With this choice the state space of the vacuum Bianchi type II solutions is a half-ellipsoid whose boundary coincides with the Kasner circle. The closure of the Bianchi type II subset is compact. The projection of each solution to the (Σ_+, Σ_-)-plane lies on a straight line and there are no stationary points. Thus the qualitative nature of the dynamics can easily be determined. Any solution moves from one point of the Kasner circle over the top of the ellipsoid to another point of the Kasner circle in such a way that its projection moves on a straight line. These straight lines can be described explicitly as follows. Consider the unique equilateral triangle symmetric under reflection in the Σ_+ axis in which the Kasner circle is inscribed (cf. Fig. 5.1). Then consider the straight lines through the vertex of the triangle which lies on the Σ_+ axis. A line of this kind which intersects the Kasner circle in two points is the projection of a Bianchi type II solution. This family of straight lines is associated to the choice $N_1 \neq 0$. Choosing instead N_2 or N_3 to be non-zero leads to other families of straight lines related to the first by a rotation of $\pm 2\pi/3$. The reason for describing this geometry in some detail is that it is believed to play a fundamental role in the dynamics of more general families of spacetimes (even inhomogeneous ones) near the initial singularity.

In the examples up to now conserved quantities could be used to reduce the problem to a dynamical system of dimension one and so general tools of the theory of dynamical systems were unnecessary. A more challenging example is given by the untilted perfect fluid spacetimes of Bianchi type II. Consider a perfect fluid with equation of state $p = w\mu$ and $0 \leq w < 1$. This is a three-dimensional dynamical system. It has exactly one non-vacuum stationary point which corresponds to a self-similar spacetime, the Collins–Stewart solution. Calculating the eigenvalues of the linearization at that point shows that they all have negative real part. Thus this stationary

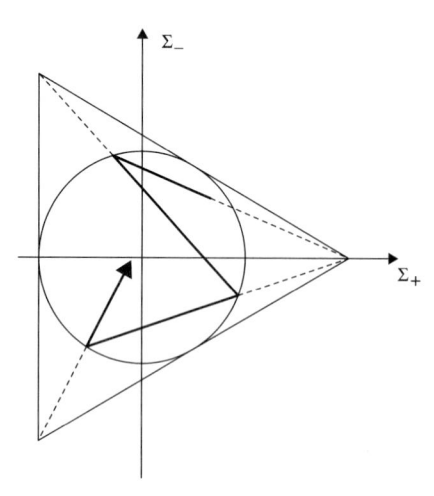

Figure 5.1 *The Kasner circle and the BKL map*

point is hyperbolic and by the Hartman–Grobman theorem all solutions which start near that point converge to it as $\tau \to \infty$. Information about the global structure of the dynamical system can be obtained by using a monotone function which is given on p. 151 of [203]. The explicit form of this montone function F is rather complicated and not surprisingly it was not found by accident – it was derived in the context of the Hamiltonian formulation of the equations. It is strictly monotone unless $\Sigma_- = 0$ and Σ_+ takes a particular value. A closer examination shows that in the latter case $\Sigma'_+ \neq 0$ except at the stationary point. It follows that except for the stationary solution the function F is monotone increasing. It vanishes on the boundary of the Bianchi type II state space. Choosing S to be this state space with the stationary solution removed, the monotonicity principle shows that the ω-limit set of any solution is the stationary point.

Next consider the dynamics of these models in the past time direction. An important role is played by the LRS solutions which are defined by the condition $\Sigma_- = 0$. For this system Σ_-^2 is strictly decreasing on the Bianchi II phase space except where it is zero. It follows that if a solution is not LRS its α-limit set must be contained in the Kasner circle. Because the Kasner circle itself consists of stationary solutions the linearization of the system about a point of the Kasner circle must have at least one zero eigenvalue and thus fail to be hyperbolic. The Hartman–Grobman theorem is not helpful in obtaining information about the local structure of the system near those points. This problem will be examined again in Section 5.7.

The LRS solutions define a two-dimensional dynamical system where the only stationary point in the interior is the Collins–Stewart solution.

On the boundary there are three stationary points, which are the two LRS Kasner solutions with $\Sigma_- = 0$ and the origin, corresponding to the spatially flat FLRW solution. The origin is hyperbolic, having one positive and one negative eigenvalue. Hence there is a one-dimensional unstable manifold and a solution which tends to the FLRW solution as $\tau \to -\infty$. In terms of the corresponding spacetime this means that the geometry becomes more and more isotropic as the singularity is approached. This kind of solution is said to have an isotropic singularity. The monotone function can be used to rule out periodic solutions. By the Poincaré–Bendixson theorem the α-limit set of any LRS solution must thus be a stationary point or a network of connecting orbits. It is in fact elementary to rule out the latter possibility and, apart from the exceptional solutions already mentioned, the α-limit set is always the Kasner solution with exponents $\frac{2}{3}, \frac{2}{3}, -\frac{1}{3}$.

The spacetimes of Bianchi type VI_0, where the signs of the N_i are $(+, -, 0)$ or a permutation thereof, are more complicated to analyse than those of type II. In the vacuum case the dynamical system is three-dimensional. Its boundary consists of the vacuum solutions of types I and II. The coordinate Σ_+ is strictly decreasing on the subsets corresponding to Bianchi VI_0 and Bianchi II. The monotonicity principle can be used to show that both the α- and ω-limit sets must lie in the Kasner circle and that they cannot contain the points with Kasner exponents $(1, 0, 0)$ and $(-\frac{1}{3}, \frac{2}{3}, \frac{2}{3})$ respectively. Consider now the Bianchi type VI_0 solutions with perfect fluid. There is a unique non-vacuum stationary point, the Collins solution. Explicit formulae for this solution can be found on p. 131 of [203]. It is hyperbolic with all eigenvalues having negative real parts. It can be shown using a monotone function that it is the ω-limit point of all solutions. It seems that the α-limit sets of general solutions have not been determined rigorously although there is a conjecture concerning them in [203].

5.6 Centre manifolds and the reduction theorem

Consider now a general stationary solution X_0 of a dynamical system. The decomposition of $\mathbb{R}^n$ associated with the linearization of a dynamical system at a hyperbolic stationary solution can be generalized to a decomposition into invariant subspaces E_+, E_c and E_- with the property that the restrictions of the linearization to these subspaces have eigenvalues whose real parts are positive, zero and negative respectively. The subspace E_c is called the centre subspace. As in the hyperbolic case there are unstable and stable manifolds which are tangent to E_+ and E_- respectively and locally unique. The centre manifold theorem says that there also exists a centre manifold which is invariant and whose tangent space at X_0 is E_c. In contrast to the stable and unstable manifolds the centre manifold need not be unique and it need not inherit the full regularity of the dynamical system. In particular it may happen for a stationary solution of a smooth dynamical system that there exists no smooth centre manifold although for any finite k there is a centre manifold which is C^k.

A further result connected with the centre manifold theorem is the reduction theorem. Before stating it, it is necessary to introduce the notion of a standard saddle. Consider a linear dynamical system $dX/dt = AX$ of dimension n for a matrix A which is hyperbolic. Thus there are k eigenvalues with positive real part and l eigenvalues with negative real part, where $k+l = n$. It can be shown that all dynamical systems of this type with fixed values of k and l are topologically equivalent. Any dynamical system which is locally topologically equivalent to a system of this type is called a standard saddle. The case where A is diagonal with k entries $+1$ and l entries -1 can be taken as a model for the standard saddles. It follows that a dynamical system is topologically equivalent to a standard saddle near any hyperbolic stationary solution. The reduction theorem generalizes this statement to non-hyperbolic stationary solutions. It says that near any stationary solution a dynamical system is locally topologically equivalent to the product of its restriction to any centre manifold and a standard saddle. This means that all the topological complications of a dynamical system near a stationary solution are already present on the centre manifold. It also means that the dynamical systems on different centre manifolds belonging to a given stationary solution are locally topologically equivalent to each other.

Suppose that the qualitative behaviour of solutions of a dynamical system near a non-hyperbolic stationary solution is to be determined. Then the reduction theorem shows that it is enough to determine the qualitative behaviour on any centre manifold. Since it is usually impossible to determine the centre manifold itself exactly this may not seem very useful. It turns out, however, that in many cases it is enough to obtain certain approximations to the centre manifold in order to identify the qualitative behaviour. This will be shown by example in Section 5.7. One particularly favourable case is that where the centre manifold is one-dimensional since the analysis of the qualitative behaviour of a one-dimensional dynamical system is very simple.

5.7 Further examples

5.7.1 Bianchi types II and VI$_0$ revisited

It will now be shown how the analysis of the non-LRS Bianchi type II solutions begun in Section 5.5 can be completed using the reduction theorem. Consider first a point of the Kasner circle on the longer open arc of the circle joining the point with Kasner exponents $(0, 1, 0)$ to that with exponents $(0, 0, 1)$. At each of the points on this arc the linearization has a two-dimensional unstable subspace. By the reduction theorem there is an open neighbourhood U such that any solution which intersects U converges to a point of the Kasner circle as $\tau \to -\infty$. It follows that if a point on this part of the Kasner circle lies in the α-limit set of a solution it is the whole limit set. Next consider a point on the shorter open arc of the Kasner circle joining the point with Kasner exponents $(0, 1, 0)$ to that with exponents $(0, 0, 1)$. The linearization at a point of this type has one positive and one negative eigenvalue. Its unstable manifold lies in the Bianchi I subset. It follows from the reduction theorem that if a point like this were in the α-limit set of a solution then there would also be vacuum Bianchi II points in that set. This contradicts what was already said in Section 5.5. Consider finally the points with Kasner exponents $(0, 1, 0)$ and $(0, 0, 1)$. Each of these has a two-dimensional centre manifold and it can be chosen to be the vacuum Bianchi II set. From what is known about the latter set it follows that apoint of this type cannot be an α-limit point. Putting all these facts together shows that the α-limit set of a Bianchi type II solution of the Einstein equations coupled to a perfect fluid with linear equation of state is always a point of the Kasner circle. Moreover this point must belong to the open arc covering two thirds of the Kasner circle mentioned above. By similar arguments it can be shown that the α-limit set of a vacuum spacetime of type VI_0 is a point of the complementary one third of the Kasner circle.

The solutions of type VI_0 do at least have the advantage that the closure of the phase space is compact. For the remaining types of Class A the phase space is non-compact and this opens up a new class of difficulties. In general very specific and complicated estimates are required. The dynamics is also genuinely more complicated than in the cases of types II and VI_0. One comforting fact is that for any Bianchi type of Class A with perfect fluid and linear equation of state there is a global monotone function. This means in particular that near the initial singularity the solution of the dynamical system approaches the sets representing simpler Bianchi types and this provides a framework for understanding the asymptotics. For instance the function $N_1 N_2 N_3$ is monotone for solutions of types VIII and IX. In the case of solutions of Bianchi Class B with perfect fluid no monotone function is known and so it is not a priori ruled out that there could be strange attractors, although there are no indications of this from numerical investigations.

This subsection will be closed by some remarks concerning more general solutions of the Wainwright–Hsu system. There is an intuitive picture of the dynamics of the most general solutions of the Wainwright–Hsu system for $\tau \to -\infty$ which has been partly confirmed by analytic results. This says that near the initial singularity a solution of Bianchi type VIII or IX follows a sequence of Bianchi II vacuum solutions. Following Bianchi type II solutions towards the past defines a mapping of the Kasner circle with the flat points removed to the Kasner circle. It is called the BKL map after Belinskii, Khalatnikov and Lifshitz and is illustrated in Fig. 5.1. It has been proved to have chaotic properties. Whether and in what sense this can be used to show that the Bianchi VIII or Bianchi IX dynamical systems, which are sometimes known as mixmaster solutions, themselves have chaotic properties is an open problem. It is known that the Bianchi IX solutions remain in a compact set and converge to the closure of the set of Bianchi II solutions as $\tau \to -\infty$. The corresponding statement for Bianchi VIII has not been proved. The most general solutions of the system corresponding to models which expand for ever are those of type VIII. They go to infinity in the space of Wainwright–Hsu variables as $\tau \to \infty$ and quite a lot of details are known concerning their asymptotics.

In the above discussion the equation of state with $w = 1$, known as stiff fluid, was excluded since in that case the solutions show qualitatively different behaviour. In the homogeneous case the stiff fluid is equivalent to a massless scalar field. For $w = 1$ the entire Bianchi type I subset of the space of Wainwright–Hsu variables consists of stationary solutions. Thus the Kasner circle is extended to a disc of stationary solutions. It turns out that all untilted perfect fluid solutions of Bianchi Class A with $w = 1$ converge to a point of the Bianchi I disc as $\tau \to -\infty$. Moreover, generic solutions converge to some point of the interior of the triangle defined by the three points representing the flat Kasner solutions in this limit.

5.7.2 The massive scalar field

In the above examples the centre manifold was explicit. Next it will be illustrated how a centre manifold which is only known approximately can be used to get information about solutions of a dynamical system. This example comes up in the study of spatially flat FLRW spacetimes with a massive scalar field as matter model. The dynamical system is

$$x_\eta = y, \qquad (5.31)$$

$$y_\eta = -x - 3y(x^2 + y^2)^{1/2}. \qquad (5.32)$$

It is taken from [23]. This is a dynamical system on the plane whose coefficients are smooth everywhere except at the origin, where they are C^1. Transforming to polar coordinates (r, θ) and introducing $\rho = r/(1 + r)$

5.7 Further examples

gives the system

$$\rho_\eta = -3\rho^2 \sin^2\theta, \tag{5.33}$$

$$\theta_\eta = -1 - 3\rho(1-\rho)^{-1} \sin\theta \cos\theta. \tag{5.34}$$

Introducing a new time coordinate τ satisfying $d\tau/d\eta = (1-\rho)^{-1}$ and setting $u = 1 - \rho$ gives

$$u_\tau = 3u(1-u)^2 \sin^2\theta, \tag{5.35}$$

$$\theta_\tau = -u - 3(1-u)\sin\theta \cos\theta. \tag{5.36}$$

Now the stationary point at the origin will be investigated. The eigenvalues of the linearization are -3 and zero. Evidently the θ-axis is invariant and in fact it is the stable manifold of the origin. The centre subspace is given by $u + 3\theta = 0$ and thus it is useful to introduce $v = u + 3\theta$ in order to study the centre manifold. This leads to the transformed system

$$u_\tau = 3u(1-u)^2 \sin^2\left(\frac{1}{3}(v-u)\right), \tag{5.37}$$

$$v_\tau = -3u\cos^2\left(\frac{1}{3}(v-u)\right) - 6u^2 \sin^2\left(\frac{1}{3}(v-u)\right) + 3u^3 \sin^2\left(\frac{1}{3}(v-u)\right)$$
$$-9\sin\left(\frac{1}{3}(v-u)\right)\cos\left(\frac{1}{3}(v-u)\right) + 9u\sin\left(\frac{1}{3}(v-u)\right)\cos\left(\frac{1}{3}(v-u)\right). \tag{5.38}$$

The centre manifold is of the form $v = \phi(u)$. Substituting this into the equations shows that $v_\tau = \phi'(u)u_\tau$. The right hand side vanishes to at least third order in u. On the other hand, on the centre manifold $v_\tau = -3(\phi(u)+u^2)+O(u^3)$. Comparing these two expressions shows that $\phi(u) = -u^2 + O(u^3)$. On the centre manifold $u_\tau = \frac{1}{3}u^3 + O(u^4)$ and hence there is a unique solution which enters the physical region.

The asymptotic form of the solution as $\tau \to -\infty$ is $u = \sqrt{3/2}(-\tau)^{-1/2} + \cdots$. It follows that $v = \frac{3}{2}\tau^{-1} + \cdots$. Hence

$$\rho = 1 - \sqrt{3/2}(-\tau)^{-1/2} + \cdots \tag{5.39}$$

and

$$\theta = -\sqrt{1/6}(-\tau)^{-1/2} + \cdots \tag{5.40}$$

The next step is to convert back to the original variables. To leading order $\eta = -\sqrt{6}(-\tau)^{1/2} + \cdots$ and so $\rho = 1 + 3/\eta + \cdots$ and $\theta = 1/\eta + \cdots$. It follows that $x = -\eta/3 + \cdots$ and $y = -\frac{1}{3} + \cdots$. The interpretation of these variables is that η, x and r are proportional to proper time, the scalar field ϕ and the mean curvature of the hypersurfaces of constant time respectively. Thus in the limit $t \to -\infty$ both ϕ and the mean curvature are proportional to t. The leading order behaviour of the scale factor for large negative times is, up to inessential constants, e^{t^2}. The Hubble parameter is decreasing but the expansion is accelerated.

5.7.3 Bianchi type III Einstein–Vlasov

Next a more complicated example of a centre manifold calculation will be carried out. The system to be studied describes an LRS Bianchi type III solution of the Einstein–Vlasov system with massive particles. When expressed in certain variables it reads

$$\dot{\Sigma}_+ = -(2-q)\Sigma_+ + M_3 + \Omega R_+ , \tag{5.41}$$

$$\dot{s} = 6s(1-s)\Sigma_+ , \tag{5.42}$$

$$\dot{z} = 2z(1-z)(1+\Sigma_+ - 3\Sigma_+ s) , \tag{5.43}$$

$$\dot{M}_3 = 2(q-\Sigma_+)M_3 , \tag{5.44}$$

where

$$\Omega = 1 - \Sigma_+^2 - M_3, \tag{5.45}$$

$$q = 2\Sigma_+^2 + \frac{1}{2}\Omega(1+R). \tag{5.46}$$

The functions R and R_+, which are the matter contributions coming from the Vlasov equation, are smooth functions of s and z which are not known explicitly. For the analysis only certain qualitative properties of these functions need to be known. The state space of the dynamical system is defined by the conditions that $s \in (0,1)$, $z \in (0,1)$, $M_3 > 0$ and $\Omega > 0$. It is assumed that the functions R and R_+, and hence the coefficients of the dynamical system, extend smoothly to the closure of the region where they were originally defined. In addition the following further assumptions on R and R_+ are made. Firstly $0 \le R \le 1$ with $R = 0$ only when $z = 1$ and $R = 1$ only when $z = 0$. Secondly $-R \le R_+ \le \frac{1}{2}R$ with $R_+ = \frac{1}{2}R$ for $s = 0$ and $R_+ = -R$ for $s = 1$.

The point $(\frac{1}{2}, 1, z_0, \frac{3}{4})$ is a stationary point for any $z_0 \in [0,1]$. Let L denote the line consisting of these points. The aim here is to determine the behaviour of solutions close to the point $(\frac{1}{2}, 1, 1, \frac{3}{4})$ in the future time direction. At this point the linearization of the dynamical system has two negative eigenvalues and two zero eigenvalues. One of the zero eigenvalues is an inevitable consequence of the presence of a line of stationary solutions

5.7 Further examples

but the other is an additional degeneracy. Now new coordinates will be introduced in order to simplify the analysis. First the stationary point with $z_0 = 1$ will be translated to the origin. Define $x = \Sigma_+ - \frac{1}{2}$, $y = 1 - s$, $w = M_3 - \frac{3}{4}$, $\tilde{z} = z - 1$. Then the transformed system is

$$\dot{x} = -\frac{1}{4} + q\left(\frac{1}{2} + x\right) - 2x + w + \Omega R_+, \tag{5.47}$$

$$\dot{y} = -6y(1-y)\left(\frac{1}{2} + x\right), \tag{5.48}$$

$$\dot{\tilde{z}} = -2(1+\tilde{z})\tilde{z}\left(-2x + \frac{3}{2}y + 3xy\right), \tag{5.49}$$

$$\dot{w} = 2\left(q - x - \frac{1}{2}\right)\left(w + \frac{3}{4}\right), \tag{5.50}$$

$$q = \frac{1}{2} + 2x + 2x^2 + \frac{1}{2}\Omega(1+R), \tag{5.51}$$

$$\Omega = -x - w - x^2. \tag{5.52}$$

Next set $Z = -\tilde{z}$, $u = x + w$ and $v = x - w$. In order to determine the behaviour of solutions of the resulting system near the origin the linearization will be examined. The centre manifold is two-dimensional. A general property of centre manifolds is that a stationary point has an open neighbourhood such that any other stationary point in that neighourhood lies on any centre manifold of the original point. It follows that in a neighbourhood of the stationary point of the system under consideration the part of the line L close to that point lies on any centre manifold. In the case of present interest the centre manifold can be defined by the equations $y = \phi(u, Z)$ and $v = \psi(u, Z)$ where ϕ and ψ vanish at least quadratically at the origin. The invariance of the centre manifold implies that

$$\dot{y} = (\partial\phi/\partial u)\dot{u} + (\partial\phi/\partial Z)\dot{Z}, \tag{5.53}$$

$$\dot{v} = (\partial\psi/\partial u)\dot{u} + (\partial\psi/\partial Z)\dot{Z}. \tag{5.54}$$

on the centre manifold. Note that $R + R_+ = O(y)$ since this function is smooth and vanishes at $y = 0$. Hence the right hand sides of these equations vanish to at least third order at the origin. On the centre manifold

$$\dot{y} = -3\phi - 3\phi(u + \psi) + 3\phi^2(1 + u + \psi). \tag{5.55}$$

Substituting (5.55) into (5.53) shows that ϕ vanishes at quadratic order. Putting this information back into the equation (5.53) shows that ϕ vanishes at third order. This can be repeated indefinitely, with the result that ϕ

vanishes to all orders. On the centre manifold

$$\dot{v} = -\frac{3}{2}\psi - \frac{3}{8}u^2 + \frac{1}{4}u\psi + \frac{5}{8}\psi^2 - \frac{1}{4}u^3 + \frac{1}{4}u^2\psi + \frac{5}{4}u\psi^2 + \frac{3}{4}\psi^3$$
$$+ \Omega\left(R_+ - \frac{1}{2}R\right) + \frac{1}{4}\Omega(1+R)(3\psi - u). \tag{5.56}$$

Since ϕ vanishes to all orders it follows from the fact that $R + R_+ = O(y)$ that R_+ and $-R$ are equal to all orders on the centre manifold. Hence to quadratic order $\psi(u, Z) = -\frac{1}{12}u^2 - r_2 uZ + \ldots$ where r_2 is the derivative of R with respect to z, evaluated at the stationary point of interest. Next the dynamics on the centre manifold will be examined.

$$\dot{u} = u^2 + \frac{11}{4}u\psi + \frac{3}{4}\psi^2 + \frac{9}{16}u^3 + \frac{15}{16}u^2\psi + \frac{3}{16}u\psi^2 - \frac{3}{16}\psi^3$$
$$+ \Omega\left[(R + R_+) + \frac{1}{4}R(3u - \psi)\right], \tag{5.57}$$

while

$$\dot{Z} = 2uZ - 2uZ^2 + 2Z\psi - 2Z^2\psi - 3Z(1-Z)\phi(1+u+\psi). \tag{5.58}$$

Discarding terms of order greater than two gives the truncated system

$$\dot{u} = u^2, \tag{5.59}$$
$$\dot{Z} = 2uZ. \tag{5.60}$$

For this system $u^{-2}Z$ is a conserved quantity and so the qualitative behaviour of the solutions is easily determined. Is the qualitative behaviour of solutions of the full system on the centre manifold similar? The terms on the right hand side of the restriction of the evolution equations to the centre manifold written out explicitly above contain a factor of u. This is not an accident and in fact the full equations are of the form

$$\dot{u} = uf(u, Z), \tag{5.61}$$
$$\dot{Z} = ug(u, Z) \tag{5.62}$$

for some functions f and g of any arbitrary finite degree of differentiability. This is because of the fact, already mentioned above, that the line L of stationary points is locally contained in the centre manifold. Thus the restriction of the system to the centre manifold has a line of stationary points with $u = 0$. This means that the vector field defining the dynamical system vanishes for $u = 0$. The existence of the functions f and g then follows from Taylor's theorem. The fact that L lies on the centre manifold also implies that $\psi(u, Z) = uh(u, Z)$ for a function h of arbitrary finite differentiability.

It is possible to get more precise information about the function f as follows. Recall that $R+R_+$ vanishes modulo the function ϕ. Since ϕ vanishes to all orders it follows that expressions containing a factor $R + R_+$ can be ignored in all perturbative calculations. Hence

$$\Omega = -u - \frac{1}{4}u^2 - \frac{1}{2}u\psi - \frac{1}{4}\psi^2. \tag{5.63}$$

It follows that every term in the evolution equation for u on the centre manifold contains a factor of u^2, either directly or via the fact that $\psi = uh$. Hence $f(0, Z) = 0$.

For the purpose of qualitative analysis the above system can be replaced by the system

$$u' = f(u, Z) + \cdots, \tag{5.64}$$
$$Z' = g(u, Z) + \cdots, \tag{5.65}$$

which has the same integral curves away from the Z-axis. Only solutions with $Z \geq 0$ are physical. The Z-axis is an invariant manifold of the rescaled dynamical system. This new system has a hyperbolic stationary point at the origin which is a source. This makes it easy to determine the phase portrait for the system near the endpoint of L at $z = 1$. What can be seen is that there is an open set of initial data in the physical region such that the corresponding solutions approach the stationary point with $z = 1$ as $\tau \to \infty$ and there is a neighbourhood of the endpoint of L at $z = 1$ such that in that neighbourhood no solution in the physical region approaches a point of L with $z < 1$.

5.8 Bifurcation theory

When studying a dynamical system it may be very difficult to directly determine its qualitative features such as the presence of stationary or periodic solutions. A method which can sometimes be used to obtain information about these is to embed the system of interest in a family of systems depending on one or more parameters. Here the discussion is restricted to the case of a single parameter λ. This area is known as bifurcation theory because it may happen that there is a qualitative change of the dynamics as some parameter value is passed. For instance a stationary point may give rise to two stationary points which move away from it as λ is varied. In a pictorial representation of this situation the set of stationary points has a bifurcation in a concrete sense.

One of the most famous types of bifurcation is the Hopf bifurcation which can be used to prove the existence of periodic solutions of a dynamical system under certain circumstances. Consider a one parameter family of dynamical systems $dX^i/dt = f(X^i, \lambda)$ which has a stationary point at the origin when $\lambda = 0$. Suppose further that $Df(0,0)$ has a pair of purely imaginary eigenvalues which are non-zero and all other eigenvalues have non-zero real parts. Then it can be shown that there is a smooth family of stationary points depending on λ near zero which passes through the origin at $\lambda = 0$. Now restrict to the case that the derivative d of the real parts of the eigenvalues which become purely imaginary at $\lambda = 0$ is non-zero at $\lambda = 0$. Under the assumption that one further number σ (the Liapunov number) is non-zero the qualitative nature of the dynamics near the origin is determined for λ small. On one side of $\lambda = 0$ there is a one-parameter family of periodic solutions (one for each value of λ) which converge to the origin as $\lambda \to 0$. Which side this is, and whether the periodic solutions are stable or unstable, is determined by the signs of d and σ. This can be made more concrete in the two dimensional case by the statement that the dynamical system is topologically equivalent to a normal form which looks particularly simple in polar coordinates. The explicit formula is

$$\dot{r} = r(\lambda + \sigma r^2), \quad \dot{\theta} = 1. \tag{5.66}$$

Here the coefficient σ can be taken to be ± 1. Of course, due to the reduction theorem, this also gives full information about the qualitative behaviour in higher dimensional cases.

It seems that there are solutions of the Einstein equations exhibiting Hopf bifurcations although this has not been confirmed in detail. The examples are tilted spatially homogeneous solutions of the Einstein–Euler equations of Bianchi class B. It is apparently easier to use other methods to prove the existence of the periodic solutions associated to this bifurcation. It should be noted that the periodicity here is obtained for self-similar variables analogous to the Wainwright–Hsu ones and so the corresponding spacetimes are discretely self-similar, i.e. there is an action of the group of integers by homotheties.

5.9 Global existence for homogeneous spacetimes

This section presents a global existence theorem for a rather general class of spatially homogeneous solutions of the Einstein–matter system. Here the spatial dimension is arbitrary. Consider those spatially homogeneous spacetimes defined by one-parameter families of left invariant metrics on a Lie group G. In the three-dimensional case these are the Bianchi models. Suppose that a solution of this type of the Einstein–matter equations is defined on an interval (t_1, t_2) of a Gaussian time coordinate t based on a homogeneous hypersurface, where t_1 and t_2 are allowed to be infinite. The metric at each time can be identified with a left invariant metric on the Lie group G. By passing to the universal cover if necessary it may be assumed that G is simply connected. This does not change the dynamics of the solutions.

Definition 5.2 *A matter model satisfies the matter continuation condition (MCC) if a spatially homogeneous solution of the corresponding Einstein–matter equations defined on an interval (t_1, t_2) with the mean curvature bounded in a neighbourhood of the finite time t_1 or t_2 can be extended to a neighbourhood of t_1 or t_2 respectively.*

It is assumed in this section that the matter satisfies the DEC. It is also assumed that there is some time t in the interval (t_1, t_2) at which $H(t) < 0$ which corresponds to the model expanding at that time.

Theorem 5.3 *Consider a spatially homogeneous solution of the Einstein–matter system in n space dimensions, defined by left invariant metrics on a Lie group G, on an interval (t_1, t_2) corresponding to data at time $t = t_0$ with a matter model satisfying the DEC and the MCC and $H(t) < 0$ at some time. Suppose that this is the maximal interval of existence. If the Lie group G admits no left invariant metrics of positive scalar curvature then $t_2 = \infty$. If $n = 3$ and the SEC is satisfied then the spacetime is future geodesically complete.*

Proof First it will be shown that tr$k < 0$ on the whole interval. Suppose that tr$k$ was zero at some time. Then by the Hamiltonian constraint and the non-positivity of the scalar curvature it follows that at that time $k_{ab} = 0$, $\rho = 0$ and $R = 0$. Since the dominant energy condition holds it follows that $\rho = 0$ everywhere, so that the solution is vacuum. There is a general result which says that if no left invariant metric with positive scalar curvature exists on a Lie group G then any left invariant metric on G with non-negative scalar curvature is flat. It follows that the initial data for the solution at the time where tr$k$ vanishes is trivial. The spatial metric is flat and the second fundamental form zero. Hence the spacetime is flat and tr$k$ is zero everywhere, a contradiction. Thus tr$k$ is negative on the whole spacetime. Combining the generalization of eqn (2.36) to n space dimensions with the Hamiltonian constraint gives in the homogeneous case

$$\partial_t (\text{tr} k) = \left[-\frac{1}{n-1} R + \frac{n}{n-1} \sigma_{ij} \sigma^{ij} + \frac{\kappa}{n-1} (n\rho + \text{tr} S) \right]. \quad (5.67)$$

Using the dominant energy condition this equation implies that tr$k$ is non-decreasing. Thus tr$k$ is bounded in the future of the initial hypersurface. It follows from the MCC that $t_2 = \infty$.

The Hamiltonian constraint, the sign of the scalar curvature and the WEC imply that $k_{ij}k^{ij} \le (\text{tr}k)^2$. If λ_i are the eigenvalues of k_{ij} with respect to g_{ij} then $\text{tr}k = \Sigma_i \lambda_i$ and $k_{ij}k^{ij} = \Sigma_i \lambda_i^2$. Thus the generalized Kasner exponents satisfy

$$p_1 + p_2 + \cdots + p_n = 1, \tag{5.68}$$

$$p_1^2 + p_2^2 + \cdots + p_n^2 \le 1. \tag{5.69}$$

Let K be the subset of $\mathbb{R}^n$ where these relations are satisfied. The minimum value of p_1 will now be determined. By symmetry this is also the minimum of any other p_i. The minimum of p_1 must occur on the boundary of K in the hyperplane $\Sigma_i p_i = 1$ and this is got by replacing the inequality above by an equality. If there were p_i and p_j with i and j not equal to one and $p_i \ne p_j$ then the minimum would be attained at interior points of K, contradicting what has already been said. For this would be the case at convex linear combinations of the original point and that obtained by interchanging the values of p_i and p_j. Hence at the minimum $p_2 = \ldots = p_n = \frac{2}{n}$ and $p_1 = -\frac{n-2}{n}$. It can be concluded that for any vector v^i tangent to the hypersurfaces of homogeneity the following holds:

$$k_{ij}v^i v^j \le -\frac{n-2}{n}(\text{tr}k)g_{ij}v^i v^j. \tag{5.70}$$

Consider now a future-directed causal geodesic and let q^i be the frame components of the projection of its tangent vector onto the hypersurfaces of constant t. Then

$$\frac{d}{dt}(g_{ij}q^i q^j) = 2k_{ij}q^i q^j$$

$$\le -\frac{2(n-2)}{n}(\text{tr}k)g_{ij}q^i q^j. \tag{5.71}$$

In the case $n=3$ and assuming the strong energy condition the inequality $\partial_t(\text{tr}k) \ge \frac{1}{3}(\text{tr}k)^2$ holds. It follows that $\text{tr}k(t) \ge -3/(C+t)$ for some positive constant C. Combining this with equation (5.71) gives

$$\frac{d}{dt}(\log g_{ij}q^i q^j) \le 2/(C+t). \tag{5.72}$$

Hence

$$(g_{ij}q^i q^j)^{-1/2} \ge C(1+t)^{-1}. \tag{5.73}$$

The affine parameter length of the geodesic up to a certain point is given by integral of $(\epsilon + g_{ij}q^i q^j)^{-1/2}$ with respect to t, where ϵ is 1 for a timelike geodesic parametrized by arc length and zero for a null geodesic. Future geodesic completeness is equivalent to the divergence of this integral as the upper limit tends to infinity. This is guaranteed by (5.73). □

In three dimensions the only simply connected Lie group which does not satisfy the curvature condition of the theorem is $SU(2)$. Hence the theorem covers all Bianchi types except type IX. In higher dimensions there is a general result which says that a simply connected Lie group admits no left invariant metrics of positive scalar curvature if and only if it is diffeomorphic to a Euclidean space. The argument in the proof establishing geodesic completeness does not work if $n \geq 4$ or the SEC is violated. It is an open problem to prove geodesic completeness in either of those cases.

There is a variant of the MCC which is sometimes useful and will now be defined.

Definition 5.4 *A matter model satisfies the weak matter continuation condition (WMCC) if a spatially homogeneous solution of the corresponding Einstein–matter equations defined on an interval (t_1, t_2) with g_{ij}, $(\det g)^{-1}$, k_{ij} bounded and R bounded above in a neighbourhood of the finite time t_1 or t_2 can be extended to a neighbourhood of t_1 or t_2 respectively.*

The MCC obviously implies the WMCC. It will now be shown that the converse is also true provided the scalar curvature is bounded above. Suppose for definiteness that the endpoint of the interval under consideration is $t = t_2$ – the argument in the case $t = t_1$ is almost identical. When tr$k$ is bounded in a neighbourhood of t_2 it follows from the Hamiltonian constraint that $k_{ij}k^{ij}$ is bounded on that neighbourhood. The evolution equation for the volume element of the metric implies that $\det g$ and its inverse are bounded. Since the mean curvature is bounded its evolution equation implies that the integral of $k_{ij}k^{ij}$ over the interval $[t_0, t_2)$ is bounded. Before proceeding some facts from linear algebra must be recalled. Let A and B be $n \times n$ symmetric matrices with A positive definite. Define the relative norm $\|B\|_A$ by

$$\|B\|_A = \sup\{\|Bx\|/\|Ax\|, x \neq 0\}. \tag{5.74}$$

When A is the identity this coincides with the usual matrix norm $\|B\|$. The inequality $\|B\| \leq \|B\|_A \|A\|$ is immediate. A fact which is needed here is that

$$\|B\|_A \leq \sqrt{\operatorname{tr}(A^{-1}BA^{-1}B)}. \tag{5.75}$$

This can be proved by noting that there is a basis z_i and constants λ_i with the property that $Bz_i = \lambda_i A z_i$ for each i. The left hand side of (5.75) is given by the maximum modulus of any λ_i. On the other hand the eigenvalues of $A^{-1}BA^{-1}B$ are $\{\lambda_i^2\}$ so that the right hand side of (5.75) is equal to $(\sum_i \lambda_i^2)^{1/2}$. Let $\|g\|$ and $\|k\|$ denote the norms of the matrices with entries g_{ij} and k_{ij} respectively. Let $\|k\|_g$ be the relative norm of the matrix with entries k_{ij} with respect to the matrix with entries g_{ij}. Then

$$\|g(t)\| \leq \|g(t_0)\| + 2\int_{t_0}^t \|k(s)\| ds \qquad (5.76)$$

$$\leq \|g(0)\| + 2\int_{t_0}^t \|k(s)\|_{g(s)} \|g(s)\| ds \qquad (5.77)$$

$$\leq \|g(0)\| + 2\int_{t_0}^t (k^{ij}k_{ij})^{1/2}(s) \|g(s)\| ds. \qquad (5.78)$$

Applying Gronwall's inequality gives

$$\|g(t)\| \leq \|g(0)\| \exp\left[2\int_{t_0}^t (k^{ij}k_{ij})^{1/2}(s)\right]. \qquad (5.79)$$

Using the fact that t_2 is finite it can be concluded that the components g_{ij} are bounded on $[t_0, t_2)$. Using the inequality $\|k(s)\| \leq (k^{ij}k_{ij})^{1/2}\|g(s)\|$ then shows that the components k_{ij} are bounded. This completes the proof that under the upper bound on the scalar curvature the WMCC implies the MCC.

There remains the question which matter models satisfy the WMCC. The straightforward answer is that any reasonable matter model should satisfy this condition. This has actually been proved (in four-dimensions) in the case of a perfect fluid (with some mild restrictions on the equation of state), for a mixture of non-interacting perfect fluids and for collisionless matter [169]. This is not a difficult task and it comes down to verifying that the matter does not form singularities as long as the hypotheses in the WMCC are satisfied. For most matter models this is an ODE problem and the standard continuation criterion for ODE plays a central role. For kinetic matter it is necessary to do some more work by hand.

5.10 An application to surface symmetry

In surface symmetric spacetimes the hyperbolic aspects of the Einstein equations are suppressed. In physics language this can be expressed by saying that there are no gravitational waves in surface symmetry. These statements are only heuristic in nature but it is worth trying to understand them a little better before coming to the mathematical analysis. In a surface symmetric spacetime there is only one spatial direction in which the

5.10 An application to surface symmetry

unknowns vary. If a gravitational wave were present it would have to propagate in that direction. The oscillations which make up the wave are in the directions orthogonal to the direction of propagation. The changes in the two-dimensional metric in these directions due to the oscillations should be traceless. There is, however, an isotropy group which causes these degrees of freedom to vanish – a two by two matrix which is invariant under rotations must be proportional to the identity. In Gowdy spacetimes, for instance, the isotropy is absent and gravitational waves are possible. The effect of the lack of waves on the form of the Einstein equations is that coordinates can be chosen so that the essential field equations are ordinary differential equations in t or r. These equations are coupled together so that in total the Einstein equations are a system of partial differential equations. However the dynamics can be analysed by studying solutions of ordinary differential equations, which is why this subject is discussed in this chapter.

Because of the absence of gravitational waves the only surface symmetric vacuum spacetimes are spatially homogeneous. This is proved below. Thus in order to get a class of spacetimes which are genuinely inhomogeneous some matter must be introduced. Among the simplest choices are a scalar field or collisionless matter. Here the latter choice is made and the model is given by the Einstein–Vlasov system. For simplicity only the plane symmetric case is considered. A cosmological constant is included. The metric is (4.27) with $\kappa = 0$. To get an explicit form of the Vlasov equation it is necessary to choose a way of parametrizing the momenta. Here the following variables are used:

$$w = e^\lambda p^1, \quad F = t^4((p^2)^2 + (p^3)^2). \tag{5.80}$$

The quantity F is a conserved quantity along the characteristics of the Vlasov equation and may be thought of as angular momentum. (For the various uses of the word 'characteristic' see Chapter 7.)

With these assumptions the equations to be solved look as follows. The Einstein equations are

$$e^{-2\mu}(2t\dot{\lambda} + 1) - \Lambda t^2 = 8\pi t^2 \rho, \tag{5.81}$$

$$e^{-2\mu}(2t\dot{\mu} - 1) + \Lambda t^2 = 8\pi t^2 p, \tag{5.82}$$

$$\mu' = -4\pi t e^{\lambda+\mu} j, \tag{5.83}$$

$$e^{-2\lambda}(\mu'' + \mu'(\mu' - \lambda')) - e^{-2\mu}\left(\ddot{\lambda} + (\dot{\lambda} - \dot{\mu})\left(\dot{\lambda} + \frac{1}{t}\right)\right) + \Lambda = 4\pi q. \tag{5.84}$$

The last of these equations is a PDE and this might seem to contradict what was said above. It turns out, however, that this last equation is satisfied automatically if all the other Einstein equations and the Vlasov equation are satisfied. Thus it plays a subordinate role. The matter quantities on the right hand side of the Einstein equations are defined by

$$\rho(t,r) = \frac{\pi}{t^2} \int_{-\infty}^{\infty} \int_0^{\infty} \sqrt{1 + w^2 + F/t^2} f(t,r,w,F) dF dw, \tag{5.85}$$

$$p(t,r) = \frac{\pi}{t^2} \int_{-\infty}^{\infty} \int_0^{\infty} \frac{w^2}{\sqrt{1 + w^2 + F/t^2}} f(t,r,w,F) dF dw, \tag{5.86}$$

$$j(t,r) = \frac{\pi}{t^2} \int_{-\infty}^{\infty} \int_0^{\infty} w f(t,r,w,F) dF dw, \tag{5.87}$$

$$q(t,r) = \frac{\pi}{t^4} \int_{-\infty}^{\infty} \int_0^{\infty} \frac{F}{\sqrt{1 + w^2 + F/t^2}} f(t,r,w,F) dF dw. \tag{5.88}$$

The Vlasov equation takes the form

$$\partial_t f + \frac{e^{\mu - \lambda} w}{\sqrt{1 + w^2 + F/t^2}} \partial_r f - (\dot\lambda w + e^{\mu - \lambda} \mu' \sqrt{1 + w^2 + F/t^2} \partial_w f) = 0. \tag{5.89}$$

Before continuing, the vacuum case will be analysed. The equation for μ' shows that μ depends only on t. The equation for λ reduces to

$$\dot\lambda = -\frac{1}{2t}(1 - \Lambda t^2 e^{2\mu}). \tag{5.90}$$

Hence

$$\lambda(t,r) = \lambda(t_0, r) - \int_{t_0}^{t} \frac{1}{2s}(1 - \Lambda s^2 e^{2\mu(s,r)}) ds. \tag{5.91}$$

By reparametrizing r it can be arranged that $\lambda(t_0, r)$ is constant and then after this transformation λ only depends on t and the spacetime is manifestly spatially homogeneous. Note that a priori the coordinate r can be chosen freely on a given Cauchy hypersurface. It is this which leads to the possibility and the necessity of doing the transformation in this proof. In the case $\Lambda = 0$ the equations can be integrated explicitly. Transforming to a Gaussian time coordinate then reveals that the spacetime is the Kasner solution with exponents $(-\frac{1}{3}, \frac{2}{3}, \frac{2}{3})$.

The aim of this section is to study solutions of the above equations with a prescribed initial datum f_0 on a hypersurface where t is constant. The first step is to prove a local existence theorem with a continuation criterion. The main idea is to define a sequence of iterates by solving the individual equations sequentially and then show that on a suitable time interval these iterates form a Cauchy sequence. This is basically the same strategy as in the proof of the local existence theorem for ordinary differential equations. There are, however, some serious difficulties to be overcome. With the obvious bounds the estimates do not close. From a PDE perspective this means that there is some loss of differentiability during the iteration. Two main ideas are used to overcome these difficulties. One is to replace μ' in one of the field equations by an a priori independent quantity $\tilde{\mu}$ and then show at the end of the proof that $\mu' = \tilde{\mu}$. Another is to use the geometrical background of the problem when estimating first derivatives of the solution of the characteristic system. These first derivatives can be thought of as the variation vectors of congruences of timelike geodesics and can thus, by a suitable choice of variables, be thought of as Jacobi fields. The evolution equations for these depend only on the curvature and not on arbitrary first derivatives of the Christoffel symbols. This observation can be used to avoid the apparent loss of derivatives.

Once a local existence theorem has been proved the maximal interval of existence can be considered for a solution determined by an initial datum prescribed at $t = t_0$. It is a subset (t_-, t_+) of the interval $(0, \infty)$. Let $P(t)$ be the supremum of w on the set where f is non-zero at time t. The proof of the local existence theorem shows that if $P(t)$ is bounded on the interval $(t_-, t_0]$ then $t_- = 0$. It also shows that if $P(t)$ is bounded on $[t_0, t_+)$ and if, in addition, the maximum value of μ in space is bounded as a function of time there then $t_+ = \infty$. Using a cancellation property in the characteristic system it can be shown that if the maximum value of μ is bounded $P(t)$ is also bounded.

In the case $\Lambda < 0$ it is not hard to show that the solutions do not always exist globally to the future. This can be seen explicitly in the vacuum case. At the same time it follows from an argument given in Chapter 3 that the solution is not geodesically complete in either time direction. There it was shown that there is an upper bound on the length of any timelike geodesic.

For $\Lambda \geq 0$, on the other hand, global existence in the future can be proved. This necessitates a series of steps which will not be recapitulated here. An identity which plays a key role in the argument will be written down since it and analogues are useful in variety of classes of spacetimes with Killing vectors. The identity reads

$$\frac{d}{dt} \int_0^1 e^{\mu+\lambda} \rho(t,r) dr = -\frac{1}{t} \int_0^1 e^{\mu+\lambda} \left[2\rho + q - \frac{\rho+p}{2}(1 - \Lambda t^2 e^{2\mu}) \right] dr. \tag{5.92}$$

It is interesting to note that although this identity was first derived for the Vlasov equation it holds for any matter model. The matter quantities

ρ, p, j, q can be defined in terms of the energy–momentum tensor for any matter model and the identity is obtained for a plane symmetric spacetime by integrating the equation $\nabla_\alpha T^{\alpha\beta} = 0$ in space.

In the case $\Lambda = 0$ not much is known about the asymptotics in the future. In particular it is not known, as might be guessed, that the spacetimes are future geodesically complete. In the case $\Lambda > 0$ more can be done. The leading terms in the asymptotic behaviour of the components of the metric are obtained.

$$\lambda = \log t[1 + o(1)], \qquad (5.93)$$
$$\mu = -\log t[1 + o(1)]. \qquad (5.94)$$

Taking only the leading terms and discarding the remainders leads to the de Sitter space. Hence at late times all of these spacetimes resemble the de Sitter solution asymptotically. The proof of this result requires some delicate estimates for the solutions of the characteristic system. A term involving the cosmological constant dominates everything else in the equation at sufficiently late times. The spacetimes are future geodesically complete.

5.11 Further reading

A thorough introduction to ordinary differential equations is [98]. See also [3]. A lot of intuition on ODE and dynamical systems can be obtained from [107]. The article [14] is full of information, in a rather compressed form. The book [77] is probably still the best reference for the theory of asymptotic expansions. The best introduction to Bianchi models is the book [203]. The form of the Wainwright–Hsu equations written down here is taken from [203] and differs from the original reference [204] by some numerical factors. A good account of centre manifold theory is given in [43]. A detailed discussion of the reduction theorem can be found in [120]. The centre manifold analysis for a dynamical system arising from the Einstein–Vlasov system is based on [173]. The results on the past attractor for Bianchi type IX vacuum spacetimes were proved in [184]. The theory of the Hopf bifurcation is explained in [95]. Information about the role of the Hopf bifurcation in spatially homogeneous solutions of the Einstein equations can be found in [105]. An account of the scalar curvature of Lie groups and homogeneous spaces is given in [26]. The application to surface symmetry follows [163] and [198].

6 Functional analysis

Functional analysis plays a role in the local existence and uniqueness theorem for ODE as described in Section 5.1. In the study of partial differential equations it plays a much bigger role and so in preparation for that topic some relevant facts about functional analysis are collected in this chapter.

6.1 Abstract function spaces

Consider the set of functions on a given space with a certain regularity, for instance continuous functions. In many cases, including the example of continuous functions, adding two functions of this type gives another of the same type as does multiplying a function of this type by a constant. Abstractly this means that the set of functions of the given type forms a vector space. In general there is also some notion of convergence for these functions, e.g. uniform convergence of continuous functions. This defines a topology on the vector space and the algebraic and topological structures satisfy certain compatibility conditions. Thus, for instance, the basic algebraic operations of addition and multiplication by a scalar are continuous. Abstract function spaces are obtained by abstracting certain of these properties. It is then possible to prove theorems for large classes of spaces which can be applied in many concrete situations.

The best known type of abstract function space is the Hilbert space. (Here Hilbert spaces over the real numbers are considered.) A Hilbert space H is a vector space with an inner product $\langle\,,\,\rangle$. This is a bilinear mapping from $H \times H$ to $\mathbb{R}$ which is assumed positive definite, i.e. $\langle x, x \rangle \geq 0$ for all x and $\langle x, x \rangle = 0$ implies $x = 0$. Furthermore it is assumed that the space is complete, a notion which will be explained soon. The inner product defines a norm $\|\ \|$ by $\|x\|^2 = \langle x, x \rangle$. The norm defines a topology and a notion of convergence. A sequence $\{x_n\}$ is said to converge to $x \in H$ if $\|x - x_n\| \to 0$ as $n \to \infty$. A Cauchy sequence $\{x_n\}$ is one for which $\|x_m - x_n\| \to 0$ as m and n both tend to infinity. The property of completeness is that for any Cauchy sequence $\{x_n\}$ in H there exists $x \in H$ such that $x_n \to x$ as $n \to \infty$. Existence theorems for partial differential equations often proceed by constructing a Cauchy sequence in a suitable Hilbert space and then showing that the limit of this sequence, whose existence is guaranteed by completeness, is the desired solution.

An important generalization of the concept of a Hilbert space is given by that of a Banach space. A Banach space B is a vector space with a norm which is complete. The defining properties of a norm are as follows. It is a mapping $\| \ \|$ from B to the non-negative real numbers such that

- $\|x\| = 0$ implies $x = 0$;
- $\|ax\| = |a|\|x\|$ for $x \in B$ and $a \in \mathbb{R}$;
- $\|x + y\| \le \|x\| + \|y\|$ for any x and y in B.

Evidently a Hilbert space with the norm defined above is a Banach space. Let F be a closed subset of a Banach space B and ϕ a mapping from F to itself which satisfies $\|\phi(x) - \phi(y)\| \le K\|x - y\|$ for all x and y in F and a constant $K < 1$. A mapping with this property is called a contraction. The contraction mapping principle (Banach fixed point theorem) says that any contraction ϕ has a unique fixed point x_*, i.e. there is a unique $x_* \in F$ which satisfies $\phi(x_*) = x_*$. This statement can be used to prove local existence and uniqueness of solutions of ODE. In Section 5.1 a slightly different proof was used. An isomorphism between Banach spaces is a linear mapping which is not only one to one and onto but is also continuous together with its inverse. A linear mapping L is called bounded if there is a constant C such that $\|L(x)\| \le C\|x\|$ for all x. A linear mapping of Banach spaces is continuous if and only if it is bounded. Thus an isomorphism of Banach spaces is a linear mapping which is one to one and onto and bounded together with its inverse. In fact the continuity of the inverse is automatic – this is the open mapping theorem for Banach spaces [189].

Banach spaces share many properties with finite-dimensional vector spaces. For instance it is possible to build up a theory of ODE in a Banach space which is very similar to that for $\mathbb{R}^n$ presented in the last chapter. Local existence and uniqueness can be proved in essentially the same way. For mappings between Banach spaces derivatives can be defined much as in several variable calculus. The implicit function theorem holds. To state this let B_1, B_2 and B_3 be Banach spaces and let ϕ be a smooth mapping from an open subset U of $B_1 \times B_2$ to B_3. The derivative of ϕ at a point (x, y) of $B_1 \times B_2$ for a fixed value of the second argument is a linear mapping from B_1 to B_3. If this mapping is an isomorphism from B_1 to B_3 and $\phi(x, y) = 0$ then the implicit function theorem says that in a neighbourhood of y there is a smooth mapping ψ from B_2 to B_1 with $x = \psi(y)$ which satisfies $\phi(\psi(z), z) = 0$.

It is also important to be aware of the differences between infinite-dimensional Banach spaces and the finite-dimensional ones which are all isomorphic to Euclidean spaces. In an infinite-dimensional Banach space the unit ball is not compact. This means that there are sequences with bounded norm which have no convergent subsequence. This may be a problem when proving existence theorems. One way of getting around it is to use different topologies on the same space. If B_1 and B_2 are Banach spaces a mapping ϕ from B_1 to B_2 is called compact if $\|x_n\| \leq C$ for a sequence $\{x_n\}$ implies that $\{\phi(x_n)\}$ has a convergent subsequence. An important special case is where the underlying vector spaces of B_1 and B_2 are the same and ϕ is the identity. This is the case of two different topologies on the same space.

Another important concept for infinite-dimensional Banach spaces is that of weak convergence. Let B be a Banach space and ϕ a bounded linear functional on B, i.e. a bounded linear mapping from B to $\mathbb{R}$. Define $\|\phi\| = \sup\{\|\phi(x)\| : \|x\| = 1\}$. The set of all bounded linear functionals on B endowed with the norm just defined is a Banach space. It is called the dual space of B and is denoted by B' or B^*. In the following the latter notation is used. This is a direct generalization of the notion of the dual of a finite-dimensional vector space. There is a natural embedding Φ of B into the double dual B^{**} given by $\Phi(x)(\phi) = \phi(x)$. In general Φ is not onto and $\Phi(B)$ is strictly smaller than B^{**}. In the case that Φ is an isomorphism the space B is said to be reflexive. Every Hilbert space is reflexive but many important Banach spaces are not. A sequence $\{x_n\}$ in a Banach space is said to converge weakly if the sequence of real numbers $\{\phi(x_n)\}$ converges for any $\phi \in B^*$. A related concept is that of weak* convergence. A sequence $\{\phi_n\}$ in B^* is said to converge weak* to ϕ if $\phi_n(x)$ converges to $\phi(x)$ for any fixed x in B. Weak and weak* convergence are useful in many existence theorems since they are easier to obtain than convergence in norm. An example of a useful statement using these concepts is the Banach–Alaoglu theorem. It says that if X is a separable Banach space then any bounded sequence in X^* has a weak*-convergent subsequence. (A Banach space is said to be *separable* if it has a countable dense subset.)

The following simple lemma is important in the study of concrete function spaces.

Lemma 6.1 *Let B_1 and B_2 be Banach spaces, X a dense linear subspace of B_1 and L a linear mapping from X to B_2. If there is a constant C such that $\|Lx\|_{B_2} \leq C\|x\|_{B_1}$ for all $x \in X$ then there is a unique linear mapping $\tilde{L}$ from B_1 to B_2 such that $\tilde{L}(x) = L(x)$ for all $x \in X$ and $\|\tilde{L}x\|_{B_2} \leq C\|x\|_{B_1}$ for all $x \in B_1$.*

6.2 Distributions

Before introducing the concept of distributions it is useful to define multiindices. They provide a compact notation for higher order derivatives which is very useful in PDE theory. A multiindex of order s is a sequence of n non-negative integers $\alpha_1, \ldots, \alpha_n$ with $\alpha_1 + \cdots + \alpha_n = s$. Let $|\alpha| = s$. By definition

$$D^\alpha u = \frac{\partial^{|\alpha|} u}{(\partial x^1)^{\alpha_1} \ldots (\partial x^n)^{\alpha_n}}. \tag{6.1}$$

There are also some related definitions. If $\xi \in \mathbb{R}^n$ and α is a multiindex then $\xi^\alpha = (\xi^1)^{\alpha_1} \ldots (\xi^n)^{\alpha_n}$ while $\alpha! = \alpha_1! \ldots \alpha_n!$.

Distributions are generalizations of functions which are often useful in the theory of partial differential equations. Let f be a smooth function on $\mathbb{R}^n$. The *support* of f is the closure of the set where f is non-zero. For any function ϕ belonging to the space $\mathcal{D} = C_c^\infty(\mathbb{R}^n)$ of smooth functions on $\mathbb{R}^n$ with compact support it is possible to define $T_f(\phi) = \int f(x)\phi(x)dx$. This defines a linear functional T_f on $\mathcal{D}$ which has certain continuity properties. For instance if f_n converges uniformly to f then $T_{f_n}(\phi)$ converges to $T_f(\phi)$. The idea of distribution theory is to replace the function f by the linear functional T_f and then to generalize T_f to any linear functional on $\mathcal{D}$ which is continuous with respect to the topology of uniform convergence of functions and all their derivatives on compact subsets. The vector space of all distributions is denoted by $\mathcal{D}'$. It is the dual space of $\mathcal{D}$. There is a natural topology on $\mathcal{D}'$ which is analogous to the weak* topology on the dual of a Banach space. A sequence T_n of distributions converges to $T \in \mathcal{D}'$ iff $T_n(\phi) \to T(\phi)$ for any $\phi \in \mathcal{D}$. In this context the functions in $\mathcal{D}$ are called test functions. With this definition of convergence the image of the natural embedding of $\mathcal{D}$ in $\mathcal{D}'$ given by integration against a test function is dense. This means that any distribution can be approximated by test functions.

Not all functions are differentiable which means that for a certain function f on $\mathbb{R}^n$ there is no function which can be called $\partial f/\partial x^1$. There is, however a definition of derivative for distributions which is defined on all of $\mathcal{D}'$. Thus it is possible, for instance, to take a function which is merely continuous, associate a distribution to it by integrating it against test functions and then define a distribution which is its derivative. The general definition is as follows. If $T \in \mathcal{D}'$ and α is a multiindex let $D^\alpha T(\phi) = (-1)^{|\alpha|} T(D^\alpha \phi)$. Note that this is defined for any $T \in \mathcal{D}'$ and that if f is smooth the natural compatibility condition $T_{D^\alpha f} = D^\alpha T_f$ holds, as can be seen by integration by parts.

The most famous distribution which is not a function is the Dirac δ. By definition the Dirac δ at the point x, δ_x or $\delta(x)$, is defined by $\delta_x(\phi) = \phi(x)$.

A difficulty with distributions is that while there is an obvious way of multiplying them by test functions they cannot in general be multiplied with each other. For instance the expression δ^2 makes no sense. There is a theorem, the Schwartz impossibility theorem, which says that there is no generalization of functions satisfying some natural axioms which allows the generalized functions to be both differentiatied and multiplied freely. This problem with multiplication limits the applications of distributions to nonlinear PDE but the applications which are possible are very important.

The Schwartz space $\mathcal{S}$ is defined to be the space of smooth functions ϕ on $\mathbb{R}^n$ such that for any multiindices α and β the function $x^\alpha D^\beta f$ is bounded. (Often the Schwartz space is defined in a strictly analogous way for complex-valued functions.) For each natural number k and multiindex α a quantity $\|\phi\|_{k,\alpha} = \sup\{|x|^k |D^\alpha \phi|\}$ is defined. A sequence $\{\phi_n\}$ in $\mathcal{S}$ is said to converge to $\phi \in \mathcal{S}$ if $\|\phi_n - \phi\|_{k,\alpha} \to 0$ for all k and α. If a distribution T satisfies an inequality of the form $|T(\phi)| \le C\|\phi\|_{k,\alpha}$ for some choice of k and α and all $\phi \in \mathcal{S}$ then T extends uniquely to a linear functional on $\mathcal{S}$. In this case T is said to be a tempered distribution. The space of all tempered distributions is denoted by $\mathcal{S}'$.

If $f \in \mathcal{S}$ (complex-valued functions) then its Fourier transform $\hat{f}$ exists. Since the Fourier transform converts multiplication by the coordinate function x^j into the derivative operator $-i\partial/\partial x^j$ and vice versa it is easy to see that $\hat{f} \in \mathcal{S}$. In fact the Fourier transform defines a continuous linear mapping from $\mathcal{S}$ to itself. This mapping has a unique extension to a linear mapping from $\mathcal{S}'$ to itself defined by $\hat{T}(\phi) = T(\hat{\phi})$. This provides a definition of the Fourier transform for arbitrary tempered distributions. As a consequence of Plancherel's theorem it agrees with the usual Fourier transform on $\mathcal{S}$ considered as a subspace of $\mathcal{S}'$. For f and g belonging to $\mathcal{S}$ the convolution $f * g$ is defined by

$$(f * g)(x) = \int_{\mathbb{R}^n} f(y)g(x-y)dy. \tag{6.2}$$

It also belongs to the Schwartz space. The Fourier transform converts multiplication of functions into convolution and vice versa. This allows some more insight into the question when distributions can be multiplied and when they cannot. A way to try to define the product of two distributions is to take their Fourier transforms, do the convolution of the results and then transform back. The obstruction to carrying out these steps is then that certain integrals which come up diverge. This idea is exploited in the theory of *microlocal analysis*, which will be discussed next.

Let f be a function on $\mathbb{R}^n$ which is locally integrable, i.e. the integral of $|f|$ over any compact subset of $\mathbb{R}^n$ is finite. Then f defines a distribution by integration against test functions. The function f is said to be smooth at the point x if there is a test function ϕ with $\phi(x) \neq 0$ such that the product ϕf is smooth. In that case $\phi f \in \mathcal{S}$ and it follows that $\widehat{\phi f}$ and its derivatives of all orders fall off faster than any power of $|\xi|$. The set of points at which f is not smooth in this sense is called the singular support of f. Informally f can be said to have a singularity at the point x if x belongs to the singular support of f. In microlocal analysis this idea is refined so as to say when a function is singular at a particular point and in a particular direction. To formulate this the notion of a conic neighbourhood is required. If $\xi \in \mathbb{R}^n$ is a non-zero vector an open neighbourhood of ξ in $\mathbb{R}^n \backslash \{0\}$ is called a conic neighbourhood of ξ if it is invariant under rescalings by positive constants. Given a point x and a direction defined by a vector $\xi_0 \neq 0$ the function f is said to be microlocally smooth at (x, ξ_0) if for a test function ϕ with $\phi(x) \neq 0$ there is a conic neighbourhood of ξ_0, where the Fourier transform of ϕf and its derivatives of all orders fall off faster than any power of $|\xi|$. The set of (x, ξ) where f is not microlocally smooth is called the wave front set of f. It should be noted that the theory of the wave front set can be formulated in an invariant way on a manifold and that in that context it is a subset of the cotangent bundle. The projection of the wave front set to the base manifold is the singular support. For example the wave front set of $\delta(0)$ consists of the points $(0, \xi)$ for all $\xi \neq 0$. For the function $\theta(x)$ defined to be zero for $x^1 < 0$ and one for $x^1 \geq 0$ the wave front set consists of all pairs (x, ξ) with $x^1 = 0$ and ξ having components $(\xi^1, 0, \ldots, 0)$. If f and g are tempered distributions such that the wave front set of f does not intersect the set obtained from the wave front set of g by replacing ξ by $-\xi$ then the product of f and g can be defined by means of the Fourier transform and convolution.

6.3 Concrete function spaces

Let Ω be an open subset of $\mathbb{R}^n$ with compact closure. Denote the set of all continuous real-valued functions on $\bar{\Omega}$ by $C^0(\bar{\Omega})$. It is a vector space with the operations of pointwise addition and multiplication by real numbers. The norm defined by $\|f\|_{C^0} = \sup\{|f(x)| : x \in \bar{\Omega}\}$ makes it into a Banach space. Similarly, using the norm $\|f\|_{C^k} = \sum_{i=0}^{k} \sup\{|D^i f(x)| : x \in \bar{\Omega}\}$ makes the set of C^k functions on $\bar{\Omega}$ into a Banach space for any natural number k. Here $D^i f$ denotes the collection of all partial derivatives of f of order i. There is no natural way to make the set of C^∞ functions into a Banach space. It is possible define a topology on the C^∞ functions using all C^k norms together but the result, instead of being a Banach space, is a more general type of space known as a Fréchet space. In Fréchet spaces there is no simple existence and uniqueness theory for ODE and no simple inverse function theorem. There is a replacement for the latter, the Nash–Moser theorem, but it is much harder to use.

An example of a compact embedding involving C^k spaces is provided by the Arzela–Ascoli theorem. The statement of this requires the definition of equicontinuity which will now be given. A set F of functions on a subset of $\mathbb{R}^n$ with values in $\mathbb{R}^k$ is called *uniformly equicontinuous* if given $\epsilon > 0$ there exists $\delta > 0$ such that $|x - y| < \delta$ implies $|f(x) - f(y)| < \epsilon$ for all $f \in F$. The important point here is that δ is independent of the function f. A version of the Arzela–Ascoli theorem says that if a sequence of functions on a compact set is uniformly bounded and equicontinuous then it has a uniformly convergent subsequence. This implies, using the mean value theorem, that if Ω is a compact domain with smooth boundary the embedding of $C^1(\bar{\Omega})$ into $C^0(\bar{\Omega})$ given by the identity is compact. There is also a generalization to infinite dimensions which has applications in PDE theory. Let $X_1 = C^1([0, T], B_1)$ be the space of continuously differentiable functions from $[0, T]$ into the Banach space B_1 and $X_2 = C^0([0, T], B_2)$ be the space of continuous functions from $[0, T]$ to the Banach space B_2. Suppose that there is a compact embedding of B_2 into B_1. Then any bounded sequence in $X_1 \cap X_2$ has a subsequence which converges in $C^0([0, T], B_1)$.

As will be explained later the C^k spaces are not convenient for the study of elliptic equations. Better are the Hölder spaces, which will now be introduced. A function f on $\mathbb{R}^n$ is called Hölder continuous of order α for a real number α belonging to the interval $(0, 1)$ if there is a constant C such that for any points x and y the inequality $|f(x) - f(y)| \leq C|x - y|^\alpha$ holds. For a bounded and uniformly Hölder continuous function f on an open set Ω it is possible to define the Hölder norm

$$\|f\|_{C^{k,\alpha}} = \sup\{|f(x)| : x \in \Omega\}$$
$$+ \sup\{|f(x) - f(y)|/|x - y|^\alpha : x, y \in \Omega, x \neq y\}. \qquad (6.3)$$

Using this a space C^α of Hölder continuous functions can be defined. Similarly it is possible to define a space $C^{k,\alpha}$ where the norm is the sum of the C^k norm and the Hölder norm of the derivative of f of order k. These spaces are Banach spaces. They are better adapted to elliptic PDE than C^k spaces.

It will be seen later that while Hölder spaces are well adapted to elliptic equations they are not well adapted to hyperbolic PDE. There spaces of a different kind are appropriate. These are the Sobolev spaces which will now be introduced. A locally integrable function f on $\mathbb{R}^n$ is said to belong to L^p for some real number p in the interval $[1, \infty)$ if $\int_{\mathbb{R}^n} |f|^p < \infty$. It is said to belong to L^p_{loc} if the corresponding integral over any compact subset of $\mathbb{R}^n$ is finite. A measurable function is said to belong to L^∞ if there is a positive real number M such that the measure of the set where $|f|$ is greater than

M is zero. The smallest such number is called the essential supremum of f and denoted ess sup$f$. For $p < \infty$ define $\|f\|_{L^p} = [\int_{\mathbb{R}^n} |f|^p]^{1/p}$. This defines a norm on the space of L^p functions which makes it into a Banach space. Strictly speaking it is necessary to quotient by the functions which vanish except on a set of measure zero but this is of little practical importance and will be ignored here. The essential supremum makes the L^∞ functions into a Banach space. The case $p = 2$ is special since L^2 is in fact a Hilbert space with the inner product $\langle f, g \rangle = \int_{\mathbb{R}^n} fg$. For $1 \leq p < \infty$ the dual space of L^p is L^q where q is the conjugate exponent to p, i.e. the unique number satisfying $1/p + 1/q = 1$. However the dual space of L^∞ is not L^1. The fact that L^∞ is the dual of the separable Banach space L^1 means that the Banach–Alaoglu theorem applies to it. It can also be applied to more complicated spaces like $L^\infty([0,T], L^2(\mathbb{R}^n))$ and this fact plays a role in Chapter 8.

Spaces analogous to the L^p spaces can be defined which contain derivatives and these are the Sobolev spaces $W^{m,p}$. Let

$$\|f\|_{W^{m,p}} = \sum_{|\alpha| \leq m} \|D^\alpha f\|_{L^p}. \tag{6.4}$$

The Sobolev space $W^{m,p}$ consists of those functions whose distributional derivatives up to order m are in L^p with the norm just defined. It is a Banach space. It is possible to define local versions of these spaces by restricting the integrals (for $p < \infty$) or the suprema (for $p = \infty$) to compact subsets. Sobolev spaces can be defined on an open subset of $\mathbb{R}^n$ in a way strictly analogous to the case of the whole of $\mathbb{R}^n$. Sobolev spaces can also be defined on manifolds. There are two basic approaches to doing this. In one of these a function f is written as a sum of pieces f_n each of which has compact support in a coordinate domain. Then the Sobolev norm of f is defined as the sum of the Sobolev norms of the functions $\tilde{f}_n$ obtained by transporting the functions f_n to $\mathbb{R}^n$ using the coordinates. The latter norms have already been defined since they concern functions on open subsets of Euclidean space. In the second approach Sobolev norms are defined using a smooth Riemannian metric on the manifold. The metric provides a volume form for integration and allows covariant derivatives to be used in the definitions of the norms. On a compact manifold all these definitions determine the same class of functions and the norms for difference choices of decompositions into pieces or Riemannian metrics are equivalent. This means that if $\|\ \|_1$ and $\|\ \|_2$ are two norms defined by different choices then there is a positive constant C such that $C^{-1}\|f\|_1 \leq \|f\|_2 \leq C\|f\|_1$ for all functions f in the space.

For finite $p > 1$ the Sobolev spaces are well adapted to the regularity properties of elliptic PDE. This is not so for $p = \infty$ or $p = 1$. Hyperbolic equations are more delicate and in that case only the choice $p = 2$ leads to optimal regularity results.

For elliptic equations on Euclidean space the question of solutions which tend to zero at infinity is important. In general relativity this is of relevance for the study of initial data for asymptotically flat spacetimes. In this context the ordinary Sobolev (or Hölder) spaces have certain disadvantages. It is sometimes appropriate to replace them by certain weighted Sobolev spaces. Let $\sigma(x) = (1+|x|^2)^{1/2}$. The weighted L^p norm is defined for $1 \leq p < \infty$ by

$$\|f\|_{L^p_\delta} = \left[\int_{\mathbb{R}^3} \sigma^{-\delta p - n} f^p\right]^{1/p}. \tag{6.5}$$

This coincides with the ordinary L^p space for $\delta = -n/p$ and gets larger with increasing δ. The weighted L^∞ norm is

$$\|f\|_{L^\infty_\delta} = \text{ess sup}(\sigma^{-\delta}|f|). \tag{6.6}$$

The weighted Sobolev spaces are defined by the norms

$$\|f\|_{W^{m,p}_\delta} = \sum_{|\alpha| \leq m} \|D^\alpha f\|_{L^p_{\delta - |\alpha|}}. \tag{6.7}$$

Define $H^k_\delta = W^{k,2}_\delta$. The intuitive picture is that functions in one of these spaces with weight δ grow or decay like r^δ at infinity, where $r = |x|$. A function which belongs to a weighted Sobolev space has the property that taking a derivative gives extra decay to the extent of one power of r. The advantages of the weighted Sobolev spaces are illustrated by the following example. If $\Delta \phi = 4\pi\rho$ for a non-negative smooth function ρ of compact support then, unless ρ is identically zero, ϕ does not belong to L^2 and hence also not to any unweighted Sobolev space based on L^2. This is because the leading order behaviour of the solution is given by m/r where $m = \int \rho(x) dx$ is the total mass. On the other hand the solution ϕ does belong to the weighted Sobolev spaces H^k_δ for $\delta > -1$.

In PDE theory it often happens that it is possible to obtain information about solutions in Sobolev spaces but in the end information on the pointwise properties of the solutions is required. In order to get this information it is necessary to relate different kinds of function spaces. This is done by means of the Sobolev embedding theorems [1]. The general form of these theorems is as follows. There are two Banach spaces B_1 and B_2 of functions on $\mathbb{R}^n$ and the theorem says that a function f which belongs to B_1 also belongs to B_2 and satisfies the inequality $\|f\|_{B_2} \leq C\|f\|_{B_1}$ for a constant C not depending on f. In other words there is an embedding of B_1 into B_2. In many cases this embedding is actually compact. Some Sobolev embedding theorems will now be collected.

Let a (non-standard) concept 'effective regularity' be defined as follows. For a function in the space $W^{k,p}_{\text{loc}}$ in n dimensions with p finite the effective regularity is defined to be $k - n/p$. For the space $C^{k,\alpha}_{\text{loc}}$ the effective regularity is defined to be $k+\alpha$. Then the following Sobolev embedding principle holds. If $f \in W^{k,p}_{\text{loc}}$ then f belongs to any local Sobolev or Hölder space whose effective regularity is no greater. To give an important example, if $f \in W^{k,p}_{\text{loc}}$ in n dimensions with $k > n/p$ then f is continuous. On a compact manifold M there are corresponding embedding theorems. For instance, if $k - n/p > l - n/q$ then any function $f \in W^{k,p}(M)$ belongs to $W^{l,q}(M)$ and there is a constant C such that $\|f\|_{L^{l,q}} \leq C\|f\|_{W^{k,p}}$ for all f in $W^{k,p}(M)$. Similarly, if $k-n/p > l+\alpha$ with $0 < \alpha < 1$ then there is a constant C such that $\|f\|_{C^{l,\alpha}} \leq C\|f\|_{W^{k,p}}$. In these inequalities if the effective regularity corresponding to the space on the left is strictly smaller than that corresponding to the space on the right then the embedding is compact. In the case of equal effective regularity compactness fails. There are suggestive analogues of these results for C^k spaces (assigned the effective regularity k) which are unfortunately false. For instance on $\mathbb{R}^2$ a function in H^1_{loc} need not be continuous or even locally bounded. In the case of Sobolev spaces on the whole of $\mathbb{R}^n$ it is necessary to consider the asymptotic behaviour as well as the local regularity. To get correct results for the embedding theorem relating two Sobolev spaces it is enough to add the requirement that $q \geq p$ in order that the local results on existence and continuity of embeddings remain valid. A well-known example is the embedding of H^1 into L^6 in $\mathbb{R}^3$. Compactness of the embedding for a strict decrease of the effective regularity does not hold for the case of $\mathbb{R}^n$. There are also embedding theorems for weighted Sobolev and Hölder spaces [18].

When proving Sobolev embedding theorems it is often possible to apply Lemma 6.1. If the smooth functions of compact support are dense in the space of interest (which is true for the Sobolev spaces with $p < \infty$) then it is enough to prove the desired inequality for smooth functions of compact support. Then the lemma guarantees that an embedding theorem holds and that the inequality holds for all functions in the Sobolev space.

In the process of proving existence and uniqueness theorems for nonlinear PDE in the setting of Sobolev spaces it is important that functions in Sobolev spaces including a sufficient number of derivatives can be multiplied and composed with smooth functions (or even with each other) in a manner which is continuous in the topology of the Sobolev space. For instance for $s > n/2$ the following estimates hold for functions on $\mathbb{R}^n$:

$$\|fg\|_{H^s} \leq C\|f\|_{H^s}\|g\|_{H^s}, \tag{6.8}$$

$$\|F(f)\|_{H^s} \leq \Phi(\|f\|_{H^s}). \tag{6.9}$$

6.3 Concrete function spaces

Here C is a constant, F a smooth function and Φ a continuous function. These are central estimates which go into the existence proofs for solutions of quasilinear hyperbolic equations. There are also other, stronger, estimates which can be used to give better continuation criteria than those just mentioned for nonlinear hyperbolic equations. These are the following Moser estimates:

$$\|D^\alpha(fg)\|_{L^2} \leq C(\|D^s f\|_{L^2}\|g\|_{L^\infty} + \|f\|_{L^\infty}\|D^s g\|_{L^2}), \quad (6.10)$$

$$\|D^\alpha(fg) - f D^\alpha g\|_{L^2} \leq C(\|D^s f\|_{L^2}\|g\|_{L^\infty} + \|Df\|_{L^\infty}\|D^{s-1} g\|_{L^2}), \quad (6.11)$$

$$\|D^\alpha(F(f))\|_{L^2} \leq \Phi(\|f\|_{L^\infty})\|D^s f\|_{L^2}. \quad (6.12)$$

Here $s = |\alpha|$ and F is a smooth function with $F(0) = 0$. Combining them with the Sobolev embedding theorem gives the previous set of estimates but they are stronger because of the fact that the norms with the largest number of derivatives (s) occur linearly on the right hand side. An estimate with this property is sometimes called tame.

A linear mapping from a Banach space B_1 to a Banach space B_2 is known as a linear operator. Thus, for instance, the Laplacian defines a linear operator from $C^2(T^n)$ to $C^0(T^n)$. In functional analysis the notion of an unbounded operator is often used. In general an unbounded operator from B_1 to B_2 does not just mean a linear mapping from B_1 to B_2 which happens not to be bounded. It often means that the linear mapping is only defined on a linear subspace X of B_1, called the domain of the operator. It is usually assumed that X is dense so that any element of B_1 can be approximated by elements of X in the norm of B_1. This kind of situation often arises in the context of concrete function spaces due to the loss of differentiability caused by differential operators. For instance, the Laplacian can be thought of as an unbounded operator from $L^2(T^n)$ to $L^2(T^n)$ with domain $H^2(T^n)$. The wave equation can be written formally as $d^2 u/dt^2 = Lu$ where L is the unbounded operator just defined. Thus the wave equation looks like an ODE on the Banach space $L^2(T^n)$. This is a dangerous point of view since L is not defined globally on $L^2(T^n)$. As it stands it is not helpful in proving existence theorems. Great caution is need when thinking of a PDE as an ODE in an infinite-dimensional space. It is often quite unhelpful. On the other hand it can be a valuable source of intuition when applied in the right way.

It is possible to extend the definition of the Sobolev spaces $W^{k,p}(\mathbb{R}^n)$ to negative values of the index k. Let B be the Banach space $W^{k,p}(\mathbb{R}^n)$ with $1 < p < \infty$ and k a non-negative integer. Since any test function belongs to B it is possible to interpret an element of B^* as a distribution. Call the space of distributions of this type $W^{-k,q}$. This definition leads to a consistent interpretation of the notation $W^{k,p}$ for any integer k and $1 < p < \infty$. The fact that L^p is the dual space of L^q avoids a conflict in the case $k = 0$.

It is also possible to define Sobolev spaces for non-integer values of k and since these are important in some applications (e.g. elliptic boundary value problems) this will be discussed briefly here. A concept which provides one route to defining these spaces is that of interpolation spaces. Let B_1 and B_2 be two Banach spaces and $\theta \in (0, 1)$. Then it is possible to define Banach spaces $[B_1, B_2]_\theta$. The intuitive idea is that these represent a continuous deformation of spaces from B_1 to B_2. As an example $[L^p(\mathbb{R}^n), L^q(\mathbb{R}^n)]_\theta = L^r(\mathbb{R}^n)$ where $r^{-1} = (1-\theta)p^{-1} + \theta q^{-1}$. Not all similarly neat looking relations involving interpolation spaces are true. It may be hard work to identify interpolation spaces with already known spaces. It should also be noted that there are two different kinds of interpolation, called the real and complex methods. A key property of interpolation spaces is as follows. If L_1 and L_2 are bounded linear operators from spaces $B_{1,0}$ to $B_{2,0}$ and $B_{1,1}$ to $B_{2,1}$ respectively then there exist bounded linear operators L_θ from $B_{1,\theta}$ to $B_{2,\theta}$ where $B_{i,\theta} = [B_{i,0}, B_{i,1}]_\theta$ for $\theta \in (0,1)$ and $i = 1, 2$. It is possible to define Sobolev spaces having a differentiability index which is a general real number by the relation $W^{\theta k, p}(\mathbb{R}^n) = [W^{k,p}(\mathbb{R}^n), L^p(\mathbb{R}^n)]_\theta$ with k an integer. Once the definition has been made the relation is true for all real k.

In the study of boundary value problems it is often necessary to restrict functions to smooth submanifolds. The functional analytic properties of this process can already be seen in the context of restricting functions on $\mathbb{R}^n$ to a linear subspace $\mathbb{R}^{n-m}$. The restriction of smooth functions from $\mathbb{R}^n$ to $\mathbb{R}^{n-m}$ extends uniquely to a continuous linear mapping from $W^{k,p}_{\text{loc}}(\mathbb{R}^n)$ to $W^{k-m/p, p}_{\text{loc}}(\mathbb{R}^{n-m})$. Intuitively it can be said that restricting a function in a Sobolev space of type L^p to a submanifold of codimension m leads to a loss of m/p derivatives. Conversely, if f is a function belonging to $W^{k-m/p,p}_{\text{loc}}(\mathbb{R}^{n-m})$ there is a function in $W^{k,p}_{\text{loc}}(\mathbb{R}^n)$ whose restriction is equal to f.

There are classes of functions intermediate between C^∞ and analytic (also called C^ω). These are the Gevrey spaces. They are defined by an inequality on their Taylor coefficients related to that characterizing analytic functions stated in the one-dimensional case in Section 5.3. The Gevrey class G^s is defined by the condition that on a neighbourhood of the origin in $\mathbb{R}^n$ a function f should satisfy the inequalities

$$|D^\alpha f(x)| \leq C^{|\alpha|+1}(\alpha!)^s \qquad (6.13)$$

for some constants C and $s > 1$ and all multiindices α. For $s = 1$ this coincides with the analytic functions and in some sense the set of C^∞ functions is the limit of the G^s functions for $s \to \infty$. Hyperbolic equations can be solved in Gevrey spaces and there are equations which are solvable in Gevrey spaces although they are not solvable in the class of C^∞ functions.

An important difference between Gevrey functions and analytic functions is that the former do not satisfy unique continuation. An analytic function defined on $\mathbb{R}^n$ which vanishes on an open subset vanishes everywhere. In contrast Gevrey functions on $\mathbb{R}^n$ with compact support are abundant. Note, for example, that the function which is equal to $e^{-1/x}$ for $x > 0$ and identically zero for $x \leq 0$, which is often used in constructing smooth cut-off functions, belongs to the class G^2.

6.4 Littlewood–Paley theory

A very powerful technique for discussing the regularity of functions with applications to PDE is Littlewood–Paley theory. Let u be a tempered distribution on $\mathbb{R}^n$ and $\hat{u}$ its Fourier transform. Consider now a sequence of smooth functions $\{\phi_n, n \geq 0\}$ with the following properties. The function ϕ_0 is supported in the unit ball about the origin. The function ϕ_1 is supported in the region $1/2 \leq |x| \leq 2$. For $n \geq 1$ the relation $\phi_{n+1}(x) = \phi_n(x/2)$ holds. Finally $\sum_{n=0}^{\infty} \phi_n(x) = 1$ for all $x \in \mathbb{R}^n$. Note that the sum makes sense since only a finite number of summands are non-vanishing at any given point of $\mathbb{R}^n$. A sequence of functions of this kind is called a *dyadic partition of unity*. It is not hard to show that such a thing exists. Define u_n to be the inverse Fourier transform of $\phi_n \hat{u}$. Heuristically it can be said that u has been decomposed into a sum of an infinite number of pieces although the sense in which the sum of pieces is to be interpreted is clearer for the Fourier transforms of the functions involved than for the functions themselves. It can be said that u_n contains only a part of u with frequencies of the order of 2^n. The Fourier transform of each u_n has compact support and so it follows from the Paley–Wiener theorem that each u_n is analytic. All trace of non-smoothness of u has disappeared from the individual dyadic pieces u_n and is encoded in their growth or decay as $n \to \infty$.

Because Littlewood–Paley theory makes such essential use of the Fourier transform it appears tied to Euclidean space (or the torus) and far from the possibility of an invariant formulation on general Riemannian manifolds. There is, however, an alternative approach which allows at least major parts of Littlewood–Paley theory to be reformulated in a geometric way. For this purpose the function to be decomposed is taken as an initial datum for the heat equation on a Riemannian manifold. The decomposition is then extracted from the late-time asymptotics of the solution. For the study of solutions of nonlinear partial differential equations it may be preferable to replace the heat equation by another geometric parabolic equation adapted to the problem at hand.

Littlewood–Paley theory can be used to unify the definitions of most of the function spaces introduced in this book up to this point and to define many more. The Besov spaces $B^s_{p,q}$ are defined by the norm

$$\|u\|_{B^s_{p,q}} = \left(\sum_{j=0}^{\infty} \left(\int_{\mathbb{R}^n} (2^{js}|u_j|)^p \right)^{q/p} \right)^{1/q}, \tag{6.14}$$

while the Lizorkin–Triebel spaces $F^s_{p,q}$ are defined by

$$\|f\|_{F^s_{p,q}} = \left(\int \left(\sum_{j=0}^{\infty} (2^{js}|u_j|)^q \right)^{p/q} \right)^{1/p}. \tag{6.15}$$

There are corresponding definitions with one or both of the indices equal to infinity where the sum or integral is replaced by the (essential) supremum, as necessary. These families contain most of the Sobolev spaces since $W^k_p = F^k_{p,2}$ for $1 < p < \infty$ and the Hölder spaces since $C^{k,\alpha} = B^{k+\alpha}_{\infty,\infty}$. This last statement prompts the question what the spaces $B^k_{\infty,\infty}$ are when k is an integer. They are called Zygmund spaces and are denoted C^k_*. The space C^1_* has an elementary characterization in terms of second order difference quotients. Solving an elliptic equation leads to a gain of two derivatives in Zygmund spaces.

What is the use of these more unusual function spaces? It often happens in PDE problems that an embedding which would be enough to solve the problem under consideration just fails in the sense that changing an index in a family of function spaces by an arbitrarily small amount leads to a valid embedding. Then it may happen that changing to a different kind of space (for instance replacing a Sobolev space by a Besov space) may allow the argument to be carried through. For instance in the refined study of the geometry of null hypersurfaces in [123] the space L^∞ in two dimensions is replaced by the Besov space $B^1_{2,1}$ which embeds into it.

6.5 Pseudodifferential operators

Consider a linear differential operator P defined by $Pu = \sum_{|\alpha| \leq s} A^\alpha D^\alpha u$ for multiindices α. If u is a tempered distribution the Fourier transform $\mathcal{F}$ can be used to write $P = \mathcal{F}^{-1} M \mathcal{F}$ where M is the operator of multiplication by the function $\sum_{|\alpha| \leq s} A^\alpha i^{|\alpha|} \xi^\alpha$ which is equal to $\sigma(P)(i\xi)$ where $\sigma(P)$ is the principal symbol of P which is defined in Section 7.1. Here it is seen how the principal symbol arises in the context of the Fourier transform.

6.5 Pseudodifferential operators

The idea of a pseudodifferential operator is to replace the symbol $\sigma(P)$ by a more general function of x and ξ which is supposed to satisfy estimates on growth or decay in $|\xi|$ for $|\xi| \to \infty$ similar to those satisfied by the symbol of a differential operator of order k. It should grow no faster than $|\xi|^k$. Each derivative with respect to ξ should reduce the order of growth by one while each derivative with respect to x should leave the order of growth unchanged. Let S^k be the set of symbols of order k. A pseudodifferential operator (ΨDO) of order k can be defined by conjugating the operator of multiplication with the symbol by the Fourier transform. Note that the order of a pseudodifferential operator can be any real number. In particular it can be negative. A symbol which satisfies the estimates required of a symbol for all real numbers k is called a symbol of order $-\infty$ and by definition $S^{-\infty} = \cap_{k \in \mathbb{R}} S^k$. The corresponding pseudodifferential operators of order $-\infty$ play an important role in the theory. Many calculations are carried out modulo operators of order $-\infty$. The calculus of pseudodifferential operators makes use of a certain notion of asymptotic series. Let $\{a_m\}$ be a sequence of symbols where $a_m \in S^{k_m}$ and $\{k_m\}$ is a decreasing sequence of real numbers which tends to $-\infty$ as $m \to \infty$. If a is a symbol then writing $a \sim \sum_k a_{k_m}$ means by definition that $a - \sum_{m \leq l} a_{k_m} \in S^{k_{l+1}}$ for each l. Given a sequence of symbols $\{a_{k_m}\}$ as above there always exists a symbol a such that $a \sim \sum_k a_{k_m}$. This determines a up to addition of an element of $S^{-\infty}$. An operator whose order is an integer k maps the Sobolev space $W^{s,p}$ into the Sobolev space $W^{s-k,p}$. In particular, the image of an element of $W^{s,p}$ under an operator of order $-\infty$ is a C^∞ function and these operators are called smoothing operators. There is a calculus of pseudodifferential operators which allows them to be used in an effective way. If P and Q are pseudodifferential operators of orders k_1 and k_2 then PQ is a pseudodifferential operator of order $k_1 + k_2$ and the symbol $\sigma(PQ)$ is equal to $\sigma(P)\sigma(Q)$ up to a remainder in $S^{k_1+k_2-1}$. The definition of ΨDO seems to be tied to Euclidean space because of the use of the Fourier transform. It is, however, possible to define ΨDO on open subsets of $\mathbb{R}^n$ and on manifolds. In the last case the symbols are naturally defined on the cotangent bundle and the operator associated to a symbol is only defined modulo smoothing operators. The basis for this definition is the behaviour of symbols under coordinate transformations which is also a part of the calculus of pseudodifferential operators. Under certain circumstances the inverse of an elliptic operator of order k can be defined as a pseudodifferential operator of order $-k$. The symbol of the inverse of P is equal to the inverse of the symbol up to a remainder which is an operator of order $-k - 1$. When the mapping properties of ΨDO between Sobolev spaces are known this immediately gives statements about elliptic regularity discussed in the next chapter for the invertible case. The function spaces which are well adapted to elliptic equations are those for which pseudodifferential operators have good mapping properties. More generally what can be obtained instead of an inverse is a parametrix, an inverse modulo smoothing operators. In other words, if P is an operator a parametrix for P is an operator such that $PQ - I$ and $QP - I$ are smoothing operators. Parametrix constructions for hyperbolic operators are also possible but these require the theory of Fourier integral operators which is more complicated than the theory of ΨDO and will not be discussed in this book. The spaces which are well adapted to hyperbolic equations are those for which Fourier integral operators have good mapping properties.

On a compact Riemannian manifold it is possible to define a pseudodifferential operator Λ of order one which defines an isomorphism from $H^s(M)$ to $H^{s-1}(M)$. This can be used to adjust the order of an operator. For instance if P is an operator (differential or pseudodifferential) of order k then $\Lambda^{-k}P$ is a pseudodifferential operator of order zero.

6.6 Further reading

A general reference on abstract functional analysis is [190]. Good sources for many of the topics of this chapter are [2] and chapter 13 of [197]. An enlightening comparison of Banach and Fréchet spaces and an account of the Nash–Moser theorem can be found in [97]. That paper also introduces the category of tame Fréchet spaces as a framework for understanding the Nash–Moser theorem. It is related to the tame estimates mentioned in the text. A detailed treatment of Sobolev spaces on compact Riemannian manifolds is given in [15]. There are two different conventions for defining weighted Sobolev spaces. Here the convention of [18] is used where the coefficient δ corresponds directly to the rate of fall-off of the functions in that space. A text on interpolation spaces is [28]. For the geometric approach to Littlewood–Paley theory see [124] and [196]. For information about Fourier integral operators the reader is referred to [194].

7 Elliptic equations

This chapter is an introduction to elliptic equations. These often provide models for equilibrium situations in physics or other sciences. In general relativity they are also important as a tool for studying the set of solutions of the constraints. Another application is to the equations defining certain coordinate conditions.

7.1 The concept of ellipticity

In this section some basic concepts are introduced which are useful in the classification of systems of partial differential equations. They are then applied to define what elliptic equations are. Consider a system of k partial differential equations for a function $u = u(x)$ on $\mathbb{R}^n$ of the following form

$$F(x, D^s u) = 0. \tag{7.1}$$

Here $D^s u$ denotes the collection of all partial derivatives of u up to order s. This can alternatively be expressed by saying that $D^s u$ denotes the collection of derivatives $D^\alpha u$ for all multiindices with $|\alpha| \leq s$. To investigate the structure of systems of partial differential equations it is useful to linearize them. The basic idea is the same as was explained for ordinary differential equations in Section 5.4. Consider a one parameter family $u(\lambda)$ of solutions of the system (7.1). Let $v = \partial u/\partial \lambda$ evaluated at $\lambda = 0$. Then by the chain rule

$$\sum_{|\alpha| \leq s} \frac{\partial f}{\partial (D^\alpha u)}(x, D^\alpha u) D^\alpha v = 0. \tag{7.2}$$

This linear system is the linearization of (7.1) at the solution u. The reader who would like to see how linearization can be formulated in a completely invariant way on manifolds is encouraged to look at [158]. Consider now the general linear system of PDE given by

$$\sum_{|\alpha|\leq s} A^\alpha(x) D^\alpha u = 0. \tag{7.3}$$

It turns out that key properties of the equation are often determined by its principal part which is obtained by keeping only the terms in (7.3) with $|\alpha| = s$, i.e. those of the highest order. The *principal symbol* of (7.3) is the mapping from $\mathbb{R}^{2n}$ to k by k real matrices defined by

$$\sigma(x^i, \xi^i) = \sum_{|\alpha|=s} A^\alpha(x) \xi^\alpha. \tag{7.4}$$

The characteristic polynomial of (7.3) is defined to be the determinant of the principal symbol and is a real-valued function on $\mathbb{R}^{2n}$. The characteristic set is the set of all (x, ξ) with $\xi \neq 0$ for which the characteristic polynomial vanishes. A hypersurface is called characteristic if any covector which vanishes on all vectors tangent to the hypersurface belongs to the characteristic set. In a characteristic hypersurface there are preferred curves which are called bicharacteristics. They are also sometimes denoted by the overworked name 'characteristic'. This is particularly common in the case of a PDE for a single function. The system (7.3) is called elliptic if its characteristic set is empty. Of course this is equivalent to the condition that its principal symbol is invertible at any point (x, ξ). The system (7.1) is called elliptic if its linearization about any solution is elliptic. In some cases it is desirable to restrict consideration to solutions of (7.1) taking values in some specified open subset of $\mathbb{R}^k$ and only require ellipticity of the linearization for all solutions with values in that open set. Notice that the basic property defining ellipticity is a pointwise property of the coefficients. In other words, to determine whether a system is elliptic it is enough to consider the frozen coefficient problem

$$\sum_{|\alpha|=s} A^\alpha(x_0) D^\alpha u = 0. \tag{7.5}$$

for an arbitrary point x_0. This is a linear problem with constant coefficients and as such it could be approached using the Fourier transform. When this is done the principal symbol appears.

The standard example of an elliptic equation is the (flat space) Laplace equation $\Delta u = \delta^{ij} \partial_i \partial_j u = 0$. In this case the symbol is just $\delta_{ij} \xi^i \xi^j$. In a sense the class of elliptic systems can be thought of as a class which shares some of the most important properties of the Laplace equation. A more restricted class is that of strongly elliptic equations where the principal symbol is required to be not just invertible but actually uniformly positive definite on the unit ball in ξ-space.

An important issue for elliptic equations (and more generally for PDE) is that of regularity. Consider for instance the Poisson equation $\Delta u = f$, the simplest inhomogeneous elliptic equation. If u is C^2 then f is C^0. It would be nice if conversely the continuity of f implied that u was C^2. Unfortunately it is well known that this is not the case. Similarly, if f is C^k then u need not be C^{k+2}. Hölder spaces are better in this respect. If f is $C^{k,\alpha}$ then u is $C^{k+2,\alpha}$. Thus it can be said that solving the Poisson equation leads to a gain of two derivatives in Hölder spaces.

Elliptic equations also gain two derivatives in Sobolev spaces with $1 < p < \infty$. In the example of the Poisson equation, if f belongs to $W^{m,p}_{loc}$ then $u \in W^{m+2,p}_{loc}$. This is a general property of elliptic systems with smooth coefficients. The corresponding property fails for $p = \infty$. This regularity statement holds for any real number m. In particular, it is true in cases where the right hand side is a distribution and the derivatives in the Laplacian are interpreted as distributional derivatives. This implies also that any distribution which satisfies the Laplace equation on $\mathbb{R}^n$ is in fact a function of class C^∞.

The fact that the Poisson equation does not have good regularity properties in C^k spaces may at first sight appear mysterious and so it will now be examined in some detail, restricting to the two-dimensional case for simplicity. In two dimensions the Poisson equation takes the form

$$\frac{\partial^2 u}{\partial r^2} + \frac{1}{r}\frac{\partial u}{\partial r} + \frac{1}{r^2}\frac{\partial^2 u}{\partial \theta^2} = f \qquad (7.6)$$

in polar coordinates. Consider first the case where u and f are both rotationally symmetric so that the equation reduces to an ordinary differential equation which is singular at $r = 0$. Assume further that f is continuous, so that $f(r) = f(0) + o(1)$, and that u is bounded. In this case the equation can be rewritten as $\frac{d}{dr}\left(r\frac{du}{dr}\right) = rf$. The function u is evidently C^2 for $r \neq 0$ and so it only remains to examine its behaviour at the origin. The right hand side is equal to $rf(0) + o(r)$. Hence $r\frac{du}{dr}$ is of the form $C + \frac{1}{2}f(0)r^2 + o(r^2)$ for a constant C. If C were non-zero then u would be unbounded. Thus with given assumptions $du/dr = \frac{1}{2}f(0)r + o(r)$. It follows that $u = u(0) + \frac{1}{4}f(0)r^2 + o(r^2)$. Putting this information back into the equation shows that d^2u/dr^2 extends continuously to the origin where it takes the value $\frac{1}{2}f(0)$. Thus regularity in C^k does hold in the case $k = 2$ with rotational symmetry.

To find an example where regularity fails in C^k it is necessary to consider functions with a more complicated angular dependence. Suppose that when u is expressed in polar coordinates it is of the form $U(r)\sin l\theta$ for some integer $l \geq 1$. The function u satisfies the Laplace equation if and only if $U'' + \frac{1}{r}U' - \frac{l^2}{r^2}U = 0$ and this ODE can be used to construct interesting examples. One of these is obtained by multiplying the function $u(r,\theta) = r^2(-\log r)^{1/2} \sin 2\theta$ by a smooth cut-off function which is equal to one for $r \leq \frac{1}{2}$ and vanishes for $r \geq \frac{3}{4}$. This means choosing $l = 2$ and $U(r) = r^2(-\log r)^{1/2}$. In this case

$$U'' + \frac{1}{r}U' - \frac{l^2}{r^2}U = -2(\log r)^{-1/2} + \frac{1}{4}(\log r)^{-3/2}, \tag{7.7}$$

which is continuous. However the second derivatives of u are not bounded near the origin – they diverge like $(-\log r)^{1/2}$. Thus u is not C^2 and in fact not even in $W^{2,\infty}_{\text{loc}}$. It is in $W^{2,p}_{\text{loc}}$ for any $p < \infty$ as the general theorems say it must be.

7.2 Boundary value problems

For elliptic systems it is natural to pose boundary value problems. Let Ω be a bounded open subset of $\mathbb{R}^n$ with smooth boundary. In the following the emphasis will be on boundary value problems for domains Ω of this type although generalizations to non-compact domains, manifolds and rough boundaries are possible. Consider a system of the form (7.1) with smooth coefficients defined on Ω and another operator $Q(x, D^r u)$ defined for x belonging to the boundary $\partial\Omega$ of Ω. A boundary condition is imposed by requiring that, for $x \in \partial\Omega$, Qu is equal to a given function u_0 defined on $\partial\Omega$. The key questions are whether for given u_0 there exists a solution of (7.1) satisfying the boundary condition, whether that solution is uniquely determined by u_0 and whether it depends continuously on u_0 in a suitable topology. If all these three conditions hold it is said that the boundary value problem is well posed. The standard examples of a boundary value problem are when the equation is the Laplace equation and the operator Q is given by the identity or by the unit normal derivative to the boundary. These are known as the Dirichlet and Neumann problems respectively. For a compact domain Ω with smooth boundary the Dirichlet problem is well posed. The mapping $u \mapsto (\Delta u, u|\partial\Omega)$ defines an isomorphism from $H^2(\Omega)$ to $L^2(\Omega) \times H^{3/2}(\partial\Omega)$. Here is a case where it is necessary to use Sobolev spaces with non-integer differentiability index. For the Neumann problem things are a little more complicated. The problem can only be solved if the function prescribed on the boundary has integral zero. This follows from Stokes' theorem. On the other hand adding an arbitrary constant to any solution to the problem gives another one. Thus both existence and uniqueness fail. Nevertheless the Neumann problem is not really badly behaved. It has Fredholm properties as discussed in Section 7.4.

7.3 Douglis–Nirenberg ellipticity

The concept of ellipticity of a system of PDE depends only on the principal part, i.e. on the terms containing the highest order derivatives. A generalization of this is Douglis–Nirenberg ellipticity where, in a sense which will be explained, the orders are adjusted in a certain way before determining the principal part. This allows a system consisting of equations of different orders to satisfy the modified definition of ellipticity under certain conditions. Consider a linear system $Pu = 0$ and split the components of unknown into L subsets so that u is written as a sequence of vector-valued unknowns $u_1, \ldots, u_L$. Then the system may be written as $\sum_j P_{ij} u_j = 0$ where i and j run from 1 to L. (It is assumed that the left and right hand sides of the equation belong to the same space.) Let the order of the operator P_{ij} be no greater than m_{ij}. Let σ_{ij} be the principal symbol of P_{ij} if the order of P_{ij} is equal to m_{ij} and zero otherwise. Suppose there are integers s_i and t_j such that $m_{ij} = s_i - t_j$ for all i and j. Define a matrix M consisting of the blocks $|\xi|^{-m_{ij}} \sigma_{ij}$. This matrix is homogeneous of order zero in ξ. The system is said to be elliptic in the sense of Douglis and Nirenberg if the indices s_i and t_j can be chosen so that the matrix M is invertible for all ξ. A version of the ordinary elliptic theory can be developed for these systems.

A way of seeing what this definition means intuitively is the following. Let Λ be a pseudo-differential operator which defines an isomorphism from H^s to H^{s-1} for each s. Let $v_i = \Lambda^{s_i} u_i$ and multiply the equations $P_{ij}(u_i) = 0$ on the left by Λ^{t_j}. Then the equations become $Q_{ij}(v_i) = 0$ where $Q_{ij} = \Lambda^{t_j} P_{ij} \Lambda^{-s_i}$. The principal symbol of the pseudo-differential operator Q made up of the matrix of operators Q_{ij} is invertible if and only if the matrix M introduced previously is invertible. In this case it may be said that the pseudo-differential operator Q is elliptic. The idea of Douglis–Nirenberg ellipticity was originally defined by analogy with that of Leray hyperbolicity which is discussed in Section 8.5.

7.4 Fredholm operators

Let B_1 and B_2 be Banach spaces and L a bounded linear mapping from B_1 to B_2. The operator L is called Fredholm if L has closed range im L and ker L, the kernel of L, and coker $L = B_2 / \text{im} L$ are finite-dimensional. The index of L is defined as dim ker L – dim coker L. If B_1 and B_2 are finite-dimensional then it is elementary that any linear operator is Fredholm and that the index is zero. Fredholm operators have the advantage that they share many properties of operators between finite-dimensional spaces although their index is in general non-zero. Let $L(\lambda)$ be a continuous mapping from an interval in $\mathbb{R}$ to the space $L(B_1, B_2)$ of bounded linear operators whose image consists of Fredholm operators. Then ind $L(\lambda_1) =$ ind $L(\lambda_2)$ for all λ_1 and λ_2 in the interval. An advantage of this is that if $L(\lambda_1)$ is an operator whose index can be calculated rather easily and $L(\lambda_2)$ is a more complicated operator then it allows the index of $L(\lambda_2)$ to be computed. Another useful fact for computing the index is that if L is a Fredholm operator and K a compact operator then $L + K$ is Fredholm and ind $(L + K) =$ ind L. This is of relevance for differential operators since it is often the case that if the principal part of a linear differential operator defines a bounded linear operator between two function spaces then the remainder, consisting of the lower order terms, defines a compact operator between the same spaces.

Fredholm operators play an important role in the theory of elliptic systems. The fact that an elliptic operator on a compact manifold defines a Fredholm operator between suitable Sobolev spaces can be deduced from the existence of a parametrix. The Laplace operator on Euclidean space defines a Fredholm operator between weighted Sobolev spaces provided the weight is not an integer. It is also Fredholm for a finite exceptional set of integers. The same is true for more general elliptic operators whose coefficients decay appropriately at infinity. This is for instance true for the Laplace operator defined by an asymptotically flat metric. The loss of the Fredholm property for integer weights is associated with harmonic polynomials which lead to a change in the dimension of the kernel while the dimension of the cokernel remains fixed or the other way round. This means a change of the index which is not consistent with persistence of the Fredholm property.

A bounded linear operator L from a (real) Hilbert space H to itself is called symmetric or self-adjoint if $\langle Lf, g \rangle = \langle f, Lg \rangle$ for all f and g in H. An unbounded operator is called symmetric if this relation holds for f and g belonging to a suitably defined domain of L. Consider now the linear differential operator P defined by $Pf = \sum_{|\alpha| \leq k} A^\alpha D^\alpha$. Its *formal adjoint* is the operator P^* defined by $P^*f = \sum_{|\alpha| \leq k} (-1)^{|\alpha|} D^\alpha (A^\alpha f)$. If $P = P^*$ then P is called formally self-adjoint. The differential operator P can be thought of as defining an unbounded operator L from the Hilbert space $L^2(\mathbb{R}^n)$ to itself with domain the smooth functions of compact support. It is symmetric. For unbounded operators the property of being self-adjoint is much more restrictive than that of being symmetric and will not be entered into here. It is important for the spectral theory of unbounded operators. The reason for touching on this subject here is to point out the following. If a bounded Fredholm operator is self-adjoint then its index is zero since its cokernel can be identified with the kernel of its adjoint. If a differential operator is formally self-adjoint and defines a Fredholm operator between appropriate function spaces then it is often possible to show that its index is zero.

7.4 Fredholm operators

Consider for instance a second order linear elliptic operator P defined on a compact manifold M which is formally self-adjoint. It defines a bounded linear operator from $H^2(M)$ to $L^2(M)$. It will be shown that its index is zero. If u is in the kernel of P then by elliptic regularity it is C^∞. The dimension of the cokernel of P is equal to the dimension of the orthogonal complement of its image in L^2. If u is a function in this orthogonal complement then by the definition of distributional derivatives it is a distributional solution of $P^*u = 0$ where P^* is the formal adjoint of P. However $P^* = P$ and so it follows that the orthogonal complement consists of C^∞ solutions of the equation $Pu = 0$. It follows that the kernel and cokernel have the same dimension.

Now a simple example will be worked out. In a spacetime with a compact Cauchy hypersurface a well-known coordinate condition is to choose a time coordinate whose level hypersurfaces have constant mean curvature. The value of the mean curvature typically varies monotonically from one hypersurface to the next and the mean curvature itself can be used as a time coordinate. It follows from the evolution equation (2.37) for the mean curvature that with this type of CMC time coordinate the lapse function satisfies the elliptic equation:

$$L\alpha = \Delta\alpha - [k_{ab}k^{ab} + 4\pi(\rho + \mathrm{tr}S)]\alpha = 1. \tag{7.8}$$

Suppose that the spacetime is smooth and consider L as defining a linear mapping from H^2 to L^2. Assume now that the strong energy condition is satisfied. Then the expression in square brackets in this equation is non-negative. It can be concluded that $\langle L\alpha, \alpha\rangle_{L^2} \leq 0$ with equality only when α is constant, $k_{ab} = 0$ and $\rho = 0$. Suppose that the last two conditions are not both satisfied. Then it can be concluded that the kernel of L is trivial. Let f be a function which satisfies $\langle Lu, f\rangle_{L^2} = 0$ for all $u \in H^2$. Then $Lf = 0$ in the sense of distributional derivatives. By elliptic regularity it follows that f is smooth and hence that $f = 0$. Thus the cokernel of L is trivial and L is an isomorphism from H^2 to L^2. It follows that the lapse equation (7.8) has a unique solution in H^2. By elliptic regularity the solution is smooth if the initial data defining its coefficients are.

The main property of a well behaved boundary value problem for a linear elliptic equation is that the mapping $u \mapsto (Pu, Qu)$ should be Fredholm in suitable spaces. This is true in the case of the Neumann problem for the Laplace equation. There the mapping is Fredholm with index zero and the one-dimensional kernel and cokernel have been exhibited in Section 7.2. Well-posedness is the condition that the Fredholm operator is actually an isomorphism and this usually depends on specific features of the equation and boundary conditions. In the case of a scalar equation, for instance, it is often possible to use the maximum principle to get further information.

In the simplest case this is the statement that a non-constant solution of the Laplace equation cannot have a maximum in the interior of its domain of definition. When the Fredholm property is satisfied this implies concretely that existence is obtained for all data belonging to a subspace of finite codimension and that uniqueness holds up to the addition of an arbitrary element of a space of finite dimension.

To obtain the Fredholm property for a boundary value problem it is necessary to supplement the condition on the principal symbol of the operator defining ellipticity by conditions on the principal symbol of the boundary operator known as the Lopatinskii–Shapiro conditions. The first step in discussing this problem is to linearize the equation, the boundary operators and the boundary itself. This means that the original problem is replaced by a simplified one in the hope that an understanding of the simpler problem can provide a criterion for understanding the original problem. Due to the linearization of the boundary it is enough to consider equations with boundary conditions given on a half-space in Euclidean space. For convenience the notation is chosen so that y denotes a coordinate normal to the boundary and increasing towards the region where the equation is to be solved and x denotes coordinates tangential to the boundary collectively. A further simplification is to freeze the coefficients at a given point x_0. The result is a linear system with constant coefficients

$$\sum_{j+|\alpha|\leq s} A^{\alpha,j} \partial_y^j D^\alpha u = 0, \tag{7.9}$$

where the multiindices refer to the variables x. The boundary condition can be written as

$$\sum_{j+|\alpha|\leq r} B^{\alpha,j} \partial_y^j D^\alpha u(0,x) = u_0(x). \tag{7.10}$$

A solution is sought on the region $y \geq 0$. Taking a Fourier transform in the variables x leads to the ordinary differential equation

$$\sum_{j+|\alpha|\leq s} A^{\alpha,j}(i\xi)^\alpha \partial_y^j \hat{u} = 0 \tag{7.11}$$

and the initial condition

$$\sum_{j+|\alpha|\leq r} B^{\alpha,j}(i\xi)^\alpha \partial_y^j \hat{u}(0,\xi) = \hat{u}_0(\xi). \tag{7.12}$$

Now consider the question whether the ordinary differential equation (7.11) has a unique bounded solution on the interval $[0,\infty)$ with an arbitrary initial datum satisfying (7.12). If so the boundary value problem is said to satisfy the Lopatinskii–Shapiro condition. Then the mapping $u \mapsto (P(u), Q(u))$ is Fredholm between suitable spaces. To understand this condition a bit better consider the Laplace equation $\sum_{i=1}^{n} \partial_i^2 u + \partial_y^2 u = 0$ with initial data given on $y = 0$. In this case the equation (7.11) takes the form

$$\partial_y^2 \hat{u} - |\xi|^2 \hat{u} = 0. \tag{7.13}$$

The general solution is of the form

$$\hat{u}(y) = \alpha e^{|\xi|y} + \beta e^{-|\xi|y}. \tag{7.14}$$

The only bounded solutions are those with $\alpha = 0$. Dirichlet boundary conditions lead to the problem of prescribing $\alpha + \beta$ and hence β. Thus there is a unique bounded solution with prescribed data, as desired. In the case of Neumann boundary conditions $\alpha - \beta$ should be prescribed and so once again there are unique bounded solutions corresponding to appropriate data. Thus both Dirichlet and Neumann boundary conditions satisfy the Lopatinskii–Shapiro conditions. Consider next the mixed boundary condition $n^\alpha \nabla_\alpha u + ku = 0$. This leads to prescribing $\alpha(1 + k) + \beta(-1 + k)$. The Lopatinskii–Shapiro condition is satisfied for all values of k except $k = 1$.

7.5 The Einstein constraints

The Einstein constraints are the equations which must be satisfied by initial data for the Einstein equations. As they stand they do not belong to any well-known type of differential equations but there is a standard method to reduce the solution of these equations to solving elliptic equations. It is known as the conformal method. The general idea is to choose certain parts of the solution freely and then to solve elliptic equations for the remaining parts. Recently other methods of solving the constraints have been introduced but here only the conformal method is discussed. Moreover only the case of a three-dimensional initial hypersurface is treated although there is a generalization of the procedure to other dimensions.

The starting point for the conformal method is a set of so-called free data. This consists of a Riemannian metric $\tilde{g}_{ab}$, a function τ, a symmetric tensor $\tilde{k}_{ab}$ with zero trace, a function $\tilde{\rho}$ and a vector field $\tilde{j}^a$. The data to be constructed have the following form. The metric is $g_{ab} = \phi^4 \tilde{g}_{ab}$ for a positive function ϕ. The matter quantities are $\rho = \phi^{-8}\tilde{\rho}, j^a = \phi^{-6}\tilde{j}^a$. Finally

$$k_{ab} = \frac{1}{3}\tau g_{ab} + \phi^{-2}\left(\tilde{k}_{ab} + \nabla_a X_b + \nabla_b X_a - \frac{2}{3}\nabla^c X_c g_{ab}\right) \tag{7.15}$$

for a one-form X_a. In these formulae indices are raised and lowered with the background metric $\tilde{g}_{ab}$ and the covariant derivative is that associated

7 : Elliptic equations

with $\tilde{g}_{ab}$. Substituting these expressions in the constraints gives the following linear equation for X_a

$$\nabla^a(\nabla_a X_b + \nabla_b X_a - \nabla_c X^c g_{ab}) = \frac{2}{3}\phi^6 \nabla_b \tau \qquad (7.16)$$

and a scalar equation for the function ϕ

$$\nabla^a \nabla_a \phi + \frac{1}{8}\left(-R_{\tilde{g}}\phi + \kappa^{ab}\kappa_{ab}\phi^{-7} + 2\pi\phi^{-3}\tilde{\rho} - \frac{1}{12}\tau^2\phi^5\right) = 0, \qquad (7.17)$$

where $\kappa_{ab} = \tilde{k}_{ab} + \nabla_a X_b + \nabla_b X_a - \frac{2}{3}\nabla_c X^c g_{ab}$. The equation (7.17) is known in the literature as the Lichnerowicz equation although Lichnerowicz himself only wrote it down in a special case. The equations to be solved consist of a linear elliptic system (7.16) coupled to the nonlinear scalar elliptic equation (7.17). There is a simplification in the case where τ is constant (constant mean curvature case) and that is the case which has mostly been considered in the literature. With that extra condition the system (7.16) does not depend on ϕ and can be solved on its own. It then only remains to solve the Lichnerowicz equation. The discussion here is restricted to the case of a compact initial hypersurface although other cases such as that of asymptotically flat initial data can be handled in a similar way. For a compact initial hypersurface the equation can, for instance, be solved in Sobolev spaces. The elliptic operator occurring in (7.16) is Fredholm from H^{s+2} to H^s. It is formally self-adjoint and so it is no surprise that its index is zero. To get a good solution theory it remains to obtain information about its kernel. The kernel consists of the conformal Killing vector fields of the metric $\tilde{g}_{ab}$. Thus it can be seen that the operator only fails to be an isomorphism when the background metric has a certain symmetry.

The questions of existence and uniqueness of solutions of the Lichnerowicz equation on a compact manifold depend on the topology and various features of the data. The first is the Yamabe class of the metric. The Yamabe theorem says that given a Riemannian metric g_{ab} on a compact Riemannian manifold M there is a conformally rescaled metric $\tilde{g}_{ab}$ which has constant scalar curvature. By a constant rescaling the value of the scalar curvature can be taken to belong to the set $(-1, 0, 1)$. For a given metric only one of these values is attained in its conformal class. Metrics can thus be classified as Yamabe positive, zero or negative. A manifold M may admit only Yamabe negative metrics, in which case it itself is called Yamabe negative, only Yamabe negative or zero metrics, in which case it is called Yamabe zero, or metrics of all three Yamabe classes, in which case it is called Yamabe positive. Any compact manifold belongs to one of these three classes. The proof of the Yamabe theorem is deep and has interesting connections to the positive mass theorem which is discussed below. Another important feature influencing the solvability of the constraints (assuming the mean curvature to be constant) is the sign of the mean curvature. It is also important whether the energy density and the tracefree part of the second fundamental form vanish identically or not.

A powerful tool in investigating the solvability of the Lichnerowicz equation is the method of sub- and supersolutions which will now be explained briefly. Consider an equation of the form $\Delta u = f(x, u)$ for a scalar function u where the Laplacian is that defined by a Riemannian metric on a compact manifold. A subsolution u_- of the equation is a function satisfying the inequality $\Delta u_- \geq f(x, u_-)$. A supersolution satisfies the inequality $\Delta u_+ \leq f(x, u_+)$. The existence of sub- and supersolutions u_- and u_+ with $u_- \leq u_+$ implies the existence of a solution u with $u_- \leq u \leq u_+$. This is valuable since sub- and supersolutions can often be constructed by hand. In some cases even constants suffice. Let k be a constant which is larger than the maximum of $\partial f/\partial u$ for any value of u between the minimum of u_- and the maximum of u_+. Then the existence of a solution is proved by defining a sequence of functions by the equations

$$-\Delta u_{n+1} + k u_{n+1} = -f(x, u_n) + k u_n. \tag{7.18}$$

for $n \geq 0$ and $\phi_0 = \phi_+$. The introduction of k ensures the invertibility of the linear operator on the left hand side and thus that the iteration is well-defined. The functions ϕ_n satisfy $\phi_{n+1} \leq \phi_n$ and are bounded below by ϕ_- as follows from the maximum principle. From this convergence to a solution can be deduced using elliptic estimates and the Arzela–Ascoli theorem. If f is increasing in its dependence on u, a case which often occurs in the study of the Lichnerowicz equation, then uniqueness statements can also be obtained.

Although several free functions can be chosen when solving the constraints there are limits to the conditions which can be imposed on the solution. A famous example of this is a consequence of the positive mass theorem which will now be discussed. Consider the Einstein constraint equations on $\mathbb{R}^3$ with matter sources ρ and j which satisfy the inequalities $\rho \geq 0$ and $|j| \leq \rho$. These inequalities are satisfied by any solution of the constraints for a matter model which obeys the dominant energy condition. Suppose that the data are asymptotically flat in the sense that (g_{ab}, k_{ab}) tend to the flat space values $(\delta_{ab}, 0)$ at infinity in a sufficiently strong sense. Then a geometrical invariant of the data called the ADM (Arnowitt–Deser–Misner) mass is defined. In coordinates it can be written as

$$m_{\text{ADM}} = \frac{1}{16\pi} \lim_{R \to \infty} \int_{S_R} \sum_{i,j} (\partial_i g_{ij} - \partial_j g_{ii}) N^j, \tag{7.19}$$

where S_R is the sphere of coordinate radius R and N^i is its unit normal vector. The ADM mass is sometimes also called ADM energy. The reason is that under suitable circumstances an ADM four-momentum P^α_{ADM} is defined. Strictly speaking the terminology should be that P^0 is the ADM energy while $(-P_\alpha P^\alpha)^{1/2}$ is the ADM mass. This distinction does not play an important role in this book and so the terms mass and energy are used interchangeably here. The positive mass theorem says that m_{ADM} is non-negative and that if it is zero the data are data for Minkowski space. This has the consequence that initial data for a non-flat solution of the Einstein–matter equation satisfying the dominant energy condition cannot fall off faster than r^{-1} at infinity where r is the coordinate radius. This means that asymptotically flat initial data for the Einstein equations necessarily have rather slow fall-off and this can lead to difficulties when the theory of hyperbolic equations is applied to solve the Einstein equations with this type of initial data. Remarkably, once the mass is fixed the remaining properties of the initial data become rather flexible. It can be shown that for any asymptotically flat solution of the vacuum constraints and any compact subset K_1 of the initial hypersurface there is a solution which agrees with the original one on K_1 and is exactly equal to a Schwarzschild solution on the complement of a larger compact set K_2. The proof of this requires techniques more sophisticated than the conformal method. The positive mass theorem also holds in higher dimensions and limits the rate of decay of initial data for the Einstein equations to r^{-n+2}. The statement just made applies to the case of data on $\mathbb{R}^n$. It is also possible to define asymptotically flat initial data on topologically non-trivial manifolds and data for black holes which exist already at the initial time are of this type. It is assumed that there is a compact set K such that $M \backslash K$ is diffeomorphic to a disjoint union of sets each of which is the complement of a ball in $\mathbb{R}^n$. When transported to $\mathbb{R}^n$ by these diffeomorphisms the data should satisfy the fall-off conditions which define asymptotic flatness on $\mathbb{R}^n$. It is either assumed that the metric is complete or that the data is defined on a manifold with boundary and that suitable boundary conditions are satisfied there. These boundaries could for instance represent the horizons of black holes. The positive mass theorem has been proved without any restriction on the topology for $n \leq 7$. In higher dimensions proofs published up to now require a topological restriction, the existence of a spin structure. When the positive mass theorem holds part of the conclusion in the case $m = 0$ is that the topology must be that of $\mathbb{R}^n$. In some cases with boundary the positive mass theorem can be strengthened to an inequality relating properties of the boundary components (for instance their areas) and the ADM mass. This result is known as the Penrose inequality.

If a spacetime can be foliated by a family of hypersurfaces such that the data induced on these hypersurfaces are asymptotically flat in a suitable sense then the ADM mass is independent of the hypersurface – it is a conserved quantity.

7.6 Further reading

A standard text on elliptic equations is [90]. An article on elliptic equations paying special attention the questions of interest in general relativity is [73]. The techniques required to prove the existence of solutions of the constraints which are Schwarzschild outside a compact set were introduced in [64]. An important case of the Penrose inequality was first proved in [109] where references to the original work on the positive mass theorem can also be found. The result which was originally proved in four dimensions has been extended to certain higher dimensions using a different method of proof [37]. The solvability conditions for the constraints are discussed in [19]. A nice account of the method of sub- and supersolutions is given in [144].

8 Hyperbolic equations

This chapter is an introduction to hyperbolic equations. The topic is of central importance in general relativity since the Einstein evolution equations are themselves essentially hyperbolic as are the equations of motion of many of the matter fields frequently used. The qualification 'essentially' is explained in Chapter 9.

8.1 The Cauchy problem

As a starting point consider once again the system (7.1). There is no definition of the concept 'hyperbolic' which is as simple and general as that of the concept 'elliptic'. Nevertheless the machinery introduced in the elliptic case is also useful in discussing what 'hyperbolic' means. The characteristic polynomial of a system of k equations of order n is a polynomial of order kn depending on the spacetime coordinates as parameters. Thus if it is thought of as a polynomial over the complex field it has kn roots (counting multiplicity) for each value of the parameters. The characteristic set, however, consists of the real roots and there may be less of those or none at all. A necessary condition for (7.1) to be hyperbolic is that all roots of the characteristic polynomial are real. This means roughly speaking that the characteristic set is the union of kn submanifolds of the cotangent space. This picture is complicated by the fact that these submanifolds in general have singularities and that the roots may have multiplicity greater than one. The condition on the number of real roots of the characteristic polynomial is not sufficient for hyperbolicity and in fact there are a number of different inequivalent definitions of hyperbolicity which will be discussed in the following sections.

The standard example of a hyperbolic equation is the wave equation $\partial_t^2 \phi - \delta^{ij} \partial_i \partial_j \phi = 0$. It is linear and its characteristic polynomial is $-(\xi_0)^2 + \delta^{ij} \xi_i \xi_j$. In this case the covectors which belong to the characteristic set are precisely the null covectors of the Minkowski metric and the characteristic set is the null cone.

8.1 The Cauchy problem

In the same way that it is natural to pose boundary value problems for elliptic systems the natural thing for hyperbolic equations is to pose initial value problems, also known as Cauchy problems. For example in the case of the wave equation the right thing is to specify the values of ϕ and $\partial_t \phi$ on a hypersurface such as $t = 0$. Here t is the standard timelike coordinate in Minkowski space and the initial hypersurface $t = 0$ is spacelike with respect to the Minkowski metric. It is possible to define a concept of spacelike hypersurface for more general hyperbolic systems in such a way that the spacelike hypersurfaces are the appropriate ones on which to prescribe initial data. These are in particular non-characteristic in the sense that a one-form ξ which vanishes on vectors tangent to the hypersurface does not belong to the characteristic set. For a system of order k the data which should be prescribed for a hyperbolic equation are the values of the function itself and of $\partial_t^i \phi$ for $1 \le i \le k - 1$ where t is a coordinate which is constant on the initial hypersurface.

When solving an initial value problem the data have to be chosen to belong to a function space which expresses certain regularity properties. Ideally this should lead to a solution with corresponding regularity. Here things are more difficult than in the elliptic case. Not only do C^k spaces work badly as in the case of elliptic equations but the same is true for Hölder spaces and for the Sobolev spaces $W^{k,p}$ with $p \ne 2$. The standard spaces which do work well are the spaces $W^{k,2}$ which are sometimes also denoted by H^k. The importance of Sobolev spaces of type L^2 is connected with the use of the technique of energy estimates which will be explained later. For the moment suffice it to note that for the wave equation the energy

$$E = \int_{\mathbb{R}^n} \frac{1}{2}((\partial_t \phi)^2 + |\nabla \phi|^2) \tag{8.1}$$

is time independent. For more general hyperbolic systems this equality is replaced by an inequality.

An example gives some insight as to why C^k regularity is not propagated by hyperbolic equations. Consider the wave equation in Minkowski space. Specializing to the spherically symmetric case gives the equation

$$\partial_t^2 \phi = \partial_r^2 \phi + \frac{2}{r} \partial_r \phi. \tag{8.2}$$

This may be usefully rewritten as $\partial_t^2(r\phi) = \partial_r^2(r\phi)$. The general solution is thus

$$\phi(t, r) = \frac{1}{r}(f(t - r) + g(t + r)) \tag{8.3}$$

for arbitrary functions f and g. Smoothness at $r = 0$ implies that $f' = g'$. For a solution evolving from initial data of compact support it can be concluded that $f = g$. The initial data at $t = 0$ is given by $\phi_0(r) = r^{-1}(f(-r) + f(r))$ and $\phi_1(r) = r^{-1}(f'(-r) + f'(r))$. For the example choose f to be a function of compact support with the following properties. It vanishes on the interval $(-\infty, 1]$, it is equal to $(r-1)^3$ on the interval $[1, 2]$ and it is smooth for $r \geq 1$. Then ϕ_0 is C^2 and ϕ_1 is C^1. It will now be shown that the solution is not C^2. The intuitive idea is that the irregularity in the initial data at $r = 1$ propagates along an ingoing light cone and focuses at $t = 1$. Inside the past light cone of the point $(1, 0)$ the solution is identically zero. Thus if the solution were C^2 all its second order derivatives would have to vanish there. However for r small $\phi(1, r) = r^2$ and thus $\partial_r^2 \phi(1, r) = 2$. By a more complicated construction involving data with an infinite number of irregularities of this type it is possible to see that for data where u_0 and u_1 are C^2 and C^1 respectively there may be no time interval where the corresponding solution is C^2.

8.2 Examples of ill-posed problems

An initial value problem is said to be well posed if the following three conditions are satisfied

- there exist solutions corresponding to all initial data;
- the solutions are uniquely determined by the initial data;
- the mapping from initial data to solutions is continuous.

To make this precise it is necessary to specify which regularity class is assumed for the data and solutions and with respect to which topology continuity is required. The condition of continuity is sometimes called Cauchy stability. The reason for including it is as follows. If PDE are to be applied to model phenomena in the natural world it must be remembered that measurements are never exact but always associated with some error. As a consequence it is impossible to know initial data for a problem exactly and so if the solutions depend on the initial data in an uncontrollable way the model cannot make useful predictions. Cauchy stability guarantees that this does not happen and thus represents a necessary condition for the applicability of PDE to the real world. The notion of well-posed systems and its opposite, that of ill-posed systems, were introduced by Jacques Hadamard in the early twentieth century. In particular he worked out a key example of an ill-posed problem, the Cauchy problem for the Laplace equation.

The Cauchy problem for the Laplace equation will now be discussed in the simplest possible case. Consider the Laplace equation in two space dimensions which will be written as $u_{tt} = -u_{xx}$. The reason for this unconventional notation is that it corresponds to the attempt to consider the

Laplace equation as an evolution equation, an attempt which turns out to be misguided. It is assumed that the function u is periodic with period 2π in the variable x and that $t = 0$ is chosen as initial hypersurface. The initial data are the restrictions of u and $\partial_t u$ to the initial hypersurface. Call them u_0 and u_1. Doing a Fourier transform in x leads to the ODE $d^2 \hat{u}_n/dt^2 = n^2 \hat{u}_n$ which can easily be solved. The nth Fourier coefficient $\hat{u}_n$ increases exponentially with time with the exponent n. Thus there are modes which grow at an arbitrarily large exponential rate and this prevents the problem being well posed. In fact, as shown in the next paragraph, there are C^∞ initial data for which there is no corresponding solution. Note that for the wave equation, $u_{tt} = u_{xx}$, which is well posed, the modes are bounded and so this problem does not occur. There is also no problem in posing initial data for the two-dimensional wave equation on $x = 0$ but this is an artefact of two dimensions where t and x play a symmetrical role. Consider for comparison the three-dimensional case $u_{tt} = u_{xx} + u_{yy}$. In that case prescribing data on $t = 0$ leads to a well-posed problem but prescribing data on $x = 0$ leads to the same kind of difficulties that occur for the Laplace equation.

Consider the mapping Φ from $X = C^\infty([0, T] \times S^1)$ to $Y = C^\infty(S^1) \times C^\infty(S^1)$ which takes a function u to (u_0, u_1) where u_0 and u_1 are the restrictions to $t = 0$ of u and $\partial_t u$ respectively. The spaces X and Y, with the topology of uniform convergence of derivatives of all orders, have the structure of Fréchet spaces and Φ is a continuous linear map with respect to this topology. Note that uniqueness does hold in this initial value problem. If there were two different solutions with the same initial data subtracting one from the other would give a non-zero solution with zero initial data. However the Fourier coefficients satisfy homogeneous linear ODE, in this case with zero initial data. Thus by ODE uniqueness all Fourier coefficients of the hypothetical solution vanish so that it itself vanishes, a contradiction. It follows from this uniqueness statement that Φ is injective. Existence in the initial value problem for all initial data would show that Φ was surjective. It follows from the open mapping theorem for Fréchet spaces that the inverse of Φ would be continuous. This would obviously contradict the example previously exhibited. Thus in fact there must exist at least one initial data set with no corresponding solution. With a different kind of regularity the parallel argument can be carried out in Banach spaces. It suffices to choose $X = C^0([0, T], H^s(S^1)) \cap C^1([0, T], H^{s-1}(S^1))$ and $Y = H^s(S^1) \times H^{s-1}(S^1)$. These are spaces for which there is a good existence and uniqueness theory for the wave equation.

There is an example of ill-posedness which has a direct relevance to general relativity. Consider a solution of the Euler equations in Minkowski space with linear equation of state $p = w\mu$, $w < 0$, and make the simplifying assumptions that only the component v^1 of the velocity is non-vanishing and that both μ and v^1 only depend on the coordinates t and x^1. Now linearize this system about the solution where $v^1 = 0$ and $\mu = 1$. This leads to the first order linear system

$$\partial_t(\tilde{\mu}) = -2(1+w)\partial_a \tilde{u}^a, \tag{8.4}$$

$$\partial_t \tilde{u}^a = -\frac{w}{2(1+w)} \delta^{ab} \partial_b \tilde{\mu}. \tag{8.5}$$

Combining the two equations gives $\partial_t^2(\tilde{\mu}) = w \Delta \tilde{\mu}$ which, after a simple rescaling, is just the Laplace equation. Thus these equations do not have a well-posed Cauchy problem. In view of this it is remarkable that fluids whose equation of state satisfies $w < 0$ are often used in cosmology. Since, however, it is common to consider only homogeneous solutions the inherent instability of these matter models does not become apparent. It is to be expected that the ill-posedness of the linearized Euler equations seen here implies ill-posedness of the full Einstein–Euler system with this equation of state but this has not been worked out in detail in the literature.

From the above it can be seen that it is often not hard to identify signs of ill-posedness in a system. It is generally much harder to show that the Cauchy problem for a particular system is well posed. The next sections present some of the general approaches to doing so.

8.3 Symmetric hyperbolic systems

The symmetric hyperbolic systems form a class for which a good theory of the Cauchy problem is available. It is widely applicable and relatively easy to use. Consider a first order system of PDE of the form $F(x, u, Du) = 0$. Let one of the coordinates $t = x^0$ be singled out and denote the others by x^i, $1 \le i \le n$. Thus the system becomes $F(t, x^i, u, \partial_t u, \partial_i u) = 0$. Let $v = \partial_t u$. Then

$$\frac{\partial F}{\partial_t u} \partial_t v + \frac{\partial F}{\partial_i u} \partial_i v + \frac{\partial F}{\partial u} v = -\frac{\partial F}{\partial t}, \tag{8.6}$$

$$\partial_t u = v. \tag{8.7}$$

This is a quasilinear equation for the pair (u, v) and it turns out that if the Cauchy problem can be solved for (8.6) and (8.7) it can also be solved for the original equation. It suffices to solve (8.6) and (8.7) for (u, v) with data (u_0, v_0) such that v_0 is obtained from u_0 by substituting into the original equation. For this reason it is sufficient to discuss the Cauchy problem for a quasilinear system and this is what is done in the following. This simplifies the discussion somewhat. A general quasilinear system is called symmetric hyperbolic if the corresponding quasilinear system is symmetric hyperbolic in the sense defined below. The general first order quasilinear system is

$$A^0(x,u)\partial_t u + A^i(x,u)\partial_i u + B(x,u) = 0. \tag{8.8}$$

It is called symmetric hyperbolic if for all u taking values in some open set G the matrices A^0 and A^i are symmetric and A^0 is positive definite. It is called uniformly symmetric hyperbolic for some set of values of x and u if A^0 is uniformly positive definite for those values in the sense that there is a positive constant C such that $\langle A^0 u, u\rangle \geq C|u|^2$. Note that if a symmetric hyperbolic system is linearized about any solution the result is a linear symmetric hyperbolic system. Conversely, if the linearization is always symmetric hyperbolic the same is true of the original nonlinear system. Thus the question whether a system is symmetric hyperbolic can be reduced to the linear case. In fact this can be taken even further. Symmetric hyperbolicity is a pointwise condition on the coefficients and so it can be said that all the information is contained in the constant coefficient linear systems obtained by freezing coefficients at individual points. The roots of the characteristic polynomial (which are sometimes just referred to as characteristics) of a symmetric hyperbolic system are always real. This results from applying the following fact from linear algebra to the principal symbol. If M and N are symmetric matrices with M positive definite then an eigenvector v of N with respect to M with eigenvalue λ is defined to be a (possibly complex) solution of the equation $Nv = \lambda Mv$ for some complex number λ. It is a theorem that under the assumptions just made λ is always real and that the eigenvectors are complete in the sense that they span the whole space. (See e.g. [87], Vol. 1, p. 310.) The eigenvalues need not be distinct. Correspondingly, the multiplicity of the characteristics of a symmetric hyperbolic system may vary with x and u.

The wave equation, which is the central example of a hyperbolic system, can be reduced to a symmetric hyperbolic system as follows. Starting with the equation $\partial_t^2 u = \delta^{ij}\partial_i\partial_j u$, introduce $v = \partial_t u$ and $w_i = \partial_i u$. Then

$$\partial_t v = \sum_i \partial_i w_i, \tag{8.9}$$

$$\partial_t w_i = \partial_i v. \tag{8.10}$$

This is a symmetric hyperbolic system and solving the Cauchy problem for the wave equation can be reduced to solving the Cauchy problem for this first order system. Recall that the initial data for the wave equation are the values of u and $\partial_t u$ at $t = 0$, denoted by u_0 and u_1 respectively. Now define data for the first order system by $v_0 = u_1$ and $(w_i)_0 = \partial_i u_0$. If a solution of the first order system corresponding to these initial data is given it is possible

8 : Hyperbolic equations

to define a function $u(t, x) = u_0(x) + \int_0^t v(s, x) ds$. The claim is that u satisfies the wave equation with initial data (u_0, u_1). The statement about the initial data is clear. The solution satisfies $v = \partial_t u$. To show that u satisfies the wave equation it remains to show that $w_i = \partial_i u$. This is a consequence of the equation $\partial_t(w_i - \partial_i u) = 0$, which follows from the first order system, and the choice of initial data. This reduction process could have been done in a slightly different way, keeping u as an unknown in addition to the new unknowns and adding the equation $\partial_t u = v$. There is no advantage in doing this for the standard wave equation but it is necessary for more complicated wave equations containing terms which depend on the function u itself as well as its derivatives.

This reduction process can in fact be applied just as well to the linear wave equation $\nabla^\alpha \nabla_\alpha \phi = 0$ defined by an arbitrary Lorentzian metric $g_{\alpha\beta}$. Written out explicitly in coordinates this equation takes the form

$$g^{\alpha\beta} \partial_\alpha \partial_\beta \phi - \Gamma^\alpha \partial_\alpha \phi = 0, \tag{8.11}$$

where $\Gamma^\alpha = g^{\beta\gamma} \Gamma^\alpha_{\beta\gamma}$. Exactly the same procedure can be used to reduce more general systems of quasilinear wave equations to quasilinear symmetric hyperbolic systems and it will be written down in the case of the system

$$g^{\alpha\beta} \nabla_\alpha \nabla_\beta u = F(t, x, u, Du). \tag{8.12}$$

Introduce the new variables by $u_\alpha = \partial_\alpha u$. Then

$$-g^{00} \partial_0 u_0 - 2g^{0a} \partial_a u_0 = g^{ab} \partial_a u_b + F(t, x, u, u_0, u_a), \tag{8.13}$$

$$g^{ab} \partial_0 u_a = g^{ab} \partial_a u_0, \tag{8.14}$$

$$\partial_0 u = u_0. \tag{8.15}$$

This is a symmetric hyperbolic system in the unknowns (u, u_α). For data satisfying $u_a = \partial_a u$ solutions of the symmetric hyperbolic system satisfy the system of nonlinear wave equations of interest.

There is a theory which applies directly to the wave equation without the need for reduction to first order, the theory of strictly hyperbolic systems. A system is called strictly hyperbolic if all characteristics are real and of multiplicity one. This condition is satisfied for the wave equation – it is a single equation of second order and has two distinct characteristics, the two sheets of the null cone. However the concept of strict hyperbolicity as stated is rather fragile. A system of two uncoupled wave equations

$$\partial_t^2 u_1 = \delta^{ij} \partial_i \partial_j u_1,$$

$$\partial_t^2 u_2 = \delta^{ij} \partial_i \partial_j u_2 \tag{8.16}$$

8.3 Symmetric hyperbolic systems

is not strictly hyperbolic. It has the same characteristics as the wave equation but for this system they are of multiplicity two. On the other hand reducing the two wave equations as above to get a system of first order equations leads to a symmetric hyperbolic system. This is an example of a more general phenomenon. Consider two symmetric hyperbolic systems for unknowns u_1 and u_2 respectively. Then the system for the pair (u_1, u_2) obtained by taking the equations for u_1 and u_2 together is symmetric hyperbolic.

When written in terms of the electric and magnetic fields the Maxwell evolution equations (3.26) and (3.27) are seen to form a symmetric hyperbolic system on any given spacetime. The other Maxwell equations, the constraints, say that E and B have zero divergence and are intrinsic to the initial hypersurface. It follows from the evolution equations that the constraint quantities, i.e. the quantities which vanish when the constraints hold, satisfy the system

$$\partial_t(\nabla_a E^a) - (\mathrm{tr}\,k)\nabla_a E^a = 0, \tag{8.17}$$

$$\partial_t(\nabla_a B^a) - (\mathrm{tr}\,k)\nabla_a B^a = 0 \tag{8.18}$$

in Gauss coordinates. Hence if the constraints are satisfied for $t = 0$ and the evolution equations are satisfied everywhere then the constraints are satisfied everywhere. This is described by saying that the constraints propagate. This is enough to ensure the propagation of the constraints in any coordinate system since it is possible to change to Gauss coordinates locally in a given solution of the Einstein–Maxwell system. In this way it can be seen that to solve the full system of Maxwell equations it suffices to solve the Maxwell evolution equations with appropriate initial data. The theory of symmetric hyperbolic systems provides the tool to solve the latter problem.

In the case of the Yang–Mills equations on an arbitrary spacetime it is presumably possible to proceed in a similar way, except for the fact that the components of the potential must be included as extra variables. This will be worked out here in the case of Minkowski space. Electric and magnetic parts of the field strength can be defined just as in the Maxwell case and these satisfy a system of equations whose principal part is the same as that of the Maxwell equations except for the presence of an extra index. It is symmetric hyperbolic for fixed values of the gauge potential. It is necessary to augment the system by an evolution equation for the potential. At the same time the Yang–Mills equations have a gauge invariance which causes problems for the Cauchy problem similar to those caused by diffeomorphism invariance in the case of the Einstein equation as discussed in Section 9.1. To overcome this problem a gauge condition must be imposed and the simplest one is temporal gauge where it is assumed that $A_0^I = 0$. The time derivatives of the components A_a^I can then be expressed in terms of F_{a0}^I and lower order terms. This leads to a symmetric hyperbolic system for the quantities (A_a^I, E^{aI}, B^{aI}). It must still be checked that all coordinate and gauge conditions propagate as a consequence of the Yang–Mills equations. This question will not be addressed here.

Consider now the isentropic relativistic Euler equations on an arbitrary spacetime background. Contracting the equation $\nabla_\alpha T^{\alpha\beta} = 0$ with u_β gives

$$u^\alpha \nabla_\alpha \mu + (\mu + p)\nabla_\alpha u^\alpha = 0. \tag{8.19}$$

Projecting the Euler equation with $h^\gamma{}_\beta$ gives

$$(\mu + p)h^\gamma{}_\beta u^\alpha \nabla_\alpha u^\beta + h^\gamma{}_\beta \nabla^\beta p. \tag{8.20}$$

Since u^α is a unit vector $u_\alpha \nabla_\beta u^\alpha = 0$ and $\nabla_\alpha u_\beta u^\alpha u^\beta = 0$. Combining these with (8.19) and (8.20) and using the equation of state gives

$$u^\alpha \nabla_\alpha \mu + (\mu + p)h_\beta{}^\alpha \nabla_\alpha u^\beta = 0, \tag{8.21}$$

$$(\mu + p)\bar{h}_{\beta\gamma} u^\alpha \nabla_\alpha u^\beta + f'(\mu) h^\alpha{}_\gamma \nabla_\alpha \mu = 0 \tag{8.22}$$

where $\bar{h}_{\alpha\beta} = g_{\alpha\beta} + 2u_\alpha u_\beta$ defines a positive definite quadratic form. Multiplying the first of these equations by $(\mu+p)$ and the second by $f'(\mu)$ leads to a symmetric system. It will now be shown to be symmetric hyperbolic provided $0 < f' < 1$ and μ is assumed to be everywhere strictly positive. To do this it is necessary to show that the contraction of the principal symbol with a future-directed timelike vector n^α is positive definite. This is easily seen to be the case if n^α is chosen to be the four-velocity of the fluid. The general case can then be treated by the following geometrical consideration. The characteristic set of the Euler equations consists of a cone (the sound cone) which, under the assumption $f' < 1$, lies outside the light cone and the spacelike hyperplane orthogonal to the four-velocity. The vector n^α can be deformed continuously to u^α through timelike vectors. During this deformation the vector (or rather the corresponding covector) is non-characteristic. Hence the contraction of the principal symbol with this covector never has a zero eigenvalue. Since the eigenvalues are all positive at the starting point u^α they must, by continuity, remain positive. Hence the contraction of the principal symbol with n^α is also positive definite.

To check that solving the symmetric hyperbolic system actually gives rise to solutions of the Euler equations it is necessary to check that the constraint $u_\alpha u^\alpha = -1$ propagates. In other words, it must be shown that if the initial data are such that $u_\alpha u^\alpha = -1$ then this condition is satisfied everywhere.

8.3 Symmetric hyperbolic systems

This is done by using the propagation equation

$$u^\alpha \nabla_\alpha (u_\beta u^\beta) = 0, \qquad (8.23)$$

which is obtained by contracting (8.22) with u^γ.

A fluid which is not isentropic can be treated in a similar way. It suffices to replace μ as a variable by p and rewrite the equation of state $p = f(\mu, s)$ as $\mu = g(p, s)$. The first term in (8.22) gets multiplied by $\partial g/\partial p$. The contribution involving $\partial g/\partial s$ coming from the chain rule vanishes due to the evolution equation for s. The system for the variables (p, u^α, s) can be brought into symmetric hyperbolic form by multiplying the equations with appropriate factors.

From a physical point of view it would seem desirable to treat solutions of the Einstein–Euler system where the energy density is compactly supported. This would be the case of a fluid body. Unfortunately this type of problem turns out to be very hard analytically. There is a danger that when the density of the fluid becomes zero the Euler equations lose their hyperbolicity. In general this is an unsolved problem but there is a trick going back to Makino which can be used to at least get large classes of solutions. This only works for certain equations of state. Consider for instance the equation of state $p = K\mu^\gamma$ with $\gamma > 1$ and introduce the variable $w = \frac{\sqrt{K\gamma}}{\gamma - 1} \mu^{\frac{\gamma-1}{2}}$. Then the equations can be written

$$\left(\frac{\mu}{\mu+p}\right)^2 u^\alpha \nabla_\alpha w + \frac{\mu}{\mu+p} (f'(\mu))^{1/2} h_\beta{}^\alpha \nabla_\alpha u^\beta = 0, \qquad (8.24)$$

$$\bar{h}_{\beta\delta} u^\alpha \nabla_\alpha u^\beta + \frac{\mu}{\mu+p} (f'(\mu))^{1/2} h^\alpha{}_\delta \nabla_\alpha w = 0. \qquad (8.25)$$

Here the loss of symmetric hyperbolicity when $\mu = 0$ has been removed in the sense that A^0 is uniformly positive definite even when μ becomes zero and that A^a remains bounded when that happens. Note also that the coefficients of the system are smooth functions of w, whatever the value of γ. A disadvantage of this system is the following. To be able to apply the usual theory of symmetric hyperbolic systems it is necessary to have initial data for w which is sufficiently differentiable. Unfortunately this translates into a strong and physically undesirable condition on the initial data for μ. It can be shown that any solution constructed using the variable w in this way has the property that the boundary of the fluid is freely falling, i.e. it is generated by geodesics. In particular, small perturbations of static solutions cannot be treated.

Before leaving the Euler equations it will be mentioned that the corresponding non-relativistic equations can also be written as a symmetric

8: Hyperbolic equations

hyperbolic system, which is as follows (in the isentropic case):

$$\partial_t \mu + u^a \partial_a \mu + \mu \partial_a u^a = 0, \tag{8.26}$$

$$\mu \delta_{ab}(\partial_t u^b + u^a \partial_a u^b) + f'(\mu) \partial_b \mu = 0. \tag{8.27}$$

It is possible to incorporate entropy and to introduce a Makino variable very much as in the relativistic case.

Under suitable assumptions on the energy density μ the equations of motion of an elastic solid can be written as a symmetric hyperbolic system for the basic unknown f and its first derivatives.

A generalization of symmetric hyperbolic systems which is often useful is given by the symmetrizable hyperbolic systems. Consider a system of the form (8.8) with matrices $A^0 = I$ and A^i not necessarily symmetric. A symmetrizer R for the system is a smooth matrix-valued function $R(x)$ such that R is symmetric and positive definite and the matrices RA^i are symmetric. Multiplying by R transforms the original system into a symmetric hyperbolic one. In this case the original system is said to be symmetrizable.

A conormal to a hypersurface H is a one-form defined on H which vanishes on vectors which are tangential to H. A hypersurface with conormal ξ_α is called spacelike with respect to a symmetric hyperbolic system of the form (7.1) if the matrix $A^0 \xi_0 + A^i \xi_i$ is positive definite. It follows in particular from this definition that the hypersurfaces $t = \text{const.}$ are spacelike. Let ϕ be a smooth function on $\mathbb{R}^n$ with the following properties. The region G of $\mathbb{R}^{n+1}$ defined by the inequalities $0 \le t \le \phi(x)$ is compact and the part of the boundary of this region where $t = \phi(x)$ is spacelike. Under these circumstances G is called a lens-shaped region (see Fig. 8.1).

Existence and uniqueness results for symmetric hyperbolic systems will now be discussed along with the important concept of the domain of dependence. Consider first a C^1 solution u of a homogeneous linear symmetric hyperbolic system of the form $A^0 \partial_t u + A^i \partial_i u + Bu = 0$ where the coefficients A^0, A^i and B are smooth functions of x. Let G be a lens-shaped region with respect to this equation. It will now be shown that if u vanishes on the part of the hypersurface $t = 0$ contained in the lens-shaped region then it vanishes on the whole lens-shaped region. This is done by means

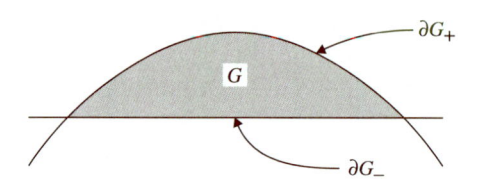

Figure 8.1 *A lens-shaped region*

8.3 Symmetric hyperbolic systems

of an inequality which will now be derived. It is an example of an energy estimate. Denote the part of the boundary ∂G in the hypersurface $t = 0$ by ∂G_- and the part in the hypersurface $t = \phi(x)$ by ∂G_+. Let k be a positive real number. Then the following identity is obtained by taking the inner product of the equation with $e^{-kt}u$, integrating the result over G and applying Stokes' theorem:

$$\int_{\partial G_+} e^{-kt}\langle (A^0 - A^i\partial_i\phi)u, u\rangle$$

$$= \int_{\partial G_-} \langle A^0 u, u\rangle + \int_G e^{-kt}\langle u, (\partial_t A^0 + \partial_i A^i)u\rangle$$

$$- k\int_G e^{-kt}\langle u, A^0 u\rangle - 2\int_G e^{-kt}\langle u, B_1 u\rangle$$

$$= \int_{\partial G_-} \langle A^0 u, u\rangle + \int_G e^{-kt}\langle (\partial_t A^0 + \partial_i A^i - 2B_1 - kA^0)u, u\rangle. \quad (8.28)$$

If the restriction of u to ∂G_- is zero and u does not vanish identically on G then the right hand side of this identity can be made negative by choosing k sufficiently large. This contradicts the fact that the left hand side is manifestly non-negative. Thus the vanishing of u on ∂G_- implies its vanishing on G.

Consider two C^1 solutions u and v of a system of the form (7.1) which is symmetric hyperbolic. Suppose that the coefficients A^0, A^i and B are smooth. The following sharp version of the mean value theorem will be needed. If F is a smooth function on an open subset U of $\mathbb{R}^n$ then there is a smooth function M on $U \times U$ such that $F(x) - F(y) = M(x, y)(x - y)$. Subtracting the equation for v from that for u and applying this result gives

$$A^0(u)\partial_t(u - v) + A^i(u)\partial_i(u - v)$$
$$+ [M_1(u, v)\partial_t v + M_2(u, v)\partial_i v + M_3(u, v)](u - v) = 0 \quad (8.29)$$

for suitable functions M_1, M_2 and M_3. This is a homogeneous linear symmetric hyperbolic equation for the difference $u - v$ with smooth coefficients. The uniqueness theorem just proved for homogeneous linear equations implies that if u and v agree on an open subset U of the initial hypersurface there is an open neighbourhood of U in $\mathbb{R}^{n+1}$ where $u = v$. In particular, if $u = v$ on all of the hypersurface $t = 0$ then the same holds on an open neighbourhood of $t = 0$. This is a basic local uniqueness theorem for symmetric hyperbolic systems.

With the local uniqueness theorem in hand it is natural to ask to what extent it can be extended to a global statement. For this purpose it is useful to introduce the concepts 'domain of dependence' and 'domain of influence'.

The reader should be warned that the names of these two concepts are sometimes interchanged in the literature. In the terminology adopted here the definition is as follows:

Definition 8.1 *Let a solution of a symmetric hyperbolic system be given on $I \times \mathbb{R}^n$ where I is an interval. A domain of dependence for a point $(t_0, x_0) \in I \times \mathbb{R}^n$ is a subset G of the initial hypersurface $t = 0$ with the property that every smooth solution v of the system which agrees with u on G also satisfies $v(t_0, x_0) = u(t_0, x_0)$.*

On the basis of this definition there is no unique domain of dependence. For instance $G = \mathbb{R}^n$ always works. Smaller domains of dependence are more interesting but there is no theorem which provides a minimal domain of dependence for a given system and solution. For any subset E of $I \times \mathbb{R}^n$ a domain of dependence for E is a subset G of the initial hypersurface which is a domain of dependence for each point of E. Conversely there is the following definition:

Definition 8.2 *Let a solution of a symmetric hyperbolic system be given on $I \times \mathbb{R}^n$ where I is an interval. The domain of influence of a subset G of the initial hypersurface is the set of all points (t, x) for which G is a domain of dependence.*

The usefulness of the concept of domain of dependence is not limited to the case of symmetric hyperbolic systems. It can also be applied to other types of hyperbolic equation. The main property is that the solution is determined at a given point by data on a proper subset of the initial hypersurface so that in a certain sense the process of solving the Cauchy problem can be localized in space. This is a characteristic property of hyperbolic equations which distinguishes them from other types of evolution equations such as the heat equation or the Schrödinger equation.

A consequence of the existence of a domain of dependence for symmetric hyperbolic equations is the property of finite speed of propagation for their solutions. Consider a solution of a symmetric hyperbolic system corresponding to an initial datum which has compact support on the hypersurface $t = 0$ in $\mathbb{R}^n$. Assuming that the coefficients of the system are smooth the system is uniformly symmetric hyperbolic. It will now be shown that if the solution exists on $\mathbb{R}^n \times [0, t_1)$ for some $t_1 > 0$ then the support of the restriction of the solution to any hypersurface $t = t_0$ with $t_0 \in (0, t_1)$ is compact. To see this, introduce regions $L_{\alpha,\beta}$ for positive constants α and β which are defined by the inequalities $0 \leq t \leq \alpha - \beta |x|^2$. There is some $\beta_0 > 0$ such that $L_{\alpha,\beta}$ is a lens-shaped region for any $\beta \leq \beta_0$. Using the uniqueness statement for solutions on a lens-shaped region it can be concluded that if the support of the initial datum is contained in the ball of radius R about the origin then the restriction of the solution to the hypersurface $t = t_0$ vanishes outside the ball of radius $R + \beta_0 t_0^2$ about the origin. This shows that the restriction of the solution to later time slices is compact and gives a crude estimate of how large it is.

Next the domain of dependence will be examined more closely in the case of the wave equation on Minkowski space. To do this consider the function $\phi(x) = t_0 - (\tau^2 + |x - x_0|^2)^{1/2}$ where τ is a positive constant. This defines a lens-shaped region for each value of τ. Hence the region $|x - x_0| \leq (t_0^2 - \tau^2)^{1/2}$ is always a domain of dependence for the point $(t_0 - \tau, x_0)$. Taking the limit $\tau \to 0$ and using the continuity of the solution shows that $|x - x_0| \leq t_0$ is a domain of dependence for the point (t_0, x_0). This argument does not depend on the dimension. The question of whether there exists a smaller domain of dependence for the wave equation is more subtle. For n odd the set $|x - x_0| = t_0$ is a domain of dependence for (t_0, x_0). This fact is associated to the name Huygens principle. For n even the corresponding statement does not hold. It is often said that the propagation of solutions of the wave equations in odd space dimensions is sharp (only on the light cone) while in even dimensions there is a tail (some residual influence inside the light cone). The fact that the domain of dependence could be pinned down as far as it could in any dimension by simple arguments follows from the fact that the geometry of the characteristics is so simple and well understood in the case of the wave equation in flat space. The analysis for nonlinear wave equations and wave maps in flat space is identical and that for the Maxwell equations very similar. For the wave equation in curved space the characteristic hyperplanes are null hyperplanes and the boundary of the domain of dependence can be expected to consist of null hypersurfaces. Complications arise due to the fact that the null cone is in general not smooth. It may contain singularities (caustics) or self-intersections. For a quasilinear system like the Euler equations things are even more complicated since the characteristics, and hence the domain of dependence, are determined by the solution considered.

It will now be sketched how local existence is proved for symmetric hyperbolic systems. Since everything can be localized using the domain of dependence the discussion will be restricted without loss of generality to the case of initial data prescribed on the hypersurface $t = 0$ in $\mathbb{R} \times \mathbb{R}^n$ which has compact support. Moreover it is assumed that the coefficients of the system other than A^0 have compact support and that A^0 itself is equal to the identity outside a compact set. The aim is to show the existence of a solution on an interval $[0, T]$. The initial data are assumed to belong to a Sobolev space H^k with $k > n/2 + 1$. This ensures, by the Sobolev embedding theorem, that the data are continuous. A strategy which can be used to prove the existence of a solution of eqn (8.8) with a prescribed initial datum is to set up an iteration by solving the equation

$$A^0(x, u_m)\partial_t u_{m+1} + A^i(x, u_m)\partial_i u_{m+1} + B(x, u_m) = 0. \tag{8.30}$$

More precisely, the initial datum u^0 is extended to a function u_0 on $\mathbb{R}^n \times [0, T]$ so as to be independent of t. Once the iterates with index less than or equal to m have been defined the iterate u_{m+1} is defined by solving eqn (8.30) with initial datum u^0. Of course in order to do this an existence theorem for linear symmetric hyperbolic systems is required. Note for the moment that if the sequence of iterates can be defined on a common time interval $[0, T]$ and if the sequence converges in a strong enough sense to allow a passage to the limit in (8.30) then the limit will be the desired solution of (8.8).

There are different approaches to proving an existence theorem for a linear hyperbolic system. One is to discretize the equation in space and time, replacing the partial derivatives by difference operators. Suppose that the coefficients in the equation are smooth. Solving the resulting system of difference equations is then an algebraic problem for which it is not difficult to obtain an existence theorem. What is harder is to obtain bounds for the solution of the difference equations for different step lengths h in the discretization which are uniform in h. This is done by means of energy estimates. The role of integration by parts in the derivation of energy estimates for differential equations is taken over by a discrete analogue, summation by parts. It is also necessary to use similar methods to estimate higher order difference quotients. Once all this has been done it can be shown that the solutions of the difference equations converge on a dense subset to a smooth function and that the limiting function satisfies the original linear symmetric hyperbolic system. The result of this proof is a global existence theorem for a linear symmetric hyperbolic system on any interval where its coefficients are well-defined and smooth. There are also other methods to reach the same goal. For instance an argument using techniques of functional analysis can be used to prove the existence of a weak solution and it can then be shown that the weak solution is in fact smooth.

Consider once again the problem of proving a local existence theorem for a symmetric hyperbolic system. It is convenient to slightly change the definition of the iteration as follows. The initial datum u^0 is approximated by a sequence u_m^0 of smooth data with compact support converging to it in H^k. The first iterate u_0 is defined by extending u_0^0 to be independent of t while in the definition of u_{m+1} the initial datum u_{m+1}^0 is used instead of u^0. The advantage of this is that all iterates are then C^∞ so that there is no problem in justifying operations which need to be carried out such as partial integration. The main idea is now to show that a subsequence of the sequence $\{u_m\}$ of iterates converges in a suitable sense. This is done

8.3 Symmetric hyperbolic systems

by using energy estimates. In general the iterate u_{m+1} may only exist on an interval $[0, T_{m+1})$ since the iterate u_m may leave the domain of definition G of the coefficients of (8.8) after that time. One of the conditions that the estimates on the iterates must fulfil is that they show that there is a common positive lower bound T for all the T_m. The basic energy estimate is just like the estimate (8.28) whose derivation has already been presented except that in the present situation the integrals can be carried out on the whole of $[0, T_m] \times \mathbb{R}^n$. The next step is to do higher energy estimates. This means that the basic equation is first acted on by a derivative D^α for some multiindex α and the energy estimate is carried out for the resulting equation. The expressions which come up can be controlled using the Moser estimates and using the basic equation to substitute for the time derivative of u_m when it occurs. The end result of all this is the following differential inequality:

$$[U^{m,s}(t)]^2 \leq \left\{ [U^{m,s}(0)]^2 + \int_0^t (1 + N^m(t'))^2 dt' \right\}$$
$$\times \exp\left\{ C \int_0^t (1 + N^m(t'))^2 dt' \right\}, \quad (8.31)$$

where

$$[U^{m,s}(t)]^2 = \sup_{0 \leq m' \leq m} \sum_{\|\alpha\| \leq s} \int_{\mathbb{R}^n} \langle A^0 D^\alpha u_{m'}, D^\alpha u_{m'} \rangle, \quad (8.32)$$

$$N^m(t) = \sup_{0 \leq m' \leq m} \|u_m(t)\|_{C^1}. \quad (8.33)$$

Note that since A^0 is bounded and uniformly positive definite the right hand side of (8.32) is equivalent to the H^s norm. As will be shown later, this inequality can be used to get a uniform bound on the iterates in the Banach space $C^0([0, T], H^s(\mathbb{R}^n))$ for T sufficiently small.

At this point it is tempting to try to use energy estimates for the differences $u_m - u_{m'}$ to show that the sequence $\{u_m\}$ is a Cauchy sequence in the Banach space where it has been shown to be bounded. Unfortunately there is a problem. The energy estimates only allow the derivatives up to order $s - 1$ of the differences of the iterates to be bounded. This difficulty is particular to quasilinear hyperbolic equations and does not occur in the semilinear case. To see where it comes from, consider the simple case of the equation $\partial_t u + u \partial_x u = 0$. The equation for the iterates is $\partial_t u_{m+1} + u_m \partial_x u_{m+1} = 0$. Subtracting these equations for two different values of the index gives

$$\partial_t(u_m - u_{m'}) + u_m \partial_x(u_m - u_{m'}) = (u_{m'} - u_m)\partial_x u_{m'}. \quad (8.34)$$

The problem arises from the first derivatives on the right hand side of this equation. For this reason another route must be taken. The differences of iterates are estimated only in L^2. Let $V^m(t) = \|u_m - u_{m-1}\|_{L^2}$. Then a suitable energy estimate gives

$$[V^m(t)]^2 \leq [V_m(0)]^2 + CT' \sup_{0 \leq t \leq T'} (1 + U^{m,s}(t)) \sup_{0 \leq t \leq T'} [V^{m-1}(t)]^2 \quad (8.35)$$

for any $T' \leq T$.

For the proof of the existence theorem it is useful to replace the unknown u in the equation by the unknown $u - u_0^0$. The new equation is of the same general form as the old one and, since u_0^0 is C^∞, the coefficients of the transformed equation are smooth. After the transformation the initial data for the iterates is small for m large. The Sobolev embedding theorem shows that for $s > \frac{n}{2} + 1$ there is a constant C such that $N^m(t) \leq CU^{m,s}(t)$. Combining this with (8.31) gives the following integral inequality:

$$[U^{m,s}(t)]^2 \leq C\left[\|u_0\|_{H^s}^2 + \int_0^t (1 + [U^{m,s}(t')]^2)^2 dt'\right]. \quad (8.36)$$

The quantity $U^{m,s}$ can be estimated by comparing it with a solution of the integral equation corresponding to the integral inequality (8.36) as explained in Chapter 5. The latter is a solution of the differential equation $df/dt = C(1 + f)^2$ with initial value $C\|u_0\|_{H^s}^2$. There is a constant $T > 0$ such that the solution of this equation is smaller than $2C\|u_0\|_{H^s}^2$ on the interval $[0, T]$. This can be used to show that on the interval $[0, T]$ all the iterates u_m stay in the region G as long as they are defined. The definition of T_m then implies that $T_m \geq T$ for all m. Moreover, the H^s norms of the iterates are uniformly bounded on this time interval. This means that the sequence $\{u_m\}$ is bounded in the space $C^0([0, T], H^s(\mathbb{R}^n))$.

The constant T' in the inequality (8.35) can be chosen so small that $CT' < 1$. It may be assumed that the initial data are approximated in such a way that $V^m(0) \leq 2^{-m}$. Then it follows that

$$\sup_{0 \leq t \leq T'} [V^m(t)]^2 \leq 2^{-2m} + K \sup_{0 \leq t \leq T'} [V^{m-1}(t)]^2. \quad (8.37)$$

Summing this from 1 to M gives

$$\sum_{m=0}^{N} \left[\sup_{0 \leq t \leq T'} [V^m(t)]^2\right] \leq 1 + a_0 + K \sum_{j=0}^{N} \left[\sup_{0 \leq t \leq T'} [V^m(t)]^2\right]. \quad (8.38)$$

This implies that the infinite sum converges and that u_m is a Cauchy sequence in the space $C^0([0, T], L^2(\mathbb{R}^n))$. The interpolation inequality

$\|u\|_{H^{s'}} \leq \|u\|_{H^s}^{\frac{s'}{s}} \|u\|_{L^2}^{1-\frac{s'}{s}}$ which holds for functions on $\mathbb{R}^n$ can now be applied. It follows that $\{u_m\}$ is a Cauchy sequence in $C^0([0,T], H^{s'}(\mathbb{R}^n))$ for all $s' < s$. Substituting this into the equation shows that $\partial_t u_m$ is a Cauchy sequence in $C^0([0,T], H^{s'-1}(\mathbb{R}^n))$. Since $s > n/2 + 1$ the number s' can be chosen to be greater than $n/2 + 1$. This shows, with the help of the Sobolev embedding theorem, that u_m converges to a limit in $C^1([0,T] \times \mathbb{R}^n)$. This limit is a classical solution of the symmetric hyperbolic system and has the prescribed initial datum. It belongs to the space $C^0([0,T], H^{s'}(\mathbb{R}^n)) \cap C^1([0,T], H^{s'-1}(\mathbb{R}^n))$. Some additional functional analytic arguments allow s' to be replaced by s in this statement. An important step is to apply the Banach–Alaoglu theorem to the bounded sequence $\{u_m\}$ in $L^\infty([0,T], H^s(\mathbb{R}^n))$. This implies that there is a subsequence which converges weak* in $L^\infty([0,T], H^s(\mathbb{R}^n))$. It must converge to the solution which was already constructed. The proof of this result also shows that for data which take values in a relatively compact subset of the open set G where the coefficients of the equation are defined the time of existence of the solution obtained only depends on the H^s norm of the initial datum.

It is possible to pass to the limit in the energy estimates for the iterates u_m to get the inequality

$$\|u(t)\|_{H^s}^2 \leq C \left[\|u_0\|_{H^s}^2 + \int_0^t (1 + \|u(t')\|_{C^1} + \|\partial_t u\|_{C^0})(1 + \|u(t')\|_{H^s}^2) dt' \right]. \tag{8.39}$$

This leads to a continuation criterion which says that if a solution exists on an interval $[0, T)$ with $T < \infty$ and if $\|u(t)\|_{C^1} + \|\partial_t u(t)\|_{C^0}$ is bounded on that interval then the solution extends to an interval $[0, T')$ with $T' > T$. This can be seen as follows. The differential inequality implies that under the given boundedness assumption $\|u(t)\|_{H^s}$ is bounded on the interval $[0, T)$. The local existence theorem then shows that there exists a solution on $[t, t + \epsilon)$ which agrees with the given solution at t where ϵ is a positive number independent of $t \in [0, T)$. By the standard uniqueness statement all these solutions agree on the intersections of their domains of definition. Hence they define a solution on the interval $[0, T + \epsilon)$ and it is possible to set $T' = T + \epsilon$. This result also has the consequence that the solution corresponding to C^∞ initial data is itself C^∞. A priori it could have been the case that the H^s solution exists on a smaller and smaller time interval as s increases so that there would have been no non-empty intersection of the domains of definition for all s. The continuation criterion shows that this problem does not occur.

The existence theorem just presented for symmetric hyperbolic systems requires data which possesses a number of derivatives in L^2 which is greater than $n/2 + 1$. At first glance this number might seem unreasonably large. It will now be shown by example that if $s < n/2 + 1$ then symmetric hyperbolic equations are not always well posed in H^s. Consider the (inviscid) Burgers' equation $\partial_t u + u \partial_x u = 0$. In this case $n = 1$. Let u_0 be a smooth initial datum of compact support which is not identically zero. Differentiating the equation with respect to x shows that $v = \partial_x u$ satisfies $\partial_t v + u \partial_x v = -v^2$. The characteristic through the point x_0 is given by the straight line $x = x_0 + u_0(x_0)t$ and u is constant along that line. If D/Dt denotes the directional derivative along that line then $Dv/Dt = -v^2$. It follows that if $v_0 = \partial_x u_0(x_0) < 0$ then v blows up after time v_0^{-1} if the solution exists that long. This shows that if V is the maximum of $-\partial_x u_0$ over all points where this quantity is positive then $V > 0$ and the future time of existence of the solution corresponding to the initial datum u_0 is less than V^{-1}. Let $\tilde{u}_0$ be the rescaled initial datum given by $\tilde{u}_0(x) = a^{-p} u_0(ax)$ for some $p < 1$. Then $\tilde{V} = a^{-p+1} V$. It follows that as $a \to \infty$ the time of existence goes to zero. To show that the initial value problem is ill-posed in H^s it then suffices to show that the H^s norm of $\tilde{u}_0$ is bounded for all $a \geq 0$. Now

$$\|\partial_x^k \tilde{u}_0 / \partial x^k\|_{L^2} = a^{-p+(k-1)/2} \|\partial_x^k u_0 / \partial x^k\|_{L^2}. \tag{8.40}$$

Hence $\|\tilde{u}_0\|_{H^2} \leq a^{-p+3/2} C \|u_0\|_{H^2}$. By the interpolation inequality mentioned above it follows that for $s \in [0, 2]$

$$\|\tilde{u}_0\|_{H^s} \leq \|\tilde{u}_0\|_{H^2}^{s/2} \|\tilde{u}_0\|_{L^2}^{1-s/2} \leq a^{s-p-1/2} \|u_0\|_{H^2}^{s/2} \|u_0\|_{L^2}^{1-s/2}. \tag{8.41}$$

It follows that the desired bound is obtained provided $p \geq s - 1/2$. For $s < 3/2$ it is possible to choose p to satisfy this constraint.

8.4 Strong hyperbolicity

A generalization of the concept of symmetric hyperbolicity is that of strong hyperbolicity. In fact strong hyperbolicity is closely related to the idea of symmetrizable systems introduced in the last section. If a point x of space is fixed then a symmetrizer R must be such that $R(x)A^i(x, \xi)$ is symmetric for all ξ. Thus one matrix must satisfy an infinite number of conditions and it is not surprising that this is difficult to achieve. In the definition of strong hyperbolicity these constraints are weakened by allowing R to be a pseudodifferential operator of order zero instead of requiring it to be a multiplication operator. This gives much more freedom. What is needed to

8.4 Strong hyperbolicity

allow this are conditions on the principal symbol. It must be diagonalizable with all eigenvalues real and it must be possible to diagonalize it by a linear transformation depending smoothly on x and ξ. Under these conditions a smooth pseudodifferential symmetrizer exists.

Let the principal symbol of the operator under consideration be denoted by $a(x, \xi, u)$ to avoid any possible confusion with the operator itself. The algebraic condition on the symbol mentioned above is equivalent to the condition that there is a symbol $b(x, \xi, u)$ which satisfies

$$b^{-1}(x, \xi, u) a(x, \xi, u) b(x, \xi, u) = d(x, \xi, u), \tag{8.42}$$

where d is diagonal. Let $r(x, \xi, u) = [b(x, \xi, u) b^T(x, \xi, u)]^{-1}$ where the superscript T denotes the transpose. Then r is the desired symmetrizer. For $b^{-1}(b^{-1})^T$ is symmetric and positive definite while $ra = (b^{-1})^T d b^{-1}$ is also symmetric. The one part of the criterion for strong hyperbolicity which is not purely algebraic is the smoothness condition. It is, however, often not hard to check once the other conditions have been verified.

For strongly hyperbolic systems a local existence theorem can be proved for data on a torus which is rather similar to that for symmetric hyperbolic systems. One disadvantage is that, because the constructions with pseudodifferential operators are non-local in space a domain of dependence is not immediately obtained and it is not possible to localize in space and transport solutions from one manifold to another as in the case of differential operators. There are ways of obtaining information about the domain of dependence by the following additional procedure. For this purpose it is necessary to introduce a suitable notion of a lens-shaped region. Suppose that in certain coordinates the hypersurfaces $t = $ const. are such that a given system of PDE is strongly hyperbolic. Then there are hypersurfaces S of the form $t = \phi(x)$ such that the region consisting of points (t, x) satisfying the condition $0 \leq t \leq \phi(x)$ is compact and if the equation is rewritten in terms of the coordinate $t' = t - f(x)$ the transformed system is strongly hyperbolic. In this case ϕ will be said to define a lens-shaped region for the system and it will be shown that solutions are determined uniquely by initial data on a region of this type. As in the case of a symmetric hyperbolic system it is enough to show that vanishing initial data implies the vanishing of the solution on this region for a linear homogeneous equation. It suffices to consider the equation satisfied by the difference of two solutions. Since for a given system a neighbourhood of any initial hypersurface can be covered by lens-shaped regions in the sense just defined a local uniqueness theorem for strongly hyperbolic systems results.

The basic uniqueness statement for a lens-shaped region is proved by a method related to the proof of Holmgren's theorem, a uniqueness theorem for PDE with analytic coefficients. Let G denote the lens-shaped region and ∂G_- and ∂G_+ denote the parts of its boundary with $t = 0$ and $t = \phi(x)$ respectively. If an operator is strongly hyperbolic the same is true of its formal adjoint. It is possible to get a local existence theorem for the formal adjoint by transporting the data to T^n. Let u be a solution of the linear strongly hyperbolic system $Pu = 0$ which vanishes on ∂G_- and let f be a smooth function on g. Let P^* be the formal adjoint of P and let v be a solution of the equation $P^*v = f$ which vanishes on ∂G_+. Then

$$\int_G uf = \int_G u(P^*v) = \int_G (Pu)v + \text{boundary terms}. \quad (8.43)$$

The first term on the right vanishes. So do the boundary terms, since that on ∂G_- contains a factor u and that on ∂G_+ contains a factor v. Hence $\int_G uf = 0$ for any smooth function f, which implies that u vanishes on G.

8.5 Leray hyperbolicity

Another notion of hyperbolicity is Leray hyperbolicity [132]. It allows operators with principal parts of different orders in the same system and may be compared to Douglis–Nirenberg ellipticity. The discussion here uses powers of an operator Λ to adjust the orders where Λ is a pseudodifferential operator which defines an isomorphism from H^s to H^{s-1} for each s. The original approach of Leray is similar but uses differential operators to do the adjustment in the orders. This involves more complicated algebra. Consider a system of PDE for an unknown u, schematically written as $P(u) = 0$. Split the components of the unknown into L subsets as in Section 7.3 and introduce operators P_{ij}. Choose indices s_i and t_j as in that case and define $Q_{ij} = \Lambda^{t_j} P_{ij} \Lambda^{-s_i}$. The order of Q_{ij} in its dependence on v_i is the order of P_{ij} in its dependence on u_i plus $t_j - s_i$. The goal is to choose the indices t_j and s_i so that each operator Q_{ij} has the same order in its dependence on v_i (say one) and lower order in its dependence on v_j for $j \neq i$. It is irrelevant for the procedure if the same amount is added to all the indices s_i and t_j simultaneously. The first condition needed for Leray hyperbolicity is that the indices can be chosen in the way just mentioned. The second is that the operators P_{ii} are strictly hyperbolic There is a third condition which makes use of the fact that the notion of a spacelike hypersurface is defined for a strictly hyperbolic system. It says that there should be a hypersurface which is spacelike for all the strictly hyperbolic operators P_{ii}. This is a condition on the characteristic sets of those operators. When all three conditions are satisfied there is a local existence theorem for the system. All components v_i

have the same degree of differentiability which means that the components u_i have different degrees of differentiability.

It should be pointed out that the method just sketched has been chosen here because of the fact that it gives an intuitive interpretation of the indices s_i and t_j and that to the author's knowledge it has never been worked out in full detail. The existence statement for Leray hyperbolic systems still rests on the original method of Leray himself.

The general concepts related to Leray hyperbolicity will now be illustrated by the concrete example of the Einstein–dust system. The explicit form of this system is as follows:

$$G_{\alpha\beta} = 8\pi \mu u_\alpha u_\beta, \tag{8.44}$$

$$\nabla_\alpha(\mu u^\alpha) = 0, \tag{8.45}$$

$$u^\alpha \nabla_\alpha u^\beta = 0. \tag{8.46}$$

The equation for the evolution of μ contains derivatives of u^α but the equation for the evolution of u^α contains no derivatives of μ. This is an obstacle to obtaining a symmetric hyperbolic system. It could be got around if the derivatives of u^α in the evolution equation for μ could be treated as not belonging to the principal part of the system. This is only possible if the derivatives of u^α are considered on the same footing as μ. The adjustment of orders occurring in the Leray theory, combined with an extra device, allows this to be achieved. In order for everything to be consistent the differentiability of the metric must be one order greater than that of u^α. Combining these two things means that the density should be two times less differentiable than the metric. This creates problems with the Einstein equations. For these are essentially (and the harmonically reduced Einstein equations are precisely) nonlinear wave equations for the metric. The solution of a system of this kind is only one degree more differentiable than the right hand side. This does not fit, since the density occurs on the right hand side. The extra device which saves the day consists in differentiating the Einstein equations once more in the direction u^α and then substituting the evolution equation for μ into the result. This gives the equation

$$u^\gamma \nabla_\gamma G_{\alpha\beta} = -8\pi \mu u_\alpha u_\beta \nabla_\gamma u^\gamma. \tag{8.47}$$

Note that the right hand side of this equation contains no derivative of μ and so is not worse than the right hand side of the undifferentiated equation in the scenario being tried here. On the other hand the left hand side is, in harmonic coordinates, a third order hyperbolic equation in $g_{\alpha\beta}$ and the solution of an equation of this type has (by the Leray theory) two more degrees of differentiability than the left hand side.

This intuitive discussion can be used to guide the choice of indices needed to make the Leray hyperbolicity of the Einstein–dust system manifest. The first thing is to decide which equations should be considered as the evolution equations for which variables and this has already been done in the above discussion. Let the unknowns $(g_{\alpha\beta}, u^\alpha, \mu)$ be denoted schematically by (u_1, u_2, u_3) to make the connection with the general discussion. The relative orders of differentiability discussed suggest choosing $s_1 = 1$, $s_2 = 2$ and $s_3 = 3$. This equalizes the expected differentiability of the different variables. In the new variables v_i the evolution equations are of orders four, three and two, respectively. Choosing $t_1 = 1$, $t_2 = 2$ and $t_3 = 1$ equalizes these. Strictly speaking the blocks into which the system must be split are not just three but fifteen, one for each component. The indices s_i and t_j are chosen to be the same for each component of one geometrical object. (It has been ignored that the variables u^α are not independent due to the normalization condition $u_\alpha u^\alpha = -1$.)

The systems which are here called Leray hyperbolic were called 'strictly hyperbolic' by Leray himself, which is not consistent with the terminology in this book. He also considered more general systems which here will be called hyperbolic in the sense of Leray–Ohya. These equations are in general not well posed in Sobolev spaces or in the class of C^∞ functions but they can be solved for data belonging to Gevrey spaces. The most prominent example of this situation connected with general relativity are the equations of relativistic magnetohydrodynamics.

8.6 The analytic Cauchy problem

It sometimes happens that for a system of equations where it is not clear whether the Cauchy problem is well-posed it is nevertheless interesting to know whether there are solutions corresponding to some restricted class of initial data. Consider the general quasilinear system (8.8) of first order with smooth coefficients. If the matrix A^0 is invertible, which is precisely the condition that the initial hypersurface $t = 0$ is non-characteristic, then it is possible to solve for $\partial_t u$ at $t = 0$ in terms of smooth initial data. Differentiating the system repeatedly with respect to t allows all time derivatives and thus all derivatives of all orders of u on the initial hypersurface to be determined. This means that there is a unique formal solution of the system with the prescribed initial data. In general there is no actual smooth solution having this formal series, as was described for the example of the Laplace equation in Section 8.2. If, on the other hand, the coefficients are analytic functions of their arguments and the data are analytic then a unique local analytic solution exists. This follows from the theorem of Cauchy–Kovalevskaya. This theorem makes no distinction between hyperbolic and elliptic equations. All that matters is that the initial hypersurface is not characteristic.

Why is the analytic theory not sufficient for a theoretical treatment of the initial value problem? One reason, already illustrated in Section 8.2, is that the solution depends on the data in a way which cannot be controlled by means of reasonable estimates. Another reason is that in the analytic set-up the important phenomenon of the domain of dependence is completely lost. This is due to the unique continuation property of analytic functions. This property is also a direct indication that the property of analyticity is an unphysical restriction. From a physical point of view there is no reason why data should not be considered which are zero in one region of space and non-zero elsewhere.

8.7 Initial boundary value problems

In many scientific problems the relevant solutions of evolution equations are determined not just by initial data but also by data on the boundary of a spatial domain. This leads to what is called an initial boundary value problem (IBVP). For example, let Ω be the unit ball in $\mathbb{R}^n$ and consider solutions $u(t,x)$, where $x \in \Omega$ and t belongs to some interval I, of the wave equation with prescribed initial data for $t = 0$ and Dirichlet data on the boundary of Ω. The latter means requiring the condition $u(t,x) = f(t,x)$ for $(t,x) \in I \times \Omega$ and a prescribed function f.

It is natural to pose an IBVP for the non-relativistic Euler equations. For a fluid in a fixed rigid container Ω can be taken to be the region whose coundary is the wall of the container. The physically relevant boundary condition is that the velocity of the fluid normal to the boundary vanishes. Similarly, an IBVP can be posed for electromagnetic waves in a cavity. If the wall of the cavity is a perfect conductor then the relevant boundary conditions are that the electric field is normal and the magnetic field tangential to the boundary. In four-dimensional language this means that the pull-back of $F_{\alpha\beta}$ to the boundary $I \times \partial\Omega$ vanishes. For the Einstein equations there is no physically natural situation analogous to those just mentioned. There is no container which can trap gravitational waves. One natural boundary occurring in general relativity, the boundary of a black hole, is a null hypersurface and so no boundary conditions are required there. The only obvious motivation coming from physics for considering an IBVP for the Einstein equations is numerical relativity. If the Einstein equations are to be solved numerically on an infinite region of space then this must be done on a finite grid. One approach is to do a transformation which maps the whole infinite region onto a finite one. This creates a boundary at infinity and may make the problem singular. There are ways of overcoming this difficulty which are also of interest for the analytical theory, as is discussed further in Section 10.2. Another alternative is to introduce an artificial boundary at a finite distance and hope that the effects of the boundary are small if the distance is large enough. Another place where an IBVP may be relevant for numerical relativity is what is known as excision. The idea is that since analytically the exterior of a black hole is not influenced by the interior it should be possible to cut out the interior in numerical calculations. This may be done using a timelike boundary slightly inside the event horizon.

Consider now a general IBVP for a symmetric hyperbolic system. By localizing in space and introducing suitable coordinates it may be assumed that initial data are given on the hypersurface $t = 0$ and boundary data on the hypersurface $y = 0$ in $\mathbb{R}^n$, where the coordinates in $\mathbb{R}^n$ are written as (x, y) as in Section 7.2 with y being a coordinate normal to the boundary. The aim is to find a criterion for well-posedness. By analogy with what was done in the case of boundary value problems and the Cauchy problem it is natural to linearize the problem at some point and restrict to the principal part of the equation. The equation is then

$$A^0 \partial_t u + A^y \partial_y u + \sum_{i=1}^{n-1} A^i \partial_i u = 0. \tag{8.48}$$

Boundary conditions are considered which are of order zero and homogeneous. Thus they are of the form $Bu(t, 0, x) = 0$ and the solution should exist on the region defined by the inequalities $t \geq 0$ and $y \geq 0$.

A method of finding a well-posedness criterion for the IBVP is to think about energy estimates. In the local existence theorem for the Cauchy problem for symmetric hyperbolic systems an integration by parts is used to get rid of the terms containing first order derivatives in the basic energy estimate. When a boundary is present boundary terms arise during this integration by parts. The estimate may nevertheless be obtained if the boundary contribution has the right sign. In other words it should correspond to dissipation of energy rather than production of energy. What this means concretely is that $\langle A^y u, u \rangle$ should be non-positive on the boundary whenever the chosen boundary condition is satisfied. In this case the boundary condition may be called dissipative. The existence theorem for the Cauchy problem proceeds by doing higher order energy estimates, i.e. energy estimates for higher order spatial derivatives. When a boundary is present this does not work any more. For derivatives tangential to the boundary the original boundary condition immediately implies a boundary condition for the

8.7 Initial boundary value problems

derivative. Unfortunately the normal derivative is not determined directly. This can lead to a loss of regularity at the boundary.

The existence proof for the IBVP will not be pursued further here. Instead one useful result which can be obtained will simply be stated. This concerns the notion of *maximally dissipative* boundary conditions. Consider the matrix A^y which by assumption is negative semidefinite when restricted to the subspace defined by the chosen boundary condition. A relatively simple case is that where the boundary is non-characteristic which means by definition that A^y is invertible. It turns out, however, that for many physically interesting problems, e.g. that for the Euler equations mentioned above, the boundary is characteristic. Suppose now that the boundary is a characteristic hypersurface of constant multiplicity, i.e. that the kernel K of A^y has constant dimension over the boundary. The boundary condition is said to be maximally dissipative if the subspace B defined by the boundary condition is such that as well as having the property that the restriction of A^y to B is negative semidefinite, the dimension of B has the maximal value compatible with this property. This maximal dimension is equal to the number of non-positive eigenvalues of A^y, counting multiplicity.

For an IBVP with maximal dissipative boundary conditions a local existence theorem for smooth solutions can be proved provided one further element is added. This is that corner conditions must be imposed. These result from the fact that for a smooth solution certain quantities can be computed at the 'corner' where $t = y = 0$ in two different ways. For instance, if a component of u is prescribed as part of the boundary condition then its time derivative at the corner can be computed by differentiating the corresponding datum on the boundary with respect to t. The same quantity can be obtained from the initial data by substituting into the differential equations. For consistency these two values must be equal. This gives a relation between initial and boundary data.

Consider the example of the system obtained by reducing the standard wave equation to first order with the new unknowns (v, w_i) as in Section 8.3 and suppose to start with that the space dimension is one. As before the region of space where the equation is to be solved is given by $y \geq 0$. In this case the matrix A^y is 2×2 with Lorentzian signature. Thus the correct number of boundary conditions for a maximal dissipative IBVP is one. This boundary condition should define a subspace which is timelike or null with respect to the quadratic form defined by the matrix A^y. For a boundary condition of the form $\alpha v + \beta w_1 = 0$ this means that the condition $\alpha\beta \geq 0$ should be satisfied. It is useful to rewrite the boundary condition as $\frac{1}{2}(\alpha + \beta)(v + w_1) + \frac{1}{2}(\alpha - \beta)(v - w_1) = 0$. Excluding the case $\alpha = -\beta$ this can be rewritten as $v + w_1 = -\frac{\alpha - \beta}{\alpha + \beta}(v - w_1)$. This expresses the ingoing null derivative at the boundary in terms of the outgoing one. In other words it describes the strength of the waves emitted from the boundary in response to incident waves coming from the interior. The maximal dissipative condition, which is equivalent to $|\frac{\alpha-\beta}{\alpha+\beta}| \le 1$, says in this case that the waves emitted from the boundary do not contain more energy than the incident waves. Suppose now that the wave equation itself is to be solved with Dirichlet boundary conditions. Then w should vanish on the boundary and this may be taken as boundary condition. It is maximal dissipative. When the first order system has been solved in this way a solution of the wave equation is obtained just as in the case of the Cauchy problem. The IBVP with Dirichlet boundary condition for the wave equation in higher dimensions can be solved in essentially the same way. The matrix A^y is augmented by a number of zeroes and vanishing data for derivatives tangential to the boundary are obtained by differentiating the Dirichlet boundary condition. In higher dimensions there are more alternative boundary conditions for the first order system available.

8.8 The null condition

This section is concerned with the question under which circumstances local solutions of a quasilinear hyperbolic system extend to global solutions. The discussion is carried out for nonlinear wave equations of the form

$$\eta^{\alpha\beta} \partial_\alpha \partial_\beta \phi = F(\phi, \partial_\alpha \phi, \partial_\alpha \partial_\beta \phi) \tag{8.49}$$

in $(n+1)$-dimensional Minkowski space. Thus the flat space wave operator acting on ϕ is equal to a nonlinear function of ϕ and its first and second order partial derivatives. It is assumed that $\phi = 0$ is a solution and that at the origin of $(\phi, \partial_\alpha \phi, \partial_\alpha \partial_\beta \phi)$-space the first derivatives of F vanish. In other words the nonlinearity is at least quadratic. It is convenient to write $F = Q + G$ where Q consists of the terms which are exactly quadratic and G of all those which vanish to at least third order at the origin.

Consider now initial data for eqn (8.49) which are small in a Sobolev space H^s with s sufficiently large and decay sufficiently fast at infinity. It is known that if the space dimension n is at least four the corresponding solution exists globally in time and decays as $t \to \infty$ in a way similar to a solution of the equation in the case $F = 0$. In the case of three space dimensions, which is of course of special interest for physics and, in particular, for general relativity things are more complicated. It is known, for instance, that if F is given by $(\partial_t \phi)^2$ global existence for small data in a Sobolev space in general fails. Global existence does hold if $Q = 0$. This last statement can be extended to the case where Q satisfies a certain restriction called the *null condition* which will now be described. The null condition says that for any null vector l^α the condition $Q(\phi, l_\alpha, l_\alpha l_\beta) = 0$ holds. If the null condition is satisfied then for all sufficiently small initial data (as measured in a suitable norm of Sobolev type) the solutions of (8.49) exist globally. Apart from the smallness condition the hypotheses of this theorem require fall-off properties which imply that the data fall off faster than a certain power of the radial coordinate r. It is necessary to assume something much stronger than decay like r^{-n+2} in n dimensions. Thus data solving the Einstein constraints do not fit into this framework.

An intuitive understanding of the meaning of the null condition can be obtained as follows. In n space dimensions with $n \geq 3$ the L^∞ norm of a solution of the linear wave equation in Minkowski space decays like $t^{-\frac{n-1}{2}}$ as $t \to \infty$. The same is true of the L^∞ norm of any partial derivative of the solution with respect to the standard Minkowski coordinates. Taking a derivative of this kind does not lead to any increase of decay. In fact the limiting factor which leads to a lower bound on the rate of decay of the L^∞ norm is the pointwise decay along null directions. Let r be a radial coordinate in $\mathbb{R}^3$. Then it is important to consider decay rates along the outgoing null rays where $v = t - r$ is constant. On those rays a decay rate in $u = t + r$ is equivalent to a decay rate in t. If ϕ is a solution of the linear wave equation then the transverse derivative $\partial_v \phi$ behaves asymptotically like t^{-1} along outgoing null rays while the derivative $\partial_u \phi$ and the derivatives in directions orthogonal to the (t, r)-plane behave like t^{-2}. In those cases taking a derivative leads to a gain of decay. Global existence theorems for small data can be obtained by global energy estimates which will not be described in detail here. It turns out to be important that if a solution of the linear equation is substituted into the quantity F the result should decay strictly faster than t^{-2} in L^∞. Thus it becomes clear why quadratic terms in three dimensions are problematic. In general they decay exactly like t^{-2}. In higher dimensions or for terms of order at least three the decay is better. Certain quadratic terms in three dimensions are better but there are also 'bad terms' like $(\partial_t \phi)^2$ which are not. The null condition serves to rule out those terms where a null derivative is multiplied by itself and by nothing else. In the case where F is given by $(\partial_t \phi)^2$, for instance, it can be rewritten as

$$(\partial_t \phi)^2 = \frac{1}{4}[(\partial_u \phi)^2 - 2\partial_u \phi \partial_v \phi + (\partial_v \phi)^2]. \tag{8.50}$$

In this case the first two terms on the right hand side are good while the third is bad. A typical example where the null condition is satisfied and global existence for small data holds is the equation $\nabla^\alpha \nabla_\alpha \phi = \eta^{\alpha\beta} \nabla_\alpha \phi \nabla_\beta \phi$. The equation for a wave map in Minkowski space has a very similar structure and also satisfies the null condition. The equation of motion for a k-essence field in Minkowski space satisfies the null condition iff the derivatives $\partial^i L/\partial \phi^i$ vanish for $i = 1, 2, 3$.

A generalization of the null condition called the weak null condition will now be written down. It only concerns the quadratic part of the nonlinearity. Suppose that $Q(\phi^i) = a^i_{jk\alpha\beta} \partial_\alpha \phi^j \partial_\beta \phi^k$ where the coefficients $a^i_{jk\alpha\beta}$ are zero unless $|\alpha| \leq |\beta| \leq 2$ and $|\beta| \geq 1$. Define a new equation for a function $\Phi(s, q)$ by

$$2\partial_s \partial_q \Phi^i = A^i_{jkmn}(\omega) \partial_q^m \Phi^j \partial_s^n \Phi^k, \tag{8.51}$$

where

$$A^i_{jkmn}(\omega) = \sum_{|\alpha|=m, |\beta|=n} a^i_{jk\alpha\beta} \hat{\omega}^\alpha \hat{\omega}^\beta \tag{8.52}$$

and $\hat{\omega} = (1, \omega)$. This asymptotic system arises from considering certain formal expansions of the solutions. The weak null condition is said to hold if all solutions of the asymptotic system with suitably decaying initial data exist globally and satisfy suitable global bounds. The null condition itself is equivalent to the condition that the coefficients A^i_{jkmn} vanish identically. The weak null condition is supposed to be a criterion for global existence of solutions of the original equations although this has not been proved in complete generality yet. The importance of this condition for the Einstein equations is explained in Section 10.3.

8.9 Global difficulties

It is typical for nonlinear hyperbolic equations that for general initial data the corresponding solutions do not exist globally in time. In this case it is said that singularities are formed. It should be kept in mind that the word singularity is used here in a different sense from that which is standard in general relativity. Most hyperbolic problems are defined on a fixed manifold, often a Euclidean space, and saying that there is a singularity simply means that the solution cannot be extended to the whole space. This is conceptually simpler than in the case of the Einstein equations where the spacetime manifold is part of the solution.

Another frequent situation for hyperbolic equations is that it is simply not known whether global existence holds for general initial data although superficially the equation appears simple. Consider for example the equation $\partial_t^2 u = \Delta u - u^{2k+1}$ on $\mathbb{R} \times \mathbb{R}^3$ for a positive integer k. In the case $k = 1$ global existence has been known for a long time. It took another thirty years to find a global existence proof for the case $k=2$. For $k>2$ the problem is still completely open – there is no global existence theorem and no example of the occurrence of a singularity. Numerical work has also failed to give useful insights for this problem. Another interesting case which gives an idea of the difficulties involved is that of the Navier–Stokes equations for an incompressible viscous fluid, which is more like a parabolic problem than a hyperbolic one. There the global existence question is open. A sociological argument indicating that the problem is hard is that the Clay Foundation has offered a prize of one million dollars to anyone who can answer this global existence question positively or negatively and that nobody has successfully claimed the prize yet.

A case where a lot more is understood about why global existence fails is that of the compressible Euler equations. There it is known that for a wide class of data shock waves are formed where the basic fluid variables remain bounded but their first derivatives blow up. There is no global existence of classical solutions. It is still possible to ask if the solutions can be extended beyond the time of shock formation so as to satisfy the equations in some generalized sense. In fact such weak solutions have been studied a lot and in recent years a good theory has been developed in the case of one space dimension. The generalized solutions are of importance in astrophysics where they come up in many situations. In space dimensions greater than one not much was known in terms of rigorous mathematics until very recently although numerical calculations can be carried out successfully. A major new development in understanding the formation of shock waves in solutions of the relativistic Euler equations without symmetry is the book of Christodoulou [58].

A problem with weak solutions is that they may not be determined uniquely by initial data and that it is unknown in many cases whether uniqueness holds. To have a hope of getting uniqueness it is often necessary to impose some extra conditions on the solutions which are physically motivated. In the case of the Euler equations these conditions are connected with the notion of entropy. For the Einstein equations no general definition of weak solutions is known. Since the Einstein equations are fundamental rather than being derived by approximation from some other underlying theory it is difficult to see where extra physical conditions on weak solutions should come from.

A standard example for obtaining some intuition about the formation of singularities in hyperbolic equations is the Burgers equation $\partial_t u + u \partial_x u = 0$. That solutions of this equation generally have a finite time of existence was shown in Section 8.3. Here a more geometric way of reaching the same conclusion will be presented. The solution is constant along the straight lines with equation $x = x_0 + tu_0$ where x_0 is a point in space and $u_0 = u(x_0)$.

This means that it is not consistent for two lines of this type with different slope to intersect if they are in the domain of existence of a classical solution. This makes it geometrically obvious that there is a singularity in at least one time direction for almost any initial data. If the initial data are such there are points x_1 and x_2 with $x_1 < x_2$ and $u(x_1) > u(x_2)$ then the corresponding solutions cannot exist for a time longer than $\frac{u(x_1)-u(x_2)}{x_2-x_1}$ in the future.

From the above discussion it should be clear that it is in general very difficult to decide whether there will be global existence for a given equation. There is, however, one type of criterion which can be used to guess whether global existence will hold and if so how difficult it is likely to be to prove it. This is based on the fact that many of the equations of interest in practice have some kind of scaling invariance and some kind of conserved quantity. In favourable cases conserved quantities can play a key role in proving global existence. Consider for example the semilinear wave equation introduced above. If $u(t,x)$ is a solution then $u_\lambda(t,x) = \lambda^{\frac{1}{k}} u(\lambda t, \lambda x)$ is also a solution. The energy

$$E = \int_{\mathbb{R}^3} \frac{1}{2}[(\partial_t u)^2 + |\nabla u|^2] + \frac{1}{2k+2} u^{2k+2} \quad (8.53)$$

is conserved. Let E_λ be the energy of the solution u_λ. Then $E_\lambda = \lambda^{\frac{2-k}{k}} E_0$. When the power in this relation is negative it means intuitively that energy can be concentrated on small scales and this indicates that a singularity may be formed. This is described by saying that the given equation is supercritical for the energy. Often global existence fails in supercritical problems and it is at any rate hard to prove. When the power is positive or zero the equation is said to be subcritical or critical for the energy respectively. Subcritical problems can be expected to be relatively easy. Critical problems are particularly interesting since they often lie exactly on the frontier of our knowledge. The classification in terms of scaling identifies the value $k = 2$ as critical for the above nonlinear wave equation. To give a couple of other examples, the Yang–Mills equations are subcritical for the energy in spacetime dimension less than five and global existence is known. The critical dimension is five and global existence is open. Wave maps are critical in spacetime dimension three and for certain target manifolds global existence is known to fail in the critical dimension. The Einstein equations are supercritical for the ADM mass in spacetime dimension four.

8.10 Comparison with parabolic equations

Parabolic equations are an important class of PDE. Since they have only had a small number of applications in general relativity they will not be discussed

8.10 Comparison with parabolic equations

here in the same detail as elliptic or hyperbolic equations. It is nevertheless useful to compare them with hyperbolic equations. There is one case, that of the Robinson–Trautman solutions, in which the vacuum Einstein equations reduce to a parabolic equation under certain special assumptions. There have also been applications to defining coordinate or gauge conditions. In particular, a parabolic equation, the inverse mean curvature flow, was used to construct a special foliation of the initial data hypersurface in the proof of the Riemannian Penrose inequality [109]. The analogous flow in a Lorentzian manifold was used to define a time coordinate in the first numerical calculation of the structure of spacetime singularities without symmetry [88].

The best known parabolic equation is the heat or diffusion equation $\partial_t u = \Delta u$. In more general linear parabolic systems the Laplace operator on the right hand side is replaced by a strongly elliptic operator. A nonlinear parabolic system can be defined as one whose linearization about any solution is parabolic. In the natural initial value problem for the heat equation the restriction of the solution u to the hypersurface $t = 0$ is prescribed. In contrast to the case of hyperbolic equations the initial value problem is only well-posed in the direction in which time increases. For general initial data there exists no corresponding solution in the direction of decreasing time. A related fact is that parabolic equations tend to be regularizing. Data with finite differentiability typically lead to solutions whose restrictions to the region $t > 0$ are smooth or even analytic, at least close to $t = 0$. This could not happen for a hyperbolic equation because in that case the evolution in both time directions is well posed.

There are similarities between hyperbolic and parabolic equations – energy estimates are a powerful tool in both cases although the way they work is somewhat different in the two cases. In some cases, in particular for a single parabolic equation with scalar unknown, there is a tool which is even more potent and which unfortunately seems to have no useful analogue in the hyperbolic case. This is the maximum principle. For the heat equation on a compact manifold such as T^n for instance it says that the maximum value of a solution is attained on the initial hypersurface and this immediately gives rather strong control on the solution. When long time evolution is considered solutions of nonlinear parabolic systems often develop singularities. This can even be a useful property as in the celebrated recent proof of the Poincaré conjecture by Perelman. This is based on the strategy developed by Hamilton of using the dynamical behaviour, including singularity formation, of solutions of a nonlinear parabolic system, the Ricci flow. It is reasonable to expect that a fruitful exchange of information between the theories of hyperbolic and parabolic equations should be possible although not very much has happened yet. In particular it should be possible to obtain a better understanding of issues like long-time asymptotics and formation of singularities.

8.11 Fuchsian methods

Fuchsian systems are a class of singular partial differential equations which have a variety of applications in general relativity. They need not be hyperbolic but they may be. The basic form of these systems is

$$t\frac{\partial u}{\partial t} + N(x)u = tf(t, x, u, \partial_i u). \tag{8.54}$$

The fact that the time derivative is multiplied by t means that the equation is singular at $t = 0$. In favourable cases a system of this type has a unique solution which tends to zero as $t \to 0$. What are the favourable circumstances? To ensure uniqueness the matrix $N(x)$ should be assumed to have some positivity properties. For instance it can be assumed that the eigenvalues of N have positive real part. In addition some regularity assumption must be made on the function f. General results are available in the case that the dependence of the coefficients on the spatial variables and the unknown is analytic. For this there is no requirement of hyperbolicity. If the dependence of the coefficients on the spatial variables and the unknown is only smooth (or of finite differentiability) the results are less developed. In that case the equations should be hyperbolic. This is natural since the Fuchsian initial value problem cannot be expected to be easier than the Cauchy problem for a regular system. To see the connection consider a system of the form

$$\partial_t v = tg(t, x, v, \partial_i v). \tag{8.55}$$

If there exists a smooth solution v of this equation with initial datum $v(0, x) = v_0(x)$ then by Taylor's theorem $v(t, x) = v_0(x) + tu(t, x)$ for some smooth function u. It follows from the equation that $\partial_t v$ vanishes for $t = 0$ and hence that u vanishes for $t = 0$. The function u satisfies the Fuchsian system (8.54) with $f(t, x, u, \partial_x u) = g(t, x, v_0 + tu, \partial_x v_0 + \partial_x u)$. Conversely, if there is a solution of (8.54) with $t = 0$ and the given choice of f then v is a solution of the original system with initial datum v_0.

The existence proof for Fuchsian systems with analytic coefficients uses an iteration which is defined by solving the equation

$$t\frac{\partial u_{j+1}}{\partial t} + N(x)u_{j+1} = tf(t, x, u_j, \partial_i u_j). \tag{8.56}$$

The solvability of this equation for a given u_j is guaranteed by the positivity assumption on N. To obtain the existence proof it must be shown that the sequence u_j converges in a suitable sense. In fact in this iteration all objects (the coefficients of the equation, the iterates, the solution) are defined on some complex domain containing the real domain where the solution is eventually defined. It is the complex analytic nature of the extended objects which gives rise to the real analyticity of the solution.

The applications of Fuchsian systems proceed as follows. To start with there is a system for an unknown v. Then some expression w with an explicit dependence on t is subtracted. It encapsulates the expected leading order asymptotic behaviour. Then, if all goes well, the difference $u = v - w$ satisfies a Fuchsian system of the form (8.54). For instance in the case of an equation which is singular at $t = 0$ the attempt is made to prescribe data at $t = 0$ for a solution u. More precisely, it is coefficients in an asymptotic expansion about $t = 0$ which are prescribed. These coefficients are used to build the function w which describes the leading order behaviour. The function v is the remainder and is expected to tend to zero for $t \to 0$. If the coefficients of the equation and the prescribed data are analytic in the spatial coordinates then the same is the case for the Fuchsian system and the more powerful analytic existence theorem can be applied. If, on the other hand, smooth data are allowed the coefficients of the Fuchsian system are only smooth and another theorem is necessary.

Systems which are not quite in the Fuchsian form with positivity of N can sometimes be brought into that form by certain transformations. For instance if instead of being of the form tf the right hand side is of the form $t^\alpha f$ for some $0 < \alpha < 1$ then introducing a new time variable by $t' = t^\alpha$ brings it into Fuchsian form. If N is not positive it may sometimes be improved by subtracting off finitely many terms in a formal asymptotic expansion. If there is a formal series $\sum_{n=1}^\infty u_n t^n$ then the new variable $u - \sum_{n=1}^k u_n t^n$ may satisfy a system of Fuchsian form where the matrix N is replaced to $N + kI$. Obviously if k is large enough then $N + kI$ is positive definite and, in particular, its eigenvalues have positive real parts.

These general remarks can be illustrated by the example of the Gowdy equations. Solutions of the Gowdy equations will be constructed which have the following form:

$$P(t, \theta) = k(\theta) \log t + \phi(\theta) + t^\epsilon u(t, \theta), \tag{8.57}$$

$$Q(t, \theta) = Q_0(\theta) + t^{2k(\theta)}(\psi(\theta) + v(t, \theta)). \tag{8.58}$$

Here ϵ is a positive constant and the functions k, ϕ, Q_0 and ψ are given. These functions are used to define the leading order asymptotics and they play the role of data at the singularity at $t = 0$. The functions u and v are remainder terms and the idea is to show that they satisfy a Fuchsian system. Since the Gowdy equations, and the equations obtained by rewriting them as equations for the unknowns u and v, are second order it is necessary to reduce them to first order in order to make contact with the theory introduced above. This is done by introducing the new unknowns

$$(u_0, u_1, u_2, v_0, v_1, v_2) = (u, t\partial_t u, t\partial_x u, v, t\partial_t v, t\partial_x v). \qquad (8.59)$$

These quantities satisfy a Fuchsian system for ϵ sufficiently small if one of two conditions are satisfied. In the first case (low velocity solutions) it is assumed that $0 < k(\theta) < 1$ while in the second (high velocity solutions) it is assumed that $\partial_x Q_0 = 0$. The significance of these conditions is discussed in Section 10.1. The theory of Fuchsian systems then shows that if the data (k, ϕ, Q_0, ψ) are analytic then there is a unique solution of the equations with u and v vanishing at $t = 0$. The functions P and Q defined in terms of u and v by (8.57) and (8.58) then satisfy the Gowdy equations. Notice that the general solution of the Gowdy equations depends on four free functions, the data for $(P, \partial_t P, Q, \partial_t Q)$ on a regular Cauchy hypersurface. In the low velocity case the solutions constructed by the Fuchsian method depend on the same number of free functions (in the analytic class) while in the high velocity case they only depend on three free functions. Thus in the sense of function counting the low velocity solutions are as general as the general solution of the Gowdy equations while the high velocity solutions are not.

In [153] there is a claim of a result using Fuchsian methods for the non-polarized Einstein–Maxwell–dilaton system with $\chi = 0$ (case A of section VI in that paper). Setting the dilaton coupling constant and the dilaton itself equal to zero defines a consistent special case and should therefore produce solutions of the Einstein–Maxwell system. Unfortunately it is already problematic to consider non-polarized solutions with $\omega \neq 0$ and $\chi = 0$. The equation of motion (4.39) for χ is in general not satisfied. It leads to a complicated integrability condition.

There are also results using Fuchsian methods for the T^3 symmetric solutions of the five-dimensional Einstein equations mentioned in Section 4.4.

It was discussed in Section 8.6 why it is not in general satisfactory to solve the Cauchy problem only for data which are analytic. The same criticism applies to the analytic Fuchsian analysis and so it is important to extend the result to the smooth case. It will now be explained how this can be done. Most of the steps involved are rather generally applicable but a final result has only been obtained in a few cases such as that of Gowdy solutions. Before explaining this method it should be pointed out that there are some results for Fuchsian systems with non-analytic coefficients which are based on other approaches. In the first of these [62] the function f is required to have some smoothness in t which means it has an asymptotic expansion in integral powers of t and the solutions obtained have the same property. The matrix $N(x)$ is only required to have no eigenvalues which are negative integers. An important element of the proof is to represent the solution in the form $u_0 + u_1$ where u_0 is polynomial in t with coefficients depending on x and u_1 is a remainder of higher order in t. Then u_1 solves a Fuchsian system where the eigenvalues of $N(x)$ have been shifted by an integer in the positive direction. If it is possible to expand to sufficiently high order then the shifted eigenvalues all have positive real parts. The only obstruction to carrying out this process is if a shifted eigenvalue becomes zero at some stage. This is prevented by the assumption on the eigenvalues of the original matrix $N(x)$. The method of [62] cannot be applied to the Gowdy system since the expansions of the coefficients contain non-integer and x-dependent powers of t and even logarithms. In the method explained in this section the technique of expanding the solution to finite order is extended to the case where terms more complicated than integral powers of t occur in the expansions.

Another theorem for Fuchsian systems with non-analytic coefficients [117] does not rely on the existence of expansions in integral powers of t but does require that the matrix N is constant, i.e. independent of x. The proofs of [62] and [117] involve the use of sophisticated techniques from functional analysis, namely semigroup theory and the Yosida approximation. These are avoided in the method presented in this section.

For a Fuchsian system with smooth coefficients the iteration used in the case of analytic coefficients can also be defined, but without the extensions into the complex domain. In this case it cannot be expected that the iteration converges (since it involves a loss of derivatives at each step) but it can be used in a different way. To see how some definitions are required. A function $u(t, x)$ is called regular if it is smooth for $t > 0$ and if it and its derivatives of any order with respect to the spatial variables x^i extend continuously to $t = 0$. (Note that this allows time derivatives of u to blow up in the limit $t \to 0$.) A formal solution of equation (8.54) of order p is a finite sequence $(u_1, \ldots, u_p)$ of regular functions which satisfies the relation

$$t\partial_t u_j + N u_j - f(t, x, u_j, \partial_i u_j) = O(t^j) \tag{8.60}$$

for each j. Here the O-symbol is to be interpreted in the sense of uniform convergence on compact subsets. With a suitable positivity assumption on N it can be shown that the above iteration produces formal solutions of any finite order,

The existence proof for Gowdy solutions proceeds by series of transformations of the Gowdy system. Introducing the new unknowns u and v and reducing to first order as in the existence proof for the analytic case produces a Fuchsian system which will be referred to as the *first reduced system*. To obtain a Fuchsian system it is necessary to fulfil the condition $\epsilon < \min(2k, 2 - 2k)$. Next the substitutions $u_2 = t\partial_x u$ and $v_2 = t\partial_x v$ are made in some places in the first reduced system. The result is called the *second reduced system*. It has the advantage that it is symmetric hyperbolic. It also has two disadvantages. The first is that the matrix N in the second reduced system has two negative eigenvalues. The second is that the equation contains the expression $t^{1+\epsilon-2k}$ which may be singular. This can be avoided subject to the restrictions already assumed on ϵ iff $k < \frac{3}{4}$. A formal solution of the first reduced system is also a formal solution of the second reduced system. Subtracting a formal solution from an actual solution and deriving an equation for the resulting quantity gives rise to a system which is both Fuchsian and hyperbolic. In the Gowdy case it will be called the *third reduced system*. If the order of the formal solution is high enough the matrix N in the third reduced system is positive definite.

To get an existence theorem for smooth initial data the first step is to approximate the smooth data by a sequence of analytic data. To each element of the sequence there is a corresponding solution, as ensured by the analytic theory. It is possible to define a third reduced system for each element of the sequence in such a way that the coefficients converge along the sequence to the corresponding coefficients for the smooth data. If it can be shown that the sequence of solutions also converges then the limit of the sequence will provide the desired solution of the Gowdy system with smooth data. The sequence is shown to be a Cauchy sequence in a Sobolev space by energy estimates. These are very similar to the standard energy estimates for regular symmetric hyperbolic systems. The main difference is that the singular term containing N contributes a term whose size cannot be controlled. However, due to the positivity of N, the sign of the large term is such that it may be discarded when doing the energy estimate. The result of this argument is that the Gowdy problem with smooth data has been handled provided $0 < k < \frac{3}{4}$. A more complicated approach, expanding the quantities to higher order explicitly, allows the range $\frac{1}{2} < k < \frac{5}{6}$ to be covered. The process can be repeated indefinitely in order to cover the whole interval $0 < k < 1$.

Even when the case of smooth data has been treated in this way one question remains open. Take the set of solutions of the equations with the given asymptotics and use them to induce initial data on a regular hypersurface $t = t_0$. Does the resulting set of initial data include any non-empty open set? There is a theorem which gives a result of this type for a class of Fuchsian equations [118] but it does not cover the case of the Gowdy equations. Its proof uses the Nash–Moser theorem.

8.12 Further reading

A nice presentation of the existence theorem for linear symmetric hyperbolic equations by discretization is given in [115]. The existence proof sketched in the nonlinear case follows the treatment in [140]. The method of Makino for treating fluid bodies was introduced for Newtonian self-gravitating fluid bodies in [141] and generalized to the relativistic case in [166]. The discussion of the Euler equations in Section 8.3 was influenced by unpublished work of U. Brauer and L. Karp. Cf. also [36]. Strong hyperbolicity is discussed in detail in [126]. Information on the applications of this concept in general relativity can be found in [182]. The original work on Leray hyperbolicity is in [132]. The example of the Einstein–dust system is taken from [46]. Concerning the Cauchy problem for relativistic magnetohydrodynamics see [133]. Theorems on formation of singularities for certain critical wave maps were proved in [188] and [127]. The Fuchsian theory for the Gowdy equations in the smooth case is taken from [172] and [195].

9 The Cauchy problem for the Einstein equations

In this chapter the general theory of hyperbolic equations is applied to the Einstein–matter system. In order not to have to confront different difficulties simultaneously the discussion in the first two sections is limited to the vacuum case. The third section describes how matter can be incorporated.

9.1 Coordinate conditions

Consider the Einstein vacuum equations. It has already been emphasized in previous sections that the principal part of the system, i.e. that containing the highest order derivatives, is of central importance. The vacuum Einstein equations are equivalent to $R_{\alpha\beta} = 0$. Now

$$R_{\alpha\beta} = g^{\gamma\delta}(\partial_\gamma \partial_\delta g_{\alpha\beta} + \partial_\alpha \partial_\beta g_{\gamma\delta} - \partial_\alpha \partial_\gamma g_{\beta\delta} - \partial_\beta \partial_\delta g_{\alpha\gamma}) + \cdots, \qquad (9.1)$$

where only the terms containing second derivatives have been written out explicitly. On the basis of these expressions the equations cannot be identified as belonging to any known type of PDE. There is a very good reason for this. The hope that the Einstein equations expressed in a fixed coordinate system might be hyperbolic is dashed by the simple observation that uniqueness in the Cauchy problem fails for these equations. Suppose that data are prescribed for $g_{\alpha\beta}$ and $\partial_t g_{\alpha\beta}$ on the hypersurface $t = 0$ and that $g_{\alpha\beta}$ is a solution with the right initial data. A new metric $g'_{\alpha\beta}$ can be obtained by acting on $g_{\alpha\beta}$ with a diffeomorphism which agrees with the identity on $t = 0$ up to first order. Then $g'_{\alpha\beta}$ induces the same initial data as $g_{\alpha\beta}$ on the initial hypersurface. Moreover, since the Einstein equations are invariant under diffeomorphisms, the metric $g'_{\alpha\beta}$ is a solution of the Einstein equations. This provides the desired example of non-uniqueness. In fact it turns out that the type of non-uniqueness just described is the only one which occurs for the Einstein equations and that the solution is unique up to diffeomorphisms in a sense which is made precise later. This property is sometimes called geometric uniqueness. The ordinary uniqueness property for a PDE is not appropriate for the coordinate form of a diffeomorphism invariant equation. It is also the case that from a physical point of view the metric should only be significant modulo diffeomorphisms. It took Einstein himself a long time to understand this issue and his discovery of the field equations of general relativity was delayed by it.

From an abstract point of view it would seem natural to prove existence and uniqueness theorems for the Einstein equations in a framework which is manifestly invariant under diffeomorphisms. Unfortunately nobody has succeeded in doing so. The only known methods make use of a special choice of coordinates or similar additional structures. Historically the first method which was found was that of harmonic coordinates or, as they are sometimes called today, wave coordinates. Here the old terminology will be kept. Very recently harmonic coordinates have seen a renaissance both in the mathematical study of the Einstein equations and in numerical relativity. A system of coordinates x^α is called harmonic for a given metric $g_{\alpha\beta}$ if the contracted Christoffel symbols $\Gamma^\alpha = g^{\beta\gamma}\Gamma^\alpha_{\beta\gamma}$ are zero. This is equivalent to the condition that the coordinates themselves satisfy the wave equation with respect to the given metric, i.e. that $\nabla_\alpha \nabla^\alpha x^\beta = 0$. Using the definition of the Christoffel symbols it is also seen to be equivalent to the condition $g^{\alpha\beta}g^{\gamma\delta}(2\partial_\gamma g_{\alpha\delta} - \partial_\alpha g_{\gamma\delta}) = 0$. When the coordinates are harmonic the expression on the right hand side of (9.1) reduces to $g^{\gamma\delta}\partial_\gamma \partial_\delta g_{\alpha\beta} + \ldots$. It follows that in harmonic coordinates the Einstein equations take the form of a system of nonlinear wave equations and are, in particular, hyperbolic. The system obtained by setting the expressions for Γ^α to zero is called the reduced Einstein equations. It can be written down in any coordinate system and is hyperbolic but is only equivalent to the Einstein equations if the coordinates happen to be harmonic.

Local harmonic coordinates can be constructed in any spacetime. For it is only necessary to solve the Cauchy problem for a hyperbolic PDE with suitable initial data. These data should be chosen so that they are functionally independent (i.e. their first derivatives are linearly independent) on the Cauchy hypersurface and hence, by continuity, in a neighbourhood of it. One natural choice is to require that on the initial hypersurface $t = 0$, $n^\alpha \partial_\alpha t = 1$, x^a are equal to some given coordinates on the initial hypersurface and $n^\alpha \partial_\alpha x^a = 0$. This means that the lapse and shift have the same values on the initial hypersurface as in a Gaussian coordinate system. Harmonic coordinates of this type are used in the next section.

9.2 The local Cauchy problem

Some care is necessary in formulating the local Cauchy problem for the Einstein equations due to the diffeomorphism invariance of the problem.

9 : The Cauchy problem for the Einstein equations

The manifold on which the solution is defined is part of the solution and thus it is not possible to prescribe the initial data on a submanifold. Instead the data are defined on an abstract manifold which is eventually identified with a submanifold of a manifold where the solution lives. An initial data set for the vacuum Einstein equations consists of a three-dimensional manifold S, a Riemannian metric g_{ab} and a symmetric tensor k_{ab} on S which satisfy the constraints (2.26) and (2.27) with $\rho = 0$ and $j^a = 0$.

Definition 9.1 *A Cauchy development of the initial data set (S, g_{ab}, k_{ab}) consists of a solution $(M, g_{\alpha\beta})$ of the vacuum Einstein equations, an embedding ϕ of S into M and a choice of unit normal vector to its image such that $\phi(S)$ is a Cauchy hypersurface and the pull-backs by ϕ of the induced metric and the second fundamental form of $\phi(S)$ corresponding to the given unit normal coincide with g_{ab} and k_{ab} respectively.*

The standard way to prove local existence of solutions of the Einstein equations with prescribed initial data is to first solve the reduced Einstein equations with appropriate initial data and then to show afterwards that the given coordinates are harmonic so that the metric which has been obtained is in fact a solution of the Einstein equations. An important step in this argument is to use the following evolution equation for the Γ^α which is satisfied by any solution of the reduced Einstein equations:

$$\nabla_\alpha \nabla^\alpha \Gamma^\beta + R^\beta{}_\alpha \Gamma^\alpha = 0. \tag{9.2}$$

This is a linear homogeneous system of hyperbolic equations. Hence if the initial data Γ^α and $\partial_t \Gamma^\alpha$ vanish for $t = 0$ these quantities vanish on a neighbourhood of $t = 0$. Suppose now that an abstract initial data set (g_{ab}, k_{ab}) for the vacuum Einstein equations is prescribed. From this a certain initial data set for the reduced Einstein equations will be constructed. The components g_{ab} are chosen to have the given values while $g_{0a} = 0$ and $g_{00} = -1$. Choose $\partial_t g_{ab} = -2k_{ab}$. The components $\partial_t g_{0\alpha}$ are fixed by requiring that $\Gamma^\alpha = 0$. It then turns out that because the abstract data satisfy the constraints the quantities $\partial_t \Gamma^\alpha$ vanish. Using the discussion of equations (9.2) above it can be concluded that a solution of the reduced Einstein equations corresponding to the data just constructed satisfies the Einstein equations on a neighbourhood of the initial hypersurface.

The reduced Einstein equations are a system of nonlinear wave equations. It follows that it is possible to prove local existence for this system by reducing it to a first order symmetric hyperbolic system as in Section 8.3 and using the fact that symmetric hyperbolic systems are known to have a well-posed local Cauchy problem. There is one possible disadvantage of this. The regularity for the initial data required for the local existence and uniqueness theorem for symmetric hyperbolic systems discussed in Section 8.1 was H^s for

$s > n/2+1$. If this is applied to the system got by deriving a first order system from the harmonically reduced Einstein equations then the initial data for the metric must be in H^s with $s > n/2+2$ in order for the theorem to apply. It is, however, known that existence theorems based on energy methods can be used to get existence and uniqueness for the reduced Einstein equations only assuming $s > n/2+1$. This is a property of quasilinear wave equations where the coefficients multiplying the second derivatives do not depend on the first derivatives. It could presumably be proved following the route sketched in Section 8.1 using the fact that the relevant symmetric hyperbolic system has a special structure. More recently techniques going beyond energy estimates have been used to weaken the condition on the initial data required to obtain local existence for the Einstein equations to $s > 2$.

The argument presented so far only suffices to give existence theorems which are local in space as well as in time. A result which is global in space can be obtained by covering the manifold on which the initial data is prescribed by a collection of local charts and piecing together the corresponding local solutions of the Cauchy problem using the domain of dependence. To be more precise, suppose that ψ_1 and ψ_2 are mappings on subsets U_1 and U_2 of the initial data manifold S with non-empty intersection which are diffeomorphisms onto open sets V_1 and V_2 of $\mathbb{R}^n$. Starting with prescribed geometric initial data, data for the reduced Einstein equations are obtained on V_1 and V_2. There are corresponding solutions on open subsets of $V_1 \times \mathbb{R}$ and $V_2 \times \mathbb{R}$. That on the subset of $V_2 \times \mathbb{R}$ can be pulled back to $V_1 \times \mathbb{R}$ by the diffeomorphism $(\psi_2 \circ \psi_1^{-1}, Id)$. It then has the same data as the solution on the open subset of $V_1 \times \mathbb{R}$ where both are defined. By the uniqueness theorem for the harmonically reduced Einstein equations the solutions must agree where they are both defined. This means that the diffeomorphism $(\psi_2 \circ \psi_1^{-1}, Id)$ can be used to glue the two solutions together consistently to get a Cauchy development of the data on $U_1 \cup U_2$. This kind of gluing can be done for the whole set of charts covering the manifold S but is rather cumbersome. An alternative procedure avoids this while using some more geometric concepts.

To describe the other approach it is necessary to introduce the concept of gauge source functions. Consider a coordinate condition which can be written abstractly in the form $P(u) = 0$. Here u are some variables related to the geometry. A new gauge condition can be defined by $P(u) = f$ while f is a function which depends on the coordinates and the solutions which is called a gauge source function. In many cases where the condition $P(u) = 0$ produces a hyperbolic form of the Einstein equations the condition $P(u) = f(x, u)$ achieves the same thing. As an example, the harmonic condition $\Gamma^\alpha = 0$ can be replaced by the condition $\Gamma^\alpha = F^\alpha$ where F^α are given functions of the coordinates. This can be rewritten in the form

$\nabla^\alpha \nabla_\alpha x^\beta = F^\beta(x^\alpha)$ which makes it clear that it is a system of semilinear wave equations for the functions x^α. The reduction of the local existence problem for the Einstein equations to that for a system of quasilinear wave equations can be done using the coordinate condition with gauge source functions in a way strictly analogous to the case where the gauge source functions are zero. The harmonic coordinate condition can be replaced by a condition using a harmonic mapping as follows. Consider the metric $g_{\alpha\beta}$ on $S \times I$ with I an open interval which is to be constructed as a solution of the Cauchy problem for the Einstein equations. Let $\bar{g}_{\alpha\beta}$ be some other Lorentz metric on the same manifold. Impose the condition that the identity should be a harmonic mapping from $g_{\alpha\beta}$ to $\bar{g}_{\alpha\beta}$. Concretely this means that $\Gamma^\alpha = \bar{\Gamma}^\alpha$ where $\bar{\Gamma}^\alpha$ are the contracted Christoffel symbols of $\bar{g}_{\alpha\beta}$. In local coordinates this is a condition involving gauge source functions. However the condition itself can be expressed in a coordinate-independent way. The problem of finding a Cauchy development is formulated as finding a metric on $S \times \mathbb{R}$ which is related to a specified metric as just described. The reduced Einstein equations obtained form a system of quasilinear hyperbolic equations which can be formulated in a coordinate invariant way. They contain the auxiliary metric $\bar{g}_{\alpha\beta}$ in a way which breaks the diffeomorphism invariance. The reduction of this system of wave equations to a symmetric hyperbolic system can be done in a coordinate invariant way. The local existence theorem for symmetric hyperbolic systems can also be done in a way which works only with geometrical objects and does not require coordinates. The result of all this is that the local existence theorem for the Einstein equations can be proved in a way which is global in space.

It remains to consider the question of uniqueness, remembering that the goal is to prove geometric uniqueness. The description involving harmonic maps will be used to do this. If two Cauchy developments of the same data with metrics $g_{\alpha\beta}$ and $g'_{\alpha\beta}$ are given then the harmonic maps relating them to $\bar{g}_{\alpha\beta}$ may be used to obtain two metrics on the same manifold satisfying the same hyperbolic system with the same initial data. They must be equal. This proves local geometric uniqueness for the solution of the Einstein equations. The harmonic maps provide the diffeomorphism occurring in the definition of geometric uniqueness.

The question of global uniqueness is more complicated. It is formulated in terms of the concept of the *maximal Cauchy development* of an initial data set. Intuitively this is the largest globally hyperbolic spacetime with the chosen initial data. The precise definition is as follows:

Definition 9.2 *Let (S, g_{ab}, k_{ab}) be an initial data set for the vacuum Einstein equations. A Cauchy development $(M, g_{\alpha\beta})$ with embedding $\phi : S \to M$ of this data set is said to be maximal if for any other Cauchy development*

9.2 The local Cauchy problem

$(M', g'_{\alpha\beta})$ *with embedding* $\phi' : S \to M'$ *there is a diffeomorphism* $\psi : M \to M'$ *with* $\psi' = \psi \circ \phi$ *and* $\psi^* g' = g$.

For any initial data set with suitable differentiability properties there exists a corresponding maximal development. That it is unique up to diffeomorphism is automatic from the definition. The existence is proved by combining local existence statements with an abstract argument which uses the axiom of choice. A more elementary proof has never been found.

A curious aspect of the usual proofs of existence and uniqueness theorems for the Einstein equations is that uniqueness is only obtained under the assumption of one more degree of differentiability than is required for the existence theorem. The reason for this is the necessity of doing coordinate transformations (or pulling back by harmonic mappings) to compare different metrics. Suppose that the initial data (g_{ab}, k_{ab}) is of the differentiability class (H^s, H^{s-1}). Then the spacetime metric is of class H^s and the Christoffel symbols are of class H^{s-1}. This means that the harmonic coordinates are of class H^s. The transformation of the components of the metric in one coordinate system to those in another involves the first derivatives of the mapping concerned. Thus the transformed metric is only guaranteed to be of class H^{s-1}. It follows that in this context uniqueness for the reduced Einstein equations in H^s with $s \geq s_0$ only implies geometric uniqueness for the Einstein equations in H^s with $s \geq s_0 + 1$.

There is one more part of well-posedness which remains to be discussed. This is the continuous dependence of the solution on the initial data. For a symmetric hyperbolic system this is easily formulated. The spaces to which the data and solutions (say on a fixed time interval $[0, T]$) belong in the statement of the existence theorem have natural topologies and it is reasonable to require continuity of the mapping from initial data to solutions with respect to these topologies. This is proved by means of energy estimates for differences of solutions. One subtlety is that, at least with the most straightforward procedures, continuity is obtained in spaces with one degree of differentiability less than those for which existence and uniqueness is obtained. The statement of Cauchy stability includes the statement that if the initial data are only varied a small amount in the topology on initial data being used then there is a common time interval of existence for all the solutions concerned. The property of continuous dependence of solutions on initial data in the Cauchy problem is called Cauchy stability. For the Einstein equations it is more difficult to define Cauchy stability because of the coordinate freedom. A simple case is where the data are defined on a compact Cauchy hypersurface and a fixed type of coordinate system (e.g. harmonic coordinates) is used to describe the solutions. Then

Cauchy stability for the Einstein equations can be taken to mean Cauchy stability for the reduced equations.

Spacetimes with symmetry play a very important role in general relativity. In this context it is important to know that initial data with symmetry lead to solutions of the Einstein equations with symmetry. Let G be a Lie group. It is not supposed that the group is connected and discrete groups are allowed as 'Lie groups of dimension zero'. Let (S, g_{ab}, k_{ab}) be an initial data set for the vacuum Einstein equations. Suppose that there is an action of G on S which leaves the data invariant. Then, as will be shown below, there is an action of G on the maximal Cauchy development $(M, g_{\alpha\beta})$ of these data which leaves the metric invariant and which restricts to an action on the image of S compatible with the original action on S. This result only covers symmetries of spacetimes which leave a given Cauchy hypersurface invariant. There are also possibilities of detecting spacetime symmetries on a Cauchy hypersurface which is not invariant but these will not be discussed here.

A symmetry of the initial data is a diffeomorphism ψ of S which leaves the data invariant. Let ϕ be the embedding of S into its maximal Cauchy development which occurs in the definition of that object. Then $\bar{\phi} = \phi \circ \psi$ also satisfies the properties of that embedding. Hence, by uniqueness up to isometry, there exists an isometry $\bar{\psi}$ of the maximal Cauchy development onto itself such that $\bar{\psi} \circ \phi = \bar{\phi}$. This implies that $\bar{\psi} \circ \phi = \phi \circ \psi$. Thus $\bar{\psi}$ is an isometry of M whose restriction to $\phi(S)$ is equal to ψ. It is thus seen that a symmetry of the initial data extends to a symmetry of the solution. It will now be shown that this extension is unique. Since a general theorem of Lorentzian (or Riemannian) geometry says that two isometries which agree on an open set agree everywhere it suffices to show that any two isometries $\bar{\psi}$ with the properties described above agree on a neighbourhood of $\phi(S)$. Let p be a point of $\phi(S)$. A neighbourhood of p can be covered by Gauss coordinates based on $\phi(S)$. An isometry preserves geodesics and orthogonality. Hence if, when expressed in Gauss coordinates, it is the identity for $t = 0$ it must be the identity everywhere. This completes the proof of the uniqueness of $\bar{\psi}$.

Now consider the situation where a Lie group G acts on S in such a way that each transformation ψ_g of S corresponding to an element g of the group is a symmetry of the initial data. Let H be the isometry group of the maximal Cauchy development and H_S the group of isometries of M which leave $\phi(S)$ invariant. The group H_S is a closed subgroup of the Lie group H and hence itself a Lie group. Each ψ_g is the restriction of a unique element $\bar{\psi}_g$ of H_S. Using uniqueness again it follows that $\bar{\psi}_{gh} = \bar{\psi}_g \circ \bar{\psi}_h$ for all elements g and h of G. This defines a homomorphism from G to H_S. This shows that there exists an action of the group G on M which extends the action on

$\phi(S)$ arising from the original action on S by means of the identification using ϕ. However this argument does not show that the action of G on M obtained in this way is smooth. To show this first consider the group H_I of all symmetries of the initial data. It is a closed subgroup of the group of isometries of the metric g_{ab} and therefore has the structure of a Lie group. The above considerations show that restriction defines an isomorphism of groups from H_S to H_I. (Here S is identified with $\phi(S)$.) If it were known that this mapping was continuous a general theorem on Lie groups [202] would show that it is also a diffeomorphism and therefore an isomorphism of Lie groups. The continuity can be seen by noting that the topology of the isometry group coming from its Lie group structure coincides with the compact open topology [125]. The continuity of the restriction mapping in the compact open topology follows directly from the definitions. It can be concluded that H_S and H_I can be identified as Lie groups. This is enough to prove the smoothness of the action.

There is another approach to proving the inheritance of symmetry by solutions from initial data which uses Killing vectors but it cannot handle discrete symmetries.

It was already remarked in Section 8.7 that most of the applications of initial boundary value problems for the Einstein equations are in the area of numerical relativity. In the case of an IBVP for the vacuum Einstein equations the setting is that there is a spacelike hypersurface S where initial data are to be prescribed and a timelike hypersurface T where boundary data are to be prescribed. The question of how to set this up in a coordinate independent way will not be addressed here. Suppose that by some means the Einstein equations have been expressed as a symmetric hyperbolic system. The questions of propagation of coordinate conditions and constraints are more difficult for an IBVP than in the case of the pure Cauchy problem. If a system is derived to show that certain quantities must vanish then vanishing boundary conditions must also be derived. The approach must be chosen very cleverly in order to achieve this.

9.3 Inclusion of matter

In the case of matter the definition of a Cauchy development must be modified to include the matter. It is then a solution of the Einstein–matter equations which induces the correct initial data for the geometry and the matter. When a suitable local existence theory is available for a given matter model there is also a maximal Cauchy development.

The Einstein–matter equations can be written in harmonic coordinates and, just as in the vacuum case, a reduced system can be defined by setting the contracted Christoffel symbols to zero. The result is a system of non-linear wave equations coupled to some matter equations. If initial data for the reduced equations are chosen as was described in the vacuum case then a corresponding solution of the reduced system will satisfy the original Einstein–matter system. The essential input for proving this is the equation (9.2) which also holds in the case with matter. The only feature of the matter which has to be used to derive this equation is that the energy–momentum tensor is divergence free.

Now the question of well-posedness of the reduced Einstein–matter equations will be considered. The reduced Einstein equations can be rewritten as a symmetric hyperbolic system with basic unknowns which, for convenience, will be denoted by u. Suppose that the equations of motion of a given matter model on any fixed background spacetime can also be written as a symmetric hyperbolic system with basic unknowns v. These two symmetric hyperbolic systems are coupled. If the coupling is such that v only occurs undifferentiated in the first system and u only undifferentiated in the second then taking both systems together gives a symmetric hyperbolic system for (u, v) and a local existence theorem for the reduced Einstein–matter system is obtained. In practice the condition on the coupling relating to the equations of motion of the matter is satisfied if these equations only depend pointwise on the metric itself and the Christoffel symbols and not, for instance, on the curvature. The condition relating to the Einstein equations is satisfied if the energy–momentum tensor only depends pointwise on the matter variables and their derivatives of order lower than the order of the equations of motion of the matter. These conditions are satisfied in very many cases. This procedure works for the ordinary nonlinear wave equation, for multiscalar fields and for wave maps. In all these cases the equations of motion are systems of nonlinear wave equations which can be reduced to symmetric hyperbolic systems as shown in Section 8.3. There are explicitly known conditions under which the equation of motion of a k-essence field is a nonlinear wave equation and in this case the procedure just described applies. It can be applied to the curvature coupled scalar field and $f(R)$ gravity through the device of writing the equations as a minimally coupled Einstein–scalar field system by means of a conformal transformation. There is also an interesting alternative to this which uses equation (3.18). A disturbing feature of the Einstein equations for a curvature coupled scalar field before conformal rescaling is the occurrence of second derivatives of ϕ on the right hand side. Introduce the first order partial derivatives of ϕ as new variables. In harmonic coordinates the curved space wave operator is identical to the flat space wave operator. Hence the first derivatives of ϕ satisfy wave equations which contain no second derivatives of the metric. Coupling these to the harmonically reduced Einstein equations leads to a system of quasilinear wave equations for the unknowns $(\phi, \nabla_\alpha \phi, g_{\alpha\beta})$ where the principal part is just the wave operator acting on each component. It can be shown that solutions of this system with data satisfying suitable compatibility conditions satisfy the Einstein equations coupled to a curvature coupled scalar field. More complicated alternative theories with Lagrangians such as $R_{\alpha\beta}R^{\alpha\beta}$ require special treatment. They have been handled using the concept of Leray hyperbolicity.

In Section 8.3 the source-free Maxwell evolution equations were written in manifestly symmetric hyperbolic form. To solve the Cauchy problem for the Einstein–Maxwell system completely it is still necessary to show that the constraints propagate. It was shown in Section 8.3 how to do this for the Maxwell constraints. Once this has been done all Maxwell equations are satisfied and the energy–momentum tensor is divergence-free. The construction of solutions of the reduced equations then ensures that the Einstein constraints are satisfied automatically. The Euler equations were also written in symmetric hyperbolic form in Section 8.3.

A case in which this procedure does not work is that of kinetic matter models. In that case the Einstein–matter equations are integrodifferential equations rather than partial differential equations in the strict sense. An existence proof for the Einstein–Boltzmann system can be constructed in close analogy with what is done in the theory of symmetric hyperbolic systems but that theory does not apply directly. There is one deficiency in the literature on this subject. A solution of the Einstein–Boltzmann system where the initial datum for the distribution function is non-negative should be such that the distribution function is non-negative everywhere but there appears to be no published proof of this. Note that to prove this it would suffice to prove the corresponding statement for a solution of the Boltzmann equation on an arbitrary background spacetime. For the special case of the Vlasov equation it is elementary to prove the desired statement.

The discussion of Cauchy stability in the last section extends without change to the case with matter. Once the existence of maximal Cauchy developments is known for a given matter model then the arguments showing the inheritance of symmetry of initial data by solutions go through as in the vacuum case. It is just necessary to require symmetries of data or solutions to preserve the matter fields as well as the purely geometric quantities as a part of the definition.

9.4 Cosmic censorship

The subject of cosmic censorship was already mentioned briefly in the introduction to the book. There are two versions of it, strong cosmic censorship and weak cosmic censorship. It is important to realize that, despite the names, the former does not imply the latter. First strong cosmic censorship will be discussed. To make it precise it is necessary to choose a class of initial data by specifying differentiability properties and some kind of boundary conditions, e.g. defined on a compact manifold or asymptotically flat.

Definition 9.3 *(Strong cosmic censorship, rough version.) Strong cosmic censorship holds if the maximal Cauchy development of any initial data for the Einstein–matter system is inextendible.*

Note that by definition the maximal Cauchy development is not extendible to a larger globally hyperbolic spacetime while keeping the same Cauchy hypersurface. However it is a priori possible that it could be extended to a spacetime which is not globally hyperbolic or which has a different Cauchy surface. A simple example of the second possibility is given by the initial data induced on the hyperboloid consisting of future-directed timelike vectors of unit length in Minkowski space. Its maximal Cauchy development is the interior of the light cone of the origin. This solution can evidently be extended.

The above formulation of strong cosmic censorship is too rough and can easily be shown to be false. This is already clear from the example of data on a hyperboloid just given. An example with data for the vacuum Einstein equations on a compact manifold is given by the flat Kasner spacetime with metric

$$-dt^2 + t^2 dx^2 + dy^2 + dz^2. \tag{9.3}$$

The coordinates x, y and z may be taken to be periodic. This is the product of a two-dimensional Lorentzian manifold on $\mathbb{R} \times S^1$, sometimes known as the Misner spacetime, with a flat torus. It is the maximal Cauchy development of data on a hypersurface of constant t. The two-dimensional space can be obtained by identifying the inside of the light cone in two-dimensional Minkowski space by a Lorentz transformation. The geometry can be extended through the light cone and the example thus violates the rough version of strong cosmic censorship. In the extension there are closed timelike curves. If it is desired to save strong cosmic censorship the way to get around this example is to say that it is extremely special and that inextendibility holds for generic data. This will be built into the improved definition of cosmic censorship given below. In general if a maximal globally hyperbolic spacetime can be extended smoothly the boundary of the original spacetime in the extension is called the Cauchy horizon. It is a null hypersurface and satisfies a Lipschitz condition. Another famous class of spacetimes which can be extended through a Cauchy horizon are the Taub–NUT spacetimes. These are the LRS Bianchi type IX solutions of the Einstein vacuum equations. It has been proved that in any other vacuum spacetime of Bianchi type IX the Kretschmann scalar is unbounded as the boundary of the maximal Cauchy development is approached. This is in perfect agreement with the idea of the improved version of cosmic censorship.

A different kind of example showing that the rough version of cosmic censorship is false involves solutions with dust. The interpretation is that this is a pathological matter model which develops singularities independently of gravity. When gravity is present the matter singularities cause singularities in the geometry. The way to save strong cosmic censorship is to say that the inextendibility of the maximal Cauchy development only holds for well-behaved matter models, for instance those which form no singularities in Minkowski space.

In general it is to be expected that finding a formulation of cosmic censorship which is true involves using both strategies just discussed. This leads to the following which requires choosing a topology on the set of initial data.

Definition 9.4 *(Strong cosmic censorship, improved version.) Strong cosmic censorship holds if the maximal Cauchy development of all initial data for the Einstein–matter system belonging to an open dense subset is inextendible provided the matter model is well-behaved.*

When cosmic censorship is mentioned in the following it is always the improved version which is meant. Because of the various choices which have to be made this is still a rather loose concept. It does, however, seem meaningful and correct to say that there is no known convincing counterexample to the improved formulation of strong cosmic censorship.

Now the relationship of strong cosmic censorship to cosmology will be considered. The properties of the FLRW models as presented in Chapter 4 lead to the following considerations. An expanding cosmological model is modelled by a spacetime with a Cauchy hypersurface whose mean curvature is negative. If the maximal Cauchy development can be covered by a foliation of Cauchy hypersurfaces all of which have negative mean curvature then the model may be called a forever expanding cosmological model. If there is a Cauchy hypersurface in the spacetime whose mean curvature is negative and another whose mean curvature is positive the spacetime may be called a recollapsing cosmological model. Consider now the simplest case where the spacetime has a compact Cauchy hypersurface. Then it can be shown that all Cauchy hypersurfaces in the spacetime are compact. If the mean curvature has a fixed sign on one of these Cauchy hypersurfaces and the timelike convergence condition is satisfied then the Hawking singularity theorem predicts geodesic incompleteness. It can be concluded that a forever expanding cosmological model has a singularity in the past while a recollapsing one has singularities both in the past and in the future. It can be shown that in a recollapsing universe there exists a compact maximal hypersurface, i.e. one whose mean curvature is zero. It has been proved that there are spacetimes with a compact Cauchy hypersurface which contain no compact hypersurface whose mean curvature has a fixed sign. It is unclear what role, if any, these have to play in cosmology. There is little understanding of their global structure.

In a forever expanding cosmological model proving strong cosmic censorship splits into two tasks. It is necessary to investigate inextendibility in the future and in the past. In the future it is reasonable to expect in many cases that the spacetime will be future geodesically complete and if this can be proved it settles the cosmic censorship question in the future. This can, however, be technically difficult and it may be practical to prove cosmic censorship in the future without proving future geodesic completeness. The only known scenario where geodesic completeness could be expected to fail for an open set of initial data is that where, intuitively, black holes are formed in an expanding universe. This statement should be seen from a physical point of view since the standard mathematical definition of a black hole is for asymptotically flat spacetimes and does not apply to the cosmological situation. Many kinds of symmetry which are often assumed to obtain tractable model problems for learning more about cosmic censorship suffice to rule out black holes and thus rule out this complication. In the case without symmetry it has to be faced. Consider now the past time direction. In the past there is a singularity and physical intuition says that this geodesic incompleteness should be accompanied by curvature blow-up. Thus a natural strategy for proving strong cosmic censorship in the past is to try to show that some curvature invariant blows up along any past-directed causal curve. In a recollapsing universe this strategy (and its time-reversed version) are appropriate to handle both time directions.

9.5 The BKL picture

The singularity theorems of Penrose and Hawking show that solutions of the Einstein equations are necessarily geodesically incomplete for large open sets of initial data. They do not, however, explain the mechanism leading to incompleteness. In other words, they do not say anything about the structure of spacetime singularities. In the context of cosmological models a general picture of the structure of spacetime singularities was developed by Belinskii, Khalatnikov and Lifshitz (henceforth abbreviated BKL). There is no proof that this picture is correct but there is a lot of evidence supporting it, consisting of heuristics, numerical calculations and mathematical proofs in special cases. The BKL picture, although itself not rigorous, has had

9.5 The BKL picture

a fundamental influence on the ways in which ODE and PDE have been applied in mathematical cosmology. It is therefore natural that it should be discussed in some detail in this book. Four main elements of this picture are the following:

1. the singularity is spacelike;
2. the evolution at different spatial points decouples;
3. the influence of matter becomes negligible near the singularity;
4. important quantities are generically oscillatory near the singularity.

None of these statements is precise and it will now be discussed how they might be understood. The first statement has to do with the causal structure of the spacetime. If two timelike geodesics hit the singularity then the intersection of their causal pasts with a small neighbourhood of the singularity should be disjoint. This is a property which is shared by a regular spacelike hypersurface but not by a regular null hypersurface. The second point is interpreted as meaning that the evolution of the solution of the PDE system (the Einstein–matter equations) can be approximated near the singularity by the solutions of a system of ordinary differential equations depending on the spatial coordinates as parameters. The system of ODE concerned is the system of spatially homogeneous Einstein equations. The third point means that spatially homogeneous solutions of the Einstein–matter equations can be approximated by solutions of the Einstein vacuum equations. This must be restricted in two ways. It is only true for generic data and not for all data – the Einstein–de Sitter solution is an example where it does not hold. It is also not true for all matter models. It is known to fail in the case that the matter is described by a massless scalar field. This last fact comes out of the BKL analysis. Thus the claim should be restricted to some 'large class' of matter models, e.g. non-stiff perfect fluids and collisionless matter. If the second and third points are accepted then the fourth point comes down to a statement about the dynamics of spatially homogeneous vacuum models and in that case rigorous results are available, as discussed in Subsection 5.7.1. The BKL analysis is done in Gaussian coordinates and assumes that there is a coordinate system of this type covering a neighbourhood of the singularity for which the singularity itself is simultaneous.

The basis of the BKL picture is to make a guess for the asymptotics of the solutions near a singularity where the solutions are parametrized by coefficients in a formal expansion and then to examine for which choices of the coefficients the guess is consistent. In the first version of this procedure due to Lifshitz and Khalatnikov it was discovered that the number of free functions which could be accommodated was one less than the number of functions in general Cauchy data. This was interpreted as saying that generic solutions had no singularities. This intepretation became untenable when the singularity theorems were discovered. The reason for the missing function is that the guess originally made implied monotone behaviour of important geometrical quantities near the singularity. Today it is believed that the generic behaviour is oscillatory and so does not fit into that picture.

The later modification of the procedure of Lifshitz and Khalatnikov which led to the BKL picture was more sophisticated. If it is assumed that a solution is behaving in a monotone way as the singularity is approached according to the old set-up then it is found that in general it must stop doing so after a finite time. Then it changes in a way which BKL described. This process repeats itself infinitely many times before the singularity is reached.

A problem with the BKL picture is that it appears to lead to the generation of small-scale structure which is not consistent with the original assumptions. The simplest example of this are the spikes which are discussed in Section 10.1. In the general case the heuristics suggest that more and more of this structure will be produced as the singularity is approached. At the same time there may be other mechanisms leading to the decay of structure and it is not clear what the final outcome will be. It is an outstanding problem to obtain rigorous results on the validity of the BKL picture for general solutions of the Einstein equations. Another challenge to the BKL picture concerns the first point on the list of properties, namely the spacelike character of the singularity. It is known that there can be null singularities in certain asymptotically flat solutions containing a black hole (see Section 11.5). Since physical intuition says that black holes can be formed in cosmological models why should null singularities not also be common there? In the context of this discussion it is important to keep in mind the following sobering fact: there is not a single example where an infinite dimensional class of solutions of the Einstein–matter equations with oscillatory behaviour near the singularity has been analysed rigorously. One reason why it is so difficult to do this is that oscillatory behaviour in inhomogeneous spacetimes always seems to be associated with an infinite amount of formation of localized spatial structure.

9.6 Further reading

For the comeback of harmonic coordinates see [161] and [137]. The method of using the reduced Einstein equations described in this chapter goes back to the original paper [81] on local existence of solutions of the Einstein equations. The optimal regularity for the existence theory using energy methods was obtained in [108]. For the improvements using other methods see [123] and references therein. The first treatment of an initial boundary value problem for the vacuum Einstein equations is in [84]. The Cauchy problem for the Einstein–Maxwell–Boltzmann system was treated in [16]. The discussion of the Cauchy problem for the conformally coupled scalar field and higher order gravity theories makes use of ideas from [154]. Generic curvature blow-up in vacuum Bianchi IX spacetimes is proved in [183]. For the original work on the BKL picture see [134], [24] and [25] and references therein.

10 Global results

10.1 Gowdy spacetimes

The Gowdy spacetimes were introduced in Section 4.4. As pointed out there the essential field equations are a system of two semilinear wave equations for functions $P(t,\theta)$, $Q(t,\theta)$ which are periodic in θ. The coefficients of the equations become singular at $t = 0$ and so if data are prescribed at $t = t_0$ for some $t_0 > 0$ the maximal interval on which a solution could be defined is $(0, \infty)$. In fact it is known that for any smooth data the corresponding solution does exist on that whole interval.

With a global existence theorem in hand the next interesting question is that of the asymptotic behaviour of the solution in the regimes $t \to \infty$ and $t \to 0$. Gowdy spacetimes defined on the interval $(0, \infty)$ are always future geodesically complete but this is a highly non-trivial result. It is proved in conjunction with other detailed information about the asymptotic behaviour. To describe the results obtained it is useful to have a geometrical point of view for describing solutions of the Gowdy equations. It is natural to think of P and Q as certain coordinates on the hyperbolic plane. Then the Gowdy equations can be thought of as equations for a mapping with values in the hyperbolic plane. They are very similar to the equations for a wave map on $(1 + 1)$-dimensional Minkowski space with values in that space. As already mentioned in Section 4.4 it is possible to go further and describe them as being exactly a wave map with symmetry on the geometry defined by the Lorentzian metric $-dt^2 + d\theta^2 + t^2 dx^2$.

A geometrical picture of a solution (P, Q) of the Gowdy equations can be obtained as follows. For each fixed value of t the solution defines a closed loop in the hyperbolic plane. Thus the whole solution can be thought of as a loop of this type which changes with time. In the case of a polarized Gowdy solution $Q = 0$ and the loop is confined to a geodesic. The late-time behaviour of a polarized solution of the Gowdy equations is that the average value of P (intuitively the centre of the loop) moves asymptotically with constant velocity with respect to the time $\tau = \log t$ along the geodesic while the diameter of the loop shrinks like $t^{-1/2}$. Since the equation for

P is linear in the polarized case it is not surprising that it makes sense to consider the average value. What is more surprising is that it also turns out to be useful to consider averages when studying the asymptotics in the non-polarized case.

By analogy with the polarized case it is possible to guess that the late-time asymptotics can be described in some sense in the following way. There is some kind of centre of the loop which moves asymptotically with constant velocity with respect to τ along a geodesic while the diameter of the loop shrinks like $t^{-1/2}$. The only differences would be that the geodesic would be general instead of a particular one and that the solution would not lie entirely on that geodesic. The reality is more complex. The functions e^P and Q correspond to Cartesian coordinates in the half-plane model of hyperbolic space. In that model geodesics are represented by circles or straight lines which meet the boundary orthogonally. For general Gowdy solutions the evolution does close in on a circle (or straight line) in the half-plane as $t \to \infty$ but this circle may meet the boundary at an angle other than a right angle or may fail to meet the boundary at all. In the last case, which represents about half of all solutions, the asymptotic behaviour is even more different from the polarized case. The solution does not settle down to any one point of the circle. Instead it continues to move around the circle for ever. The solutions have an unending oscillatory behaviour which is asymptotically periodic as a function of τ. It is interesting to note that this behaviour was not found by heuristic considerations or numerical work as is the case for so many properties of solutions of the Einstein equations but arose within a complete rigorous analysis of the asymptotics. The two types of solutions can be identified explicitly in terms of their initial data.

The distinction between the data for Gowdy spacetimes giving rise to different types of late time behaviour is based on the following conserved quantities.

$$A = \int [2Q(t\partial_t Q)e^{2P} - 2(t\partial_t P)]d\theta, \tag{10.1}$$

$$B = \int e^{2P}(t\partial_t Q)d\theta, \tag{10.2}$$

$$C = \int [(t\partial_t Q)(1 - e^{2P}Q^2) + 2Q(t\partial_t P)]d\theta. \tag{10.3}$$

The existence of these conserved quantities is associated with the fact that the Gowdy equations are invariant under an action of the group $SL(2,\mathbb{R})$, which is the identity component of the isometry group of hyperbolic space. The key diagnostic quantity is the combination $A^2 + 4BC$ which is related to a Casimir invariant of $SL(2, \mathbb{R})$. It is exactly when this invariant is negative that the solutions have oscillatory late-time behaviour.

The proofs of these statements are complicated and will not be described in detail here but some comments will be given on the general structure of the arguments. A very important tool is the energy

$$\mathcal{E} = \frac{1}{2} \int (\partial_t P)^2 + e^{2P}(\partial_t Q)^2 + (\partial_\theta P)^2 + e^{2P}(\partial_t Q)^2 d\theta, \tag{10.4}$$

which satisfies the equation

$$\frac{d\mathcal{E}}{dt} = -\frac{1}{t} \int (\partial_t P)^2 + e^{2P}(\partial_\theta Q)^2 d\theta. \tag{10.5}$$

This implies that the energy is bounded towards the future but it also provides a lot more information. In the proof of the asymptotic behaviour it is shown that for any solution the energy is eventually smaller than a certain constant η and that any solution whose energy is smaller than η has the property that $\mathcal{E}(t) \leq Ct^{-1}$ for some constant C at late times. The part of the energy containing the spatial derivatives of the unknowns has a geometric interpretation as the squared length of the loop in the hyperbolic plane. The fact that the energy decays like t^{-1} implies that the length of the curve, and hence its diameter, decays like $t^{-1/2}$.

The technique used to obtain the decay of the energy can usefully be explained in the simpler context of an ODE. This will now be done following the exposition in [185]. Consider the ODE

$$\ddot{x} + 2a\dot{x} + b^2 x = 0, \tag{10.6}$$

where a and b are constants with $a > 0$ and $b^2 > a^2$. Of course this equation can be solved explicitly but the idea here is to explain how information about the asymptotic behaviour of the solutions can be obtained without using the exact expression by a method which can be generalized to many other situations. Define an energy by

$$H = \frac{1}{2}(\dot{x}^2 + b^2 x^2). \tag{10.7}$$

Then

$$dH/dt = -2a\dot{x}^2. \tag{10.8}$$

This shows that H is non-increasing but does not immediately show that H tends to zero as $t \to \infty$ and certainly not that it decays exponentially at late times, which is the case. These facts can be proved by introducing a correction term $\Gamma = ax\dot{x}$. There are positive constants c_1 and c_2 such that $c_1 H \le H + \Gamma \le c_2 H$. On the other hand

$$d(H + \Gamma)/dt = -2a(H + \Gamma). \tag{10.9}$$

Thus estimating H is equivalent to estimating $H + \Gamma$ and the quantity $H + \Gamma$ decays exponentially with exponent $-2a$. This exponent is optimal.

A similar technique can be applied to the polarized Gowdy equation. In that case

$$\frac{d}{dt}\left(\frac{1}{2}\int (\partial_t P)^2 + (\partial_\theta P)^2 d\theta\right) = -\frac{1}{t}\int (\partial_t P)^2 d\theta. \tag{10.10}$$

A candidate for a correction term which might lead to a strong estimate is $\tilde{\Gamma} = \frac{1}{2t}\int P\partial_t P d\theta$. This does not work because of the lack of a suitable estimate of $\tilde{\Gamma}$ in terms of the energy. This comes from the zero eigenvalue of the Laplacian on the circle. This difficulty can be overcome by using the quantity

$$\Gamma = \frac{1}{2t}\int (P - \langle P \rangle)\partial_t P d\theta, \tag{10.11}$$

where $\langle P \rangle$ is the average of P over the circle. Related but more sophisticated techniques can be used to obtain information about the decay of the energy for solutions of the general Gowdy equations. For the class of solutions of the Einstein–Maxwell equations and the higher-dimensional analogues of the Gowdy solutions introduced in Section 4.4 it is possible to proceed in a similar way, introducing an energy-like quantity and proving that it tends to zero as $t \to \infty$. The more detailed information about the asymptotics of Gowdy spacetimes, and in particular their future geodesic completeness, which has been obtained requires the use of features of that special case for which generalizations to the electromagnetic or higher dimensional cases remain to be discovered.

Next the asymptotics of Gowdy spacetimes for $t \to 0$ will be considered. It is convenient to use the time coordinate $\tilde{\tau} = -\tau = -\log t$ which tends to infinity as the singularity is approached. In the picture of the solution as a moving loop a point on the loop (with a fixed value of θ) moves with a certain speed as measured by the metric on the hyperbolic plane. This speed is given by $[(\partial_{\tilde{\tau}} P)^2 + e^{-2P}(\partial_{\tilde{\tau}} Q)^2]^{1/2}$. It turns out that for many solutions this quantity converges uniformly to a function of θ as $\tilde{\tau} \to \infty$. This function is called the asymptotic velocity of the solution. There is an issue of sign here related to the linguistic distinction between speed and velocity but it will be ignored for the moment. The asymptotic velocity serves as a useful diagnostic quantity for the nature of the asymptotics as $t \to 0$. There is a large open set of initial data for which the asymptotic behaviour of the solutions has the following form:

$$P(t,\theta) = k(\theta)\log t + \phi(\theta) + o(1), \tag{10.12}$$

$$Q(t,\theta) = Q_0(\theta) + t^{2k(\theta)}(\psi(\theta) + o(1)). \tag{10.13}$$

These asymptotic statements remain true when differentiated term by term any number of times with respect to t and θ. Here k is the asymptotic velocity and is assumed to take values in the interval $(0, 1)$. These are known as low velocity solutions. As discussed in Section 8.11 it can be shown by Fuchsian techniques that the coefficient functions k, Q_0, ϕ and ψ can be prescribed freely. The result that there is an open set of initial data leading to solutions with asymptotics of this type in the approach to the singularity is proved by other more direct arguments.

Remarkably it turns out that the asymptotic velocity always exists as a pointwise limit although the convergence to that limit is not uniform. Smooth initial data may give rise to an asymptotic velocity which is discontinuous. The lack of uniform convergence which is encountered in solutions with asymptotic velocity greater than one is associated with the phenomenon of spikes. The low velocity asymptotics are such that spatial derivatives of the basic unknowns do not diverge faster than the unknowns themselves as the singularity is approached. For instance the quantity $\partial_\theta P/P$ remains bounded as the singularity is approached. In fact it converges. In the presence of spikes things look different and $\partial_\theta P/P$ typically diverges at isolated points while converging elsewhere. At these isolated points the asymptotic velocity jumps. It is the shape of the graph of P which gives rise to the name 'spike'. The spikes grow in size and become narrower as the singularity is approached. They were first observed numerically and it was rather delicate to find a convincing argument that they were not numerical artefacts. Later an analytic proof of the existence of spikes was found.

Looking at the numerical results it is easy to observe that there are two qualitatively different types of spikes. One difference between them is that the spike in P points downwards (towards lower values of P) in one case and upwards in the second. The other main difference is that in the first case Q has a kind of discontinuity while in the second case it is perfectly smooth. Spikes are shown in Figs. 10.1 and 10.2. In the second case Q is smooth and unremarkable and has therefore not been shown. For reasons to be explained later the two types of spikes are known as 'false spikes' and 'true spikes' respectively. False spikes owe their name to the fact that they are artefacts of the parametrization of the Gowdy solutions by the coordinates P and Q in the hyperbolic plane. They can be removed by a suitable explicit transformation which, in the representation of the hyperbolic space as the

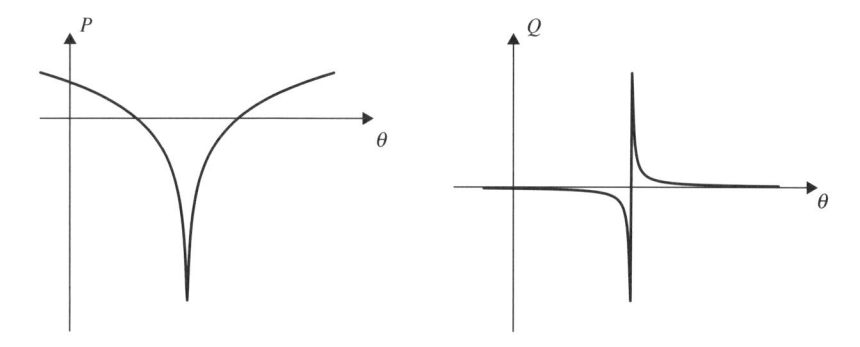

Figure 10.1 *A Gowdy false spike at fixed time*

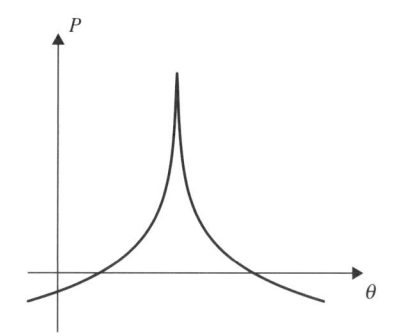

Figure 10.2 *A Gowdy true spike at fixed time*

upper half plane, is an inversion in the origin. Explicitly, new variables $\tilde{P}$ and $\tilde{Q}$ are defined by the relations

$$e^{-\tilde{P}} = \frac{e^{-P}}{Q^2 + e^{-2P}}, \tag{10.14}$$

$$\tilde{Q} = \frac{Q}{Q^2 + e^{-2P}}. \tag{10.15}$$

Since this is a coordinate effect curvature invariants such as the Kretschmann scalar show no irregular behaviour at a false spike.

True spikes are genuinely geometrical effects and they do show up in curvature invariants. They can be constructed from low velocity solutions by a certain explicit transformation (Gowdy to Ernst transformation) combined with inversion. However, unlike the inversion, the Gowdy to Ernst transformation produces a new spacetime (i.e. one which is not isometric to the original one). The proof that there exists a large class of solutions with true spikes proceeds by transforming the large class of low velocity solutions. The asymptotic velocity of the transformed solutions lies in the interval $(0, 2)$. The general analysis of the asymptotics of Gowdy spacetimes near the singularity shows that in a sense spikes are the worst thing that can happen. For generic initial data there are a finite number of spikes and away from these the solution has the simple low velocity asymptotics.

Superficially it might seem that there is no great difference between Gowdy spacetimes and vacuum spacetimes with T^2-symmetry. They are only distinguished by a single twist constant which leads to a coupling of some of the equations. This apparent similarity is very misleading. At least that is what heuristic and numerical work done up to now seems to indicate. There is almost nothing known rigorously beyond a global existence theorem and a result on the behaviour near the singularity in polarized solutions discussed below. The preliminary picture in the general case is as follows. The late-time behaviour is very different and may generically be a lot simpler than in the Gowdy case. In the approach to the initial singularity the behaviour is generically oscillatory with similarities to the mixmaster solution. It is a challenge to make these ideas rigorous.

In spacetimes with polarized T^2 symmetry the structure of the singularity seems to be rather simple and Fuchsian techniques can be used to prove the existence of a large class of solutions with simple singularities. The asymptotic expansions for these solutions are as follows.

$$U(t, \theta) = k(\theta) \log t + U_0(\theta) + o(1), \tag{10.16}$$

$$\eta = k^2(\theta) \log t + \eta_0(\theta) + o(1), \tag{10.17}$$

$$\alpha = \alpha_0(\theta) + o(1). \tag{10.18}$$

The existence theory requires the relation

$$\partial_\theta \eta_0 - 2k\partial_\theta U + \frac{\partial_\theta \alpha_0}{2\alpha_0} = 0 \tag{10.19}$$

and that the inequality $k(\theta) > 1$ be satisfied. These can be thought of as 'high velocity' solutions in the following sense. Of course it would be possible to just compare the form of the function U (called P in the Gowdy case) in the two classes of spacetimes but a more geometrical interpretation is possible. The asymptotic velocity in Gowdy spacetimes is related to the asymptotic value of one of the generalized Kasner exponents. Let λ_1 correspond to the θ direction and λ_2 and λ_3 correspond to the directions of the Killing vectors. Then k is the limiting value of the expression $\frac{1}{2}\frac{p_2 - p_3}{p_2 + p_3}$. The function k has the same geometrical interpretation in the case of T^2 symmetry. This result can be extended essentially without change to the case where a cosmological constant is present. It can also be extended to a 'half-polarized' case with one more free function provided $k < 3$.

A class of spacetimes related to the Gowdy spacetimes are those with cylindrical symmetry. As mentioned in Section 4.4 the key equations in that case define a wave map on three-dimensional Minkowski space with rotational symmetry. There are theorems on global existence and asymptotic behaviour for these spacetimes in the vacuum case. They have been generalized so as to include an electromagnetic field. In that case the target space of the wave map is a four-dimensional Riemannian manifold, the complex hyperbolic space.

10.2 Stability of de Sitter space

If an explicit solution of the Einstein equations is to be relevant to the description of reality then it must have some stability properties. The simplest explicit solution is Minkowski space. Its stability is discussed in the next section. It turns out that there is another explicit solution whose stability can be treated more easily than that of Minkowski space, namely de Sitter space which solves the vacuum Einstein equations with positive cosmological constant. This case is discussed in this section. The stability of the de Sitter solution is connected with a statement of physical interest, known as the cosmic no hair theorem.

Before coming to the example it is useful to consider the concept of stability more generally, starting with the relatively simple ODE case. Consider a dynamical system with a stationary point p. The point p is said to be stable if given any open neighbourhood U of p there exists an open neighbourhood V of p such that any solution which starts in V remains forever in U. Sometimes this notion of stability is called nonlinear stability to emphasize the difference to other concepts of stability. In this book the term stability is used as being synonymous with nonlinear stability. If p is stable and in addition there is a neighbourhood U of p such that any solution which starts in U converges to p then p is said to be asymptotically stable. The point p is said to be linearly stable if the origin, considered as a stationary point of the linearization of the original system about p, is stable. The relations between linear and nonlinear stability are not straightforward. In some cases the spectrum (set of eigenvalues) of the linearization is diagnostic for linear stability. The point p is said to be spectrally stable if the real part of each eigenvalue of the linearization is non-positive. For each of the different concepts of stability its negation is denoted by the corresponding term with stability replaced by instability. Spectral stability does not imply linear stability as can be seen by considering the matrix

$$\begin{bmatrix} 0 & 1 \\ 0 & 0 \end{bmatrix}.$$

The linear system which it defines has solutions which do not remain in any neighbourhood of the origin. On the other hand spectral instability does imply linear and nonlinear instability. This is a consequence of the existence of an unstable manifold, the Hartman–Grobman theorem and the fact that any linear system for which all eigenvalues have positive real part exhibits nonlinear instability. Finally it should be noted that linear stability does not imply nonlinear stability. A trivial example is given by the origin in the dynamical system $\dot{u} = u^2$.

There are further complications involved in notions of stability in the case of PDE. One reason is that the definition of nonlinear stability is dependent on the choice of a topology. In the finite-dimensional (ODE) case all choices are equivalent but this is no longer the case in infinite dimensions. It may happen that in the definition it is convenient to choose the topologies in which the sets U and V are open to be distinct. Spectral theory is also much more complicated in infinite dimensions and cannot be reduced to the study of eigenvalues alone. Moreover, there are no universally applicable analogues of ODE results such as that on the existence of an unstable manifold.

Now the example of de Sitter space will be considered. The formulation used here has been chosen to require the minimal amount of geometrical background while including the essential features. Consider the de Sitter space in the form

$$-dt^2 + e^{2Ht}(dx^2 + dy^2 + dz^2), \tag{10.20}$$

where $H = \sqrt{\Lambda/3}$ and think of x, y and z as periodic coordinates so that the spatial topology is that of a three-dimensional torus. The idea is to take data for this solution on a hypersurface of constant t, perturb it slightly and look at the properties of the solution which evolves from the new data. It can be shown that in Gauss coordinates the perturbed metric has the form

$$-dt^2 + e^{2Ht}(g^0_{ab} + O(e^{-2Ht}))dx^a dx^b. \tag{10.21}$$

Here the metric g^0_{ab} depends on the spatial coordinates but not on t. Its dependence on the coordinates x^a can be chosen arbitrarily, at least under the assumption of analyticity. The O-symbol is to be interpreted in the sense of uniform convergence. The asymptotic expansion remains valid when differentiated term by term an arbitrary number of times with respect to any of the coordinates. It can be seen that this means that the perturbed metric remains close to the de Sitter metric in an appropriate topology. In fact the cosmic no hair theorem demands more. It corresponds intuitively to asymptotic stability of the de Sitter solution. With the global formulation just given this is simply not true. However it is true on certain spacetime regions.

Related to this is the fact that the compactness of space is really irrelevant for this result. It can be localized in space. The reason is the nature of the domain of dependence in de Sitter space and small perturbations of it. For instance the part of de Sitter space with $t \geq 0$ inside a ball of coordinate radius H is determined by initial data on a ball of coordinate radius $4H$ on the hypersurface $t = 0$.

The relevance of this result for cosmology will now be discussed. It is a fact of observation that if the distribution of matter (galaxies) in our universe is averaged on a sufficiently large scale then it is consistent to assume that the result is homogeneous on all larger scales which can be observed. This is a justification for the use of spatially homogeneous solutions of the Einstein equations as models in cosmology. In addition the cosmic microwave background, consisting of photons which have propagated more or less unchanged except for uniform cooling since a very early epoch, is homogeneous to one part in 10^5. This raises the question whether this homogeneity can be explained as having arisen from an original inhomogeneous state by some dynamical mechanism. In a cosmological model of the kind which was standard up to ten years ago (e.g. the Einstein–de Sitter model) this is not possible for the following reason. The homogeneity includes the fact that the energy density is the same at spatial points which could never have been in causal contact since the big bang. In other words, the form of the domain of dependence in these spacetimes is such that it is easy to produce inhomogeneities in the model at the present time on arbitrarily large scales by choosing appropriate initial data at a very early time. What is needed to avoid this is a different kind of model, one where the domain of dependence has a form like that in de Sitter space. This has to do with accelerated expansion. This means that for a homogeneous and isotropic metric of the form $-dt^2 + a^2(t)d\Sigma^2$ for a fixed spatial metric $d\Sigma^2$ of constant curvature, it is necessary to have $\ddot{a} > 0$ at least some of the time. In the Einstein–de Sitter model $\ddot{a}$ is always negative. One of the Einstein equations takes the form

$$\ddot{a} = [-(4\pi/3)(\rho + 3p) + \Lambda/3]a. \quad (10.22)$$

Thus to get accelerated expansion it is necessary to have $\Lambda > 0$ or $\rho + 3p < 0$. The second of these inequalities implies that the matter must violate the strong energy condition. Under the very reasonable assumption that it satisfies the weak energy condition this means in particular that the pressure must be negative. In cosmology accelerated expansion at a very early time is known as inflation. It is believed to solve a number of other problems in cosmology. In the last ten years evidence has been accumulating that the expansion of the universe has also been accelerating at more recent times up to the present day. Exotic matter which causes accelerated cosmological expansion is called dark energy. The reason for describing it as exotic is that it violates the strong energy condition while all forms of matter normally encountered satisfy that condition. The nature of dark energy is still completely open.

As already mentioned in Chapter 1, it was observations of supernovae which led astrophysicists to believe that the expansion of the universe is accelerating. The most convincing argument for accelerated expansion today combines several types of observations to make a coherent picture known as the concordance model. In addition to the supernova data this makes use of observations of the cosmic microwave background, galaxy clustering and gravitational lensing. It should be mentioned that the arguments used lead to a lot of unsolved mathematical questions. For example an excellent exposition of the theoretical analysis of the microwave background can be found in [149]. A mathematician reading that book will see how much remains to be done to put these ideas on a rigorous mathematical basis. This is not a criticism of [149]. On the contrary, it is the clarity of that book which makes the mathematical difficulties apparent. There is also no claim that cosmology is an area of physics where mathematical foundations have been specially neglected. It is a general phenomenon that in many areas of physics there is a big gap between what can be done on a mathematically rigorous level and what is needed for the comparison with experiment and observation. This gap will not be closed in the near future but it is an exciting challenge to try and narrow it.

The discussion of dark energy can be made more concrete by giving a specific example. Consider a nonlinear scalar field with potential $e^{-\sqrt{8\pi}\lambda\phi}$ with λ a positive constant. This matter field can violate the strong energy condition. For instance it does so at any time when $\dot{\phi} = 0$. It can be shown that if $\lambda < \sqrt{2}$ then $\ddot{a} > 0$ in a homogeneous and isotropic model at all sufficiently late times. It is believed that for this model a cosmic no hair theorem holds. Thus inhomogeneous solutions with initial data close to homogeneous and isotropic can be approximated at late time by a homogeneous and isotropic model. The homogeneous and isotropic models are supposed to be asymptotically stable in a similar sense to that in which this statement holds for de Sitter space. For some values of λ this has been proved.

The proof of the stability of de Sitter space uses a system of equations called the regular conformal field equations which will now be described schematically. If $g_{\alpha\beta}$ is a solution of the vacuum Einstein equations, possibly with cosmological constant, then a conformally rescaled metric $\tilde{g}_{\alpha\beta} = \Omega^2 g_{\alpha\beta}$ is introduced. Then infinity in the original spacetime is represented by points in the conformally rescaled spacetime where $\Omega = 0$. Under suitable circumstances the metric $\tilde{g}_{\alpha\beta}$ can be extended beyond $\Omega = 0$. The asymptotic behaviour of the original spacetime is encoded in the regularity properties of $\tilde{g}_{\alpha\beta}$ at $\Omega = 0$. This is the conformal compactification of Penrose. The Einstein equations for $g_{\alpha\beta}$ can be rewritten as a system of equations for $\tilde{g}_{\alpha\beta}$ and Ω which is regular and hyperbolic even at $\Omega = 0$, the system of regular conformal field equations. This is a highly non-trivial process. Once it has been carried out certain statements about the global behaviour of spacetimes, e.g. the stability of de Sitter space, can be reduced to local statements about solutions of the conformal field equations. This procedure in its original form only works in spacetime dimension four. Recently a new version has been found which also works in higher even spacetime dimensions. The result concerning scalar fields with exponential potential mentioned above also relies on these higher-dimensional results. The scalar field problem can be related to an analogous problem for the vacuum equations with positive cosmological constant in higher dimensions by Kaluza–Klein reduction.

A disadvantage of the conformal field equations is that they are rather rigid. For instance they cannot obviously be used to prove the cosmic no hair theorem when matter is included which is not conformally invariant. This is the case for most types of phenomenological matter fields. Mathematically this rigidity can also be seen in the fact that it is not obvious how to extend the proof of the stability of de Sitter space in four dimensions to the corresponding result in higher dimensions. A variant of the conformal field equations using significantly different ideas was required to obtain the existing results on higher dimensions. In odd dimensions it seems that the conformal structure at infinity is not smooth, involving logarithmic corrections. The existence of a large class of solutions illustrating this has been proved by Fuchsian methods. These methods could presumably also be used to prove analogous existence theorems for certain Einstein–matter systems which are not conformally invariant but this has not been done yet.

There are indications that the attractor properties of the de Sitter solution extend beyond data which are close to those of the attractor solution. This can be illustrated by what happens in the spatially homogeneous case where there is a general theorem due to Wald [205].

Theorem 10.1 *(Wald) Consider a spatially homogeneous solution of the Einstein–matter system with $\Lambda > 0$ of Bianchi type I–VIII where the matter satisfies the dominant and strong energy conditions. If the solution exists for an infinite proper time in the future then as $t \to \infty$ the mean curvature* tr$k$ *converges to the value* $\sqrt{3\Lambda}$ *while the ratios* $\sigma^{ab}\sigma_{ab}/(\mathrm{tr}k)^2$, $R/(\mathrm{tr}k)^2$ *and* $\rho/(\mathrm{tr}k)^2$ *decay exponentially.*

The intuitive meaning of this result is that the influence of the shear, spatial curvature and energy density decays exponentially so that the solution resembles the de Sitter solution, where these quantities are zero, at late times. It is possible to make a more refined statement about the way in which the de Sitter solution approximates the given solution. In terms of components in a left-invariant frame on the Lie group

$$g_{ij}(t) = g_{ij}^0 e^{2Ht} + o(e^{2Ht}). \tag{10.23}$$

Note the resemblance to equation (10.21). The statement of Theorem 10.1 includes a global existence assumption. Evidently this cannot be got rid of without assuming more about the matter. A pathological matter model could spoil global existence while satisfying the energy conditions. Nevertheless it is to be expected that in the spatially homogeneous case any reasonable matter model satisfies a global existence statement. This has been proved explicitly for collisionless matter and for perfect fluids with a reasonable equation of state. What is needed is a suitable continuation criterion such as the WMCC introduced in Section 5.9. When global existence can be proved it is usually also possible to obtain a detailed description of the asymptotics of the matter fields. Wald's theorem extends without difficulty to the case of vacuum spacetimes in higher dimensions. The assumption that the symmetry should be of Bianchi type I–VIII is replaced by the requirement that the symmetry type should not be consistent with metrics of positive scalar curvature.

There are a variety of generalizations of Wald's theorem where the cosmological constant is replaced by a nonlinear scalar field whose potential has suitable properties. For instance, if the potential has a strictly positive global minimum this acts like a positive cosmological constant and late-time exponential expansion is obtained. Another case for which results are available is that where the potential is everywhere positive, tends to zero at plus infinity and satisfies $\lim_{\phi \to \infty}(V'(\phi)/V(\phi)) = 0$. The late time asymptotics can be determined by a method related to what is called the slow roll approximation. The situation where V'/V has a strictly positive limit is a borderline case. Late time accelerated expansion is only obtained if the limit is small enough. If V takes the value zero at a minimum then it is not true that there is accelerated expansion at all sufficiently late times.

The following intuitive picture can be used to organize the results on the dynamics of nonlinear scalar fields in spatially homogeneous spacetimes. The equation of motion of the scalar field

$$\ddot{\phi} + 3H\dot{\phi} = -V'(\phi) \tag{10.24}$$

is like the equation of motion of a ball which rolls on the graph of the potential V with friction determined by H. The Hubble parameter H is determined by the coupling to the Einstein equations and so there is a lot more to the problem than just the one equation for ϕ. Nevertheless the rolling ball picture is useful on the intuitive level and may also suggest strategies for proofs to confirm the heuristic conclusions. The expectation is that the ball settles down to a minimum if there is one or continues towards ever smaller values of V if there is not. In the former case the potential tends to a constant value while if the value of V at the minimum is non-zero the kinetic energy gets dissipated away and this causes H to tend to a positive constant value. This leads to the similarity of the dynamics in this situation with that in the case of a positive cosmological constant. In the case where the potential has its infimum at infinity it is more delicate to find out what happens. The slow roll situation comes about when the slope of the potential is sufficiently gentle. In that case the term $\ddot{\phi}$ in the evolution equation for ϕ can be ignored asymptotically as $t \to \infty$ and the kinetic energy can be ignored in the expression for H, leading to the relation $H = \sqrt{8\pi V/3}$. Combining these leads to the simplified equation $\dot{\phi} = -V'/\sqrt{24\pi V}$. The solutions of the original equation are approximated by solutions of the simplified equation as $t \to \infty$. Since the simplified equation is a closed equation for ϕ alone it is very useful for proving results about the details of the asymptotics. When there is a minimum which is zero the effect of the kinetic energy on the dynamics remains important for all times. In some cases a relatively simple description of its average effect can be obtained by an argument related to the virial theorem.

The conformal field equations have also been used to prove a result on the stability of anti-de Sitter space, which has the metric

$$-\cosh^2 r\, dt^2 + dr^2 + \sinh^2 r\, d\Sigma^2, \tag{10.25}$$

where $d\Sigma^2$ is the standard round metric on the two-sphere. It is a solution of the vacuum Einstein equations with $\Lambda < 0$. It is not globally hyperbolic and so it is difficult to discuss its stability in the context of the pure Cauchy problem. Introducing a suitable conformally rescaled metric makes the causal structure of this spacetime clearer. It is possible to attach a timelike boundary to it. This suggests using an IBVP to study the stability question [83]. The result is that it is possible to find perturbed anti-de Sitter spacetimes which are determined by initial and boundary data. They exist on a finite time interval in the conformally rescaled picture although the deeper meaning of this finiteness is not clear.

10.3 Stability of Minkowski space

The stability of Minkowski space is much harder to prove than that of de Sitter space and it is worth trying to understand why this is the case. One reason is that in the de Sitter case many quantities of interest decay exponentially at late times while in the Minkowski case they only decay like (small) powers of t. Looking at the discussion of the null condition in Section 8.8 suggests why this makes a big difference. The reduced Einstein equations in harmonic coordinates do not satisfy the null condition. There is also another difficulty which persists in higher dimensions where the null condition is unnecessary. This is because the decay of the initial data for the Einstein equations is limited by the constraints and, more specifically, by the positive mass theorem. The resulting slow fall-off is not enough to allow the application of standard theorems on small data global existence for nonlinear wave equations.

For a long time it was believed that it would not even be possible to prove the nonlinear stability of Minkowski space using harmonic coordinates. For this reason the first proof, due to Christodoulou and Klainerman, used a maximal time coordinate, i.e. one for which the level hypersurfaces of t satisfy $\mathrm{tr} k = 0$, and a shift zero condition. The proof involves many complicated details and the book which presents it has more than 500 pages. Later it was realized that the stability can be proved in harmonic coordinates. This requires a variety of new ideas. One of them which is of central importance is the weak null condition which was discussed in Section 8.8. The proof of the stability of Minkowski space using harmonic coordinates is much shorter than the original one. On the other hand it gives less precise information about the detailed asymptotics. It does suffice to prove the key property of geodesic completeness. In contrast to the original proof it extends in a relatively straightforward way to higher dimensions and to the Einstein–Maxwell equations [48].

The decay of asymptotically flat solutions of the Einstein equations, especially the vacuum equations, in null directions is closely related to the theory of gravitational radiation. Part of the folklore of this subject is the so-called peeling theorem. This says that if the Weyl tensor is expressed in a suitable null frame different components fall off at different rates along outgoing null geodesics. There are ten components which, as far as their fall-off behaviour is concerned, can be thought of as five pairs. These different pairs fall off like r^{-1}, r^{-2}, r^{-3}, r^{-4} and r^{-5} respectively. Since this statement concerns decay along a null direction r could equivalently be replaced by u or t. These statements follow if the spacetime possesses a sufficiently smooth conformal compactification. Unfortunately it is difficult to decide how generally a compactification of this kind exists for a solution of the Einstein equations corresponding to prescribed initial data. The original proof of the stability of Minkowski space could only confirm part of the claims of the peeling theorem. Later it was shown that under some slightly stronger assumptions on the initial data the full peeling behaviour does hold. It may well be that for the most general physically relevant initial data peeling fails.

10.3 Stability of Minkowski space

The proof of the stability of Minkowski space makes heavy use of the symmetry of Minkowski space. In the proof using harmonic coordinates it is enough to use the vector fields which have the same coordinate form as the Killing vector fields of Minkowski space. In the original proof a large part of the effort goes into constructing approximate Killing vector fields X^α which are analogues of the Killing vector fields in Minkowski space and have the property that the deformation tensor $\nabla_\alpha X_\beta + \nabla_\beta X_\alpha$ decays sufficiently fast at infinity. It then uses generalized energy estimates associated to these vector fields, as will now be explained.

The fact that the energy–momentum tensor is divergence-free leads to certain conservation laws for matter fields. If K^α is a vector field then $\nabla^\alpha(T^{\alpha\beta}K_\beta) = \frac{1}{2}T^{\alpha\beta}(\nabla_\alpha K_\beta + \nabla_\beta K_\alpha)$. This means that if K^α is a Killing vector then $P^\alpha = T^{\alpha\beta}K_\beta$ is divergence free. Hence integration over a lens-shaped region shows that the integral of $P_\alpha n^\alpha$ is the same over both components of the boundary, where n^α is the normal vector. If the energy–momentum tensor is trace-free as for instance in the case of the Maxwell field then it suffices to take a conformal Killing vector field to get the same identity. Recall that a conformal Killing vector field is one which generates diffeomorphisms which are conformal transformations and that the notion is characterized by the property that $\nabla_\alpha K_\beta + \nabla_\beta K_\alpha = f g_{\alpha\beta}$ for some function f. In this way conservation laws for matter fields can be obtained. For the gravitational field itself there is no energy–momentum tensor but there is a similar object which can be used in a similar way to get energy-like estimates. This is the Bel–Robinson tensor which is defined for a vacuum spacetime by

$$T_{\alpha\beta\gamma\delta} = C_{\alpha\rho\gamma\sigma} C_\beta{}^\rho{}_\delta{}^\sigma + C_{\alpha\rho\delta\sigma} C_\beta{}^\rho{}_\gamma{}^\sigma - \frac{1}{8} g_{\alpha\beta} g_{\gamma\delta} C_{\rho\sigma\mu\nu} C^{\rho\sigma\mu\nu}. \quad (10.26)$$

It is completely symmetric in all indices, trace-free and divergence-free. It satisfies the following analogue of the dominant energy condition. If U^α, V^α, W^α and Z^α are future-directed causal vectors then $T_{\alpha\beta\gamma\delta} U^\alpha V^\beta W^\gamma Z^\delta \geq 0$. Thus contracting the Bel–Robinson tensor with three causal conformal Killing vectors gives a positive conserved quantity. More pertinently, contracting it with three approximate conformal Killing vector fields gives a quantity which is approximately conserved.

The basic principle of the proof of the stability of Minkowski space is that of a bootstrap argument. This is an idea which is also useful in other contexts and so it will be sketched here. It is a kind of analogue of mathematical induction where the discrete index is replaced by a continuous parameter. Suppose that $u(t)$ schematically denotes a solution of an evolution equation. Let $P(t)$ denote some functional of $u(t)$ such as a weighted Sobolev norm which has the following properties:

1. $P(t)$ is continuous in t for any solution of the class being considered;
2. if $P(t)$ is bounded on a time interval $[t_0, t_1)$ where the solution is defined with $t_1 < \infty$ then the solution extends to an interval of the form $[t_0, t_1 + \epsilon)$ for some $\epsilon > 0$;
3. if a solution on an interval $[t_0, t_1)$ with sufficiently small initial data satisfies $P(t) \leq \delta$ on that interval for a sufficiently small constant δ then it satisfies $P(t) < \delta$ on that interval.

Then global existence of the solution for small data follows. For let $[t_0, T^*)$ be the maximal interval on which a solution corresponding to the prescribed initial data exists and satisfies $P(t) \leq \delta$. Here T^* may be infinite. If the data are sufficiently small then $P(t)$ is initially small. Thus by continuity (the first assumption) there is some interval on which $P(t) \leq \delta$ and T^* is well-defined as a real number or infinity. Suppose that $T^* < \infty$. Since $P(t) \leq \delta$ on the interval $[t_0, T^*)$ the solution can be extended to a longer time interval $[t_0, T^* + \epsilon)$ for some $\epsilon > 0$ by the second assumption. Since the data are sufficiently small $P(t) < \delta$ on $[0, T^*)$ by the third assumption. By continuity (first assumption) $P(t)$ remains less than δ for a short time after T^* and this contradicts the maximality of T^*. Thus in fact $T^* = \infty$. Notice that as well as global existence some information is obtained about the solution, namely that it satisfies $P(t) \leq \delta$. In a typical implementation of this scheme some decay of the solution as $t \to \infty$ is obtained for free.

The assumptions of the original theorem on the stability of Minkowski space express the idea that the initial data should be close to Minkowski data on an entire hypersurface $t = t_0$. Because of the domain of dependence it might reasonably be expected that if the data were only supposed to be defined and close to that of Minkowski space on the complement of a compact set in $\mathbb{R}^3$ then a solution would be obtained which was globally defined in the exterior of an outgoing light cone. Moreover, since asymptotically flat data are close to data for flat space on the complement of a sufficiently large compact set there should be a theorem of this kind applying to any asymptotically flat vacuum solution. It should also be possible to handle data for the Einstein–matter equations provided the support of the matter fields is compact. It is enough to just cut out a set containing the support of the matter. A result of this kind would emphasize the wide physical applicability of the theorem. Unfortunately the original proof of the stability of Minkowski space cannot easily be used to get this kind of result. The difficulty is that it uses maximal hypersurfaces, whose definition is global in space. Thus the stability of an exterior region is left open. The exterior problem was treated using a different approach in [121]. In that work the authors used a double null coordinate system to cover the exterior region.

10.4 Stability of the Milne model

The Milne model has the metric

$$-dt^2 + t^2 d\Sigma^2, \tag{10.27}$$

where $d\Sigma^2$ is a metric of constant negative curvature. This is a vacuum FLRW model with $K = -1$. It is flat and can be obtained by foliating the interior of the light cone in Minkowski space by the hypersurfaces of constant Lorentzian distance from the origin. The stability statement concerns the solution where the spatial hypersurfaces are compactified using identifications defined by a discrete group. Here the name Milne model is also used for the compactified spacetime. In higher dimensions there are analogues of the Milne model where the metric denoted by $d\Sigma^2$ above is replaced by an arbitrary Einstein metric with negative curvature, i.e. a Riemannian metric satisfying $R_{ab} = \frac{1}{n} R g_{ab}$ with R a negative constant.

Consider initial data for the (spatially compactified) Milne model on a hypersurface of constant mean curvature. In general, depending on how the compactification is done, there may be deformations of these data, depending on continuous parameters, all of whose Cauchy developments are flat. If this is not the case the data is said to be rigid. It can be proved that in the rigid case the Milne model is asymptotically stable in the following sense. Suitable dimensionless variables describing the solution converge to the values they have in the Milne model at late times. This theorem was proved using energy estimates based on the Bel–Robinson tensor in a way similar to what was done in the original proof of the nonlinear stability of Minkowski space. The argument was however simpler since the Bel–Robinson tensor only needed to be contracted with copies of a single timelike vector and not with the complicated selection of vectors used in the case of Minkowski space. This is connected with the fact that the decay of the solution in the case of the Milne model does not have the complicated direction-dependent behaviour present in the case of Minkowski space. The information obtained about the asymptotic behaviour is in particular enough to show future geodesic completeness of the spacetimes concerned. Since the Bel–Robinson tensor is particular to four dimensions a proof of stability of the higher dimensional analogues of the Milne model would have to rely on a different approach.

10.5 Stability of the flat Bianchi type III model

This section is concerned with the stability of the spacetime with metric

$$-dt^2 + t^2 d\Sigma^2 + dz^2, \tag{10.28}$$

where $d\Sigma^2$ is a two-dimensional metric of constant negative curvature. This is a spacetime with hyperbolic symmetry and it is flat. It is spatially homogeneous and of Bianchi type III. It can be obtained by foliating the interior of the light cone in three-dimensional Minkowski space by the hypersurfaces of constant Lorentzian distance from the origin and taking the product of the result with the circle. It can also be considered as a spacetime with $U(1)$ symmetry as in Section 4.5. In that framework the two-dimensional manifold is a surface of higher genus, the harmonic form is set to zero and the circle bundle is trivial. It is the solution of the Einstein–wave map equations in $2+1$ dimensions where the wave map is constant.

The stability of this solution in the future has been proved within the class of $U(1)$-symmetric spacetimes. It is expected that in the past generic solutions show oscillatory behaviour so that controlling them is out of reach of present techniques. It appears that for the analysis of the late time dynamics having a higher genus surface is advantageous in comparison to the sphere or torus. A related, but simpler, problem which played an important role in understanding the dynamics is that of solutions of the vacuum Einstein equations in $2+1$ dimensions with a compact Cauchy hypersurface. At first sight this might appear trivial since any solution of the vacuum Einstein equations in $2+1$ dimensions is flat – the Riemann tensor is determined by the Ricci tensor. However it is not so since there are global degrees of freedom. These are related to the fact that the metrics of constant negative curvature on a higher genus surface form a finite-dimensional manifold, the Teichmüller space. The Einstein equations in three dimensions define a dynamical system on Teichmüller space. It can be shown that the solutions of this dynamical system exist globally towards the future. This uses a quantity called the Dirichlet energy to obtain control of solutions. In the stability problem which is the subject of this section the evolution of Teichmüller parameters also plays an important role. For comparison note that the vacuum equations in $2+1$ dimensions can be solved explicitly when the spatial topology is that of a torus. There are no solutions where the topology is spherical, at least in the case $\Lambda = 0$. The case of a positive cosmological constant has also been studied.

In the stability proof under discussion here the time coordinate used is taken to be $t = -\tau^{-1}$ where τ is the mean curvature of spacelike surfaces of constant mean curvature in the $(2+1)$-dimensional manifold. An analytic tool of central importance for the global existence proof is the use of the following energy-like quantity

$$E(t) = \frac{1}{2}\int (|N^{-1}\partial_0 u|^2 + \nabla_a u \nabla^a u + |\tilde{k}|^2). \qquad (10.29)$$

10.5 Stability of the flat Bianchi type III model

Here u denotes the wave map occurring in the Kaluza–Klein reduction, $\tilde{k}$ is the tracefree part of the second fundamental form and the integral is taken over the two-dimensional spatial manifold. This quantity satisfies the identity

$$\frac{dE}{dt} = \frac{1}{2}\tau \int (|\tilde{k}|^2 + |N^{-1}\partial_0 u|^2). \tag{10.30}$$

Remembering that τ is negative in an expanding spacetime it is seen that $E(t)$ is a non-increasing function. Note that not all terms occurring in the integral defining the energy also occur on the right hand side of (10.30). If all terms had been present an estimate for the rate of decay of E would have been obtained. Since it is in fact not the case a more refined procedure is required. This uses the 'corrected energy' defined by

$$E_\alpha(t) = E(t) - \alpha\tau \int (u - \bar{u}), \tag{10.31}$$

where $\bar{u}$ is the mean value of u over the spatial surface. The correction term can be estimated in terms of $E(t)$ with the help of the Poincaré inequality. This means that estimating E_α is equivalent to estimating E if α is chosen appropriately. It is possible to obtain a differential inequality for the corrected energy which is what is needed for the global existence proof.

In the course of the proof estimates are obtained for higher energies containing derivatives of the quantities making up the original energy and for related corrected energies. A number of elliptic estimates have to be done to control other parts of the geometry. The harmonic energy is used to control the Teichmüller parameters. The result of all this is that differential inequalities are obtained which show that if the data are sufficiently small then the solution exists globally in time and stays small. This is in particular enough to show that the spacetime is future geodesically complete.

It may be asked why only perturbations with $U(1)$ symmetry are treated in this work while in the case of the Milne model general perturbations were allowed. From a technical point of view this has to do with the following fact. The estimates carried out are dependent on the fact that the constants in the Sobolev inequalities for the families of metrics involved are under control. If perturbations without $U(1)$ symmetry are allowed, or if the higher genus surface is replaced by a torus then it might happen that the two-dimensional geometry undergoes collapse in the sense of Cheeger and Gromov and that as a consequence the Sobolev constants behave in a way which is hard to estimate. The meaning of collapse of a sequence of Riemannian metrics is that the metric becomes singular although its curvature remains bounded. A trivial example is obtained by scaling the metric on a torus but there are also much more subtle examples. For instance it is known this kind of phenomenon plays a role in the evolution of a number of Bianchi models.

Some further remarks on the concept of collapsing will now be made. Let (M, g) be a Riemannian manifold. There is an associated distance function d which is defined by the property that if p and q are points of M then $d(p, q)$ is the infimum of the lengths of curves joining p and q. This defines a metric in the sense of the theory of metric spaces. If A is a subset of a metric space let A_r be the set of all points p such that there is a point $q \in A$ with $d(p, q) \leq r$. If A and B are two compact subsets of a metric space let $d(A, B)$, the Hausdorff distance from A to B, be the infimum of the set of numbers r such that $A \subset B_r$ and $B \subset A_r$. The Hausdorff distance gives the set of compact subsets of a fixed metric space itself the structure of a metric space in a natural way. If now A and B are abstract compact metric spaces then the Gromov–Hausdorff distance between A and B is the infimum over the Hausdorff distances between the images of all isometric embeddings of A and B into some other metric space. This makes the collection of all compact metric spaces into a metric space. In this way convergence of a sequence of Riemannian manifolds can be defined in a way which allows the limit to be a manifold of a lower dimension or some other metric space. For instance the sequence of metrics on a torus given by scaling a given flat metric by $1/n$ converges to a point in the Gromov–Hausdorff sense as $n \to \infty$. The sequence of induced metrics on the homogeneous hypersurfaces in the Taub–NUT spacetime converges to a round metric on the two-dimensional sphere as the Cauchy horizon is approached. There is also a generalization of Gromov–Hausdorff convergence to non-compact manifolds. It is called pointed Gromov–Hausdorff convergence since to make it well-defined it is necessary to choose a preferred point of each manifold in the sequence.

The pseudo-Schwarzschild solution (11.73) is included in the symmetry class treated in the stability theorem but the corresponding data need not be small. However the known evolution of that spacetime shows that the data are small at late times. This determines the asymptotic behaviour for data close to those of the pseudo-Schwarzschild solution since it suffices to use Cauchy stability to show that the solutions evolving from these data will follow the homogeneous solution into the small data regime. Thus its asymptotics is determined by the theorem for small data. If non-trivial bundles were allowed and the condition of $U(1)$ symmetry were dropped the vacuum Bianchi type VIII solutions would be included and the known complexity of the late time behaviour of those solutions gives a lower bound for the complexity of the resulting problem. If the two-dimensional spacelike manifold is taken to be a torus then the Gowdy solutions and more general solutions with T^2 symmetry are included and this gives some hints as to the difficulties to be expected.

There is a result on the structure of the past singularity in solutions of the Einstein vacuum equations with polarized $U(1)$ symmetry [113]. Solutions are obtained which depend on two free functions and this is the same number as the general polarized solution. Without the polarized condition the solutions depend on four free functions. An existence theorem for spacetimes with controlled singularity structure has also been proved in the class of half-polarized solutions which depend on three free functions. The result is proved by Fuchsian techniques and, as is typical for this technique, the data are assumed to be analytic and the generality of the solutions is measured by function counting. There is no known way of identifying the half-polarized solutions in terms of the data they induce on a regular initial hypersurface – the class is defined directly in terms of the asymptotic data at the singularity.

In what was presumably the first application of Fuchsian techniques in general relativity Moncrief proved the existence of large classes of vacuum spacetimes with $U(1)$ symmetry and compact Cauchy horizons [147], [148]. They depend on two free functions. It has been conjectured by Isenberg and Moncrief that all spacetimes with compact Cauchy horizons must possess at least one Killing vector. In the scenario they envisage the Killing vector is tangent to the null direction within the horizon, except in the case of the Kasner solutions where the space of Killing vectors is at least three dimensional. The conjecture has been proved under some additional assumptions [112]. It is related to the uniqueness theorems for black holes where there is supposed to be a Killing vector tangent to the null direction in the event horizon.

10.6 The Newtonian limit

The Newtonian theory of gravity, the predecessor of general relativity, is formulated in the following way. The gravitational interaction is described by the Newtonian potential U which satisfies the Poisson equation $\Delta U = 4\pi\rho$, where ρ is the mass density. To have a closed system of evolution equations the Poisson equation must be coupled to the equations of motion of some matter fields. Simple examples of matter fields are a perfect fluid described by the (non-relativistic) Euler equations, kinetic matter described by the (non-relativistic) Boltzmann equation or an elastic solid. The extensions of these matter models to the general relativistic context are described in Chapter 3. In the equations of motion the term describing the gravitational acceleration is given by the gradient of U.

Many textbooks on general relativity contain statements that the Newtonian theory of gravity arises as the limit of general relativity in situations where material bodies move slowly in comparison to the speed of light. This is important for physics since it underlies the fact that most physical phenomena involving gravity can be modelled rather accurately by the Newtonian theory. The arguments offered in textbooks are on a purely heuristic level and it is not at all easy to obtain precise mathematical statements on this subject. Indeed the available rigorous mathematical theory is a long way from covering the procedures which are used in applications. If the Newtonian theory is thought of as the lowest order approximation to general relativity then corresponding higher order corrections are known as post-Newtonian expansions. Unfortunately the most direct expansion does not cover the phenomenon of radiation and as soon as radiation enters the expansion breaks down. The standard lowest order approximation modelling radiation gives rise to the Einstein quadrupole formula. This seems to work very well in practice but there is no mathematical proof of the formula. It is an outstanding challenge for mathematical relativists to change this.

To approach the Newtonian limit mathematically it is convenient to proceed in the following way. Consider a family of physical systems depending on a parameter λ where typical velocities v_T in the system with parameter value λ are of order $\lambda^{1/2}$. Suppose that these systems are described using a fixed choice of physical units. Then the numerical value of the speed of light c is fixed while the numerical value of v_T tends to zero as $\lambda \to 0$. Hence $v_T/c \to 0$ as $\lambda \to 0$. It is convenient to do a transformation of units depending on λ so that after transformation the value of v_T is of order unity while the numerical value of the speed of light c expressed in these units tends to infinity as $\lambda \to 0$. In fact the choice $\lambda = c^{-2}$ will be made. In this way the Newtonian limit can be thought of as the limit $c \to \infty$.

For appropriate physical quantities f an expansion of the form

$$f(\lambda) = f_0 + \lambda f_1 + \lambda^2 f_2 + \ldots \tag{10.32}$$

is assumed. At the moment this is a purely formal expansion. The goal of the mathematical treatment of the Newtonian limit is to investigate whether and under what circumstances physical quantities derived from solutions of the Einstein–matter equations really admit asymptotic expansions with controlled errors corresponding to these formal expansions. The standard terminology is that a quantity like f_0 belongs to the Newtonian approximation, a quantity like f_1 belongs to the 1PN (first post-Newtonian) approximation, f_2 to the 2PN approximation, etc. If the formal power series is to be asymptotic in the sense of uniform convergence then it turns out that the approximation breaks down after the 2PN level. Radiation should occur at the order c^{-5}, i.e at the $2\frac{1}{2}$PN level. This raises the question why the expansion was done using only even powers of c. The only answer which will be offered here is that experience shows that it is consistent to do so up to the level where radiation enters.

What is the nature of the mathematical problem which has to be solved in the context of the Newtonian limit? The Einstein equations which are essentially hyperbolic degenerate to the Poisson equation which is elliptic as $\lambda \to 0$. This change of type of the differential equation means that this is a singular limit. It is well known that singular limits of PDE lead to obstructions to smooth parameter dependence of solutions as the limiting value is approached. The breakdown of the post-Newtonian expansions is an example of this phenomenon. Other examples are the non-relativistic (electrostatic) limit of the Maxwell equations coupled to charged matter and the incompressible limit of compressible fluids. In the first of these cases the speed of light tends to infinity while in the second it is the speed of sound. In order to come closer to understanding the difficult case of the Einstein equations it is useful to learn from what is known in these simpler cases. It should not be assumed that even the simpler cases are completely understood. Consider for example the dipole formula for electromagnetic radiation which is the analogue for the Maxwell equations of the quadrupole formula. It is only very recently that a dipole formula was proved for a self-consistent model of charged matter and even that is unlikely to be the optimal result.

Even before tackling the analytical difficulties of the Newtonian limit of the Einstein equations there are conceptual questions which have to be addressed. In general relativity the gravitational field is described by the metric which has ten components when expressed in coordinates. In Newtonian theory there is only a single function, the potential. How can the one be said to converge to the other? A framework for doing this is the following [76]. The metric in GR is replaced by two objects $g^{\alpha\beta}$ and $h_{\alpha\beta}$ depending on a parameter λ satisfying $g^{\alpha\beta}h_{\beta\gamma} = \lambda\delta^\alpha_\gamma$. These are called the time metric and space metric respectively. It is clear that they must become degenerate for $\lambda = 0$. Both $g^{\alpha\beta}$ and $h_{\alpha\beta}$ are assumed to be covariantly constant with respect to a connection $\Gamma^\alpha_{\beta\gamma}$. For $\lambda \neq 0$ this can only be the Levi-Cività connection. The field equations are written down using the Ricci tensor of the connection $\Gamma^\alpha_{\beta\gamma}$. Some care must be taken in writing the energy–momentum tensor, deciding where to use $g^{\alpha\beta}$ and where to use $h_{\alpha\beta}$. The result of this is that for each $\lambda \neq 0$ the field equations obtained are equivalent to the Einstein equations and this is just a rather redundant description of GR. For $\lambda = 0$ the equations obtained are closely related to the Newtonian theory or, more precisely, to a slight generalization of it called the Newton–Cartan theory. The connection defined by $\Gamma^\alpha_{\beta\gamma}(0)$ is called the Cartan connection and contains all the information about the Newtonian gravitational field. In suitably chosen coordinates the only non-zero components of the Cartan connection are those of the form Γ^a_{00} and these are equal to $\nabla_a U$ where U is the Newtonian potential. This set-up can be used to formulate the Newtonian limit in a precise manner. The quantities $g^{\alpha\beta}$, $h_{\alpha\beta}$ and $\Gamma^\alpha_{\beta\gamma}$ are required to have a certain degree of smoothness. The metrics $g^{\alpha\beta}$ and $h_{\alpha\beta}$ degenerate so as to have rank one and three respectively for $\lambda = 0$. Physically they are thought of as measuring time intervals and spatial distances respectively. There are some additional conditions which will not be listed here. This framework gives the optimal correspondence with the physical meaning of various quantities and helps, for instance, in finding the correct expression for the energy–momentum tensor for a specific matter model. On the other hand all objects are determined by $g^{\alpha\beta}$ for $\lambda \neq 0$ and so for the analytical theory of the Newtonian limit it is possible to formulate everything in terms of $g^{\alpha\beta}$ alone.

The Newton–Cartan theory has another application. If a mathematical problem in GR is too difficult to solve immediately it is often useful to think about what happens in the corresponding Newtonian situation. The Newton–Cartan theory can help in defining the Newtonian analogue. In ordinary Newtonian theory the potential U is defined on $\mathbb{R}^3$ and is required to vanish at infinity. Newtonian theory deals with isolated systems. Nevertheless ideas of the Newtonian theory are often applied in cosmology where a description by an isolated system in this sense does not apply. The calculations which are done are sometimes confusing since it is not clear what assumptions they are based on. The Newton–Cartan theory provides a convenient solution to these difficulties. It is possible to consider solutions of the theory on a compact manifold. It is easy to include a cosmological constant if desired. The usual calculations in the astrophysics literature can be interpreted in this framework. An example where a suitable analogue of a problem in GR can be solved is that of the cosmic no hair theorem for solutions with a positive cosmological constant (see [35]).

It has been proved that there are rather general asymptotically flat solutions of the Einstein–Vlasov system having a Newtonian limit. The limiting quantities satisfy the Vlasov–Poisson system which is discussed in Section 11.1. The coordinates used in the proof have maximal time slices and shift zero. A technical difficulty is that the deviation of the metric from the flat metric has only $1/r$ fall-off at infinity. Thus a quantity like $g_{ab} - \delta_{ab}$ does not belong to an ordinary Sobolev space. The attempt to replace ordinary Sobolev spaces by weighted ones runs into the difficulty that it does not fit with the approach to this kind of singular limit which has been applied to other problems such as the incompressible limit of compressible fluids. In the original work on the Newtonian limit of the Einstein–Vlasov system this was overcome by basing the arguments on energy estimates for the second fundamental form which does lie in an ordinary Sobolev space. Later other possibilities were discovered. While maximal slicing fits well with the post-Newtonian expansions this is not true of the shift zero condition. At the very least it makes the calculations extremely unwieldy. For this reason the work on Einstein–Vlasov already mentioned was not pushed to the 1PN level. More recently the Newtonian limit of the Einstein–Euler system has been treated. One reason for choosing the Vlasov equation as source in the original work was the absence of problems with that matter model when the density has compact support. For the Euler equations, on the other hand, there are big problems and the theorems which have been proved are confined to the case of solutions of Makino type. This work uses harmonic coordinates which fit better with the PN expansions than the shift zero condition. As a consequence it has been possible to obtain rigorous results on the first post-Newtonian approximation.

10.7 Further reading

Global existence for the Gowdy equations was first proved in [146]. The future asymptotics, including a proof of future geodesic completeness, was determined in [185]. The existence of the asymptotic velocity is a central part of [186]. Gowdy spacetimes with prescribed low velocity asymptotics were constructed in [119]. The first paper to obtain rigorous results on the existence of spikes was [181]. Spacetimes with polarized T^2 symmetry and prescribed singularity behaviour were constructed in [111]. The addition of a cosmological constant and the extension to the half-polarized case are in [63].

The cosmic no hair theorem for spacetimes with positive cosmological constant was proved in [82]. The generalization to scalar fields with certain exponential potentials was achieved in [100]. An existence theorem for spacetimes with prescribed asymptotics in the presence of a positive cosmological constant can be found in [174]. The precise statement on the asymptotics of the metric in the context of Wald's theorem is in [129]. Generalizations of Wald's theorem for various scalar field models are proved in [175], [130] and [178]. Aspects of the case of a potential with zero minimum have been discussed in [180].

The original proof of the stability of Minkowski space is the content of the book [59] while the shorter proof in harmonic coordinates is given in [137]. Peeling is obtained for a suitable class of data in [122]. The stability of the Milne model and the flat Bianchi type III model were proved in [5] and [50] and [47] respectively. The existence theory for spacetimes with prescribed behaviour near the past singularity is in [113].

A dipole formula for the Maxwell equations coupled to collisionless matter was proved in [20]. The application of the Newton–Cartan theory to cosmology is described in [35]. Theorems on the Newtonian limit in the case of collisionless matter and perfect fluids were proved in [168] and [155] respectively.

11 The Einstein–Vlasov system

The phenomenological matter model for which the most mathematical results are known for the Einstein–matter system is collisionless matter solving the Vlasov equation. A variety of existing mathematical results on this subject are surveyed in this chapter. First some of the known results for related systems are summarized.

11.1 Other kinetic equations

Before coming to the Einstein–Vlasov system itself results on some related systems will be discussed. This is motivated by the hope that understanding of the properties of solutions of the Einstein–Vlasov system may be improved by interaction with the theory of these other equations. The non-relativistic analogue of the Einstein–Vlasov system is the Vlasov–Poisson system. As stated in Section 10.6 it is possible to show rigorously that solutions of the Einstein–Vlasov system converge to solutions of the Vlasov–Poisson system as the speed of light tends to infinity. The Vlasov–Poisson system can also be used to describe charged particles. The only difference is a change of sign due to the fact that like particles repel in the electromagnetic case. The case of the Vlasov–Poisson system with an attractive force is often called the stellar dynamics case since it is used to describe galaxies. Information about astrophysical applications of this model can be found in [31]. The Vlasov–Poisson system with a repulsive force is often called the plasma physics case since that is an area of physics in which it is frequently applied.

For general C^1 initial data with compact support global existence is known for solutions of the Vlasov–Poisson system with either sign of the interaction term. The majority of the proof is the same in both cases but there is one key difference. The system has a conserved total energy which is the sum of kinetic and potential contributions. The kinetic energy is always positive but the potential energy is positive in the plasma physics case and negative in the stellar dynamics case. It follows that in the stellar dynamics case there is a possible blow-up scenario in which both potential and kinetic energies tend to infinity in modulus although their sum is constant. This is ruled out by an estimate of Horst [106]. Interestingly this estimate is essentially dimension dependent and global existence for the gravitational Vlasov–Poisson system fails in four or more space dimensions as can be proved by a contradiction argument. The case of four spacetime dimensions is critical from the point of view of scaling and in that case an explicit singular solution has been found. It is self-similar. In three space dimensions there are two very different global existence proofs. One uses clever estimates of the characteristic system while the other uses integral estimates.

The natural relativistic generalization of the Vlasov–Poisson system in the plasma physical case is the Vlasov–Maxwell system which is much simpler than the Einstein–Vlasov system. Nevertheless there is still no completely general global existence theorem for the Vlasov–Maxwell system although there is a variety of results concerning solutions with symmetry as well as a small data global existence theorem. A generalization of the gravitational Vlasov–Poisson system which is mathematically interesting, although of doubtful direct physical relevance, is the Vlasov–Nordström system. Here the Newtonian theory of gravity is replaced by a relativistic theory introduced before the birth of general relativity. Despite the fact that the Vlasov–Nordström system shows many close similarities to the Vlasov–Maxwell system there is a general global existence theorem for the former. A scalar gravitational theory has also been used as a model system for developing techniques for numerical relativity [193]. The scalar theory can be obtained from Brans–Dicke theory by discarding the field equation for the metric and replacing the metric in the other equations expressed in the Einstein frame by the flat metric.

In the so-called relativistic Vlasov–Poisson system the Vlasov equation has the relativistic form but the field equation for the matter remains non-relativistic. In the case of an attractive force solutions of this system develop singularities in three space dimensions, even in the spherically symmetric case. For a repulsive force global existence is known in the spherically symmetric case which is in fact equivalent to the spherically symmetric case of the Vlasov–Maxwell system. Global existence for solutions without symmetry is not known.

11.2 Small data global existence

11.2.1 Schwarzschild coordinates

Consider initial data for the Einstein–Vlasov system with vanishing cosmological constant which are spherically symmetric and asymptotically flat.

The metric will be written in the form (4.26) with $k = 1$. These coordinates are known as Schwarzschild coordinates. The boundary conditions are those of asymptotic flatness and require that $\mu(t, r) \to 0$ as $r \to \infty$ for each fixed t and that $\lambda(t, 0) = 0$ for all t. For a spherically symmetric metric written in polar coordinates the latter condition ensures that the geometry is smooth at the centre of symmetry instead of having a conical singularity. Initial data for the distribution function is assumed to have compact support. Then the restriction of any solution to any hypersurface of constant t has compact support. The field equations imply that $\lambda \to 0$ as $r \to \infty$. There is a close similarity between this problem and the example treated in Section 5.10. In the present case the momenta are parametrized by their spatial components v^i in an orthonormal frame. The Einstein equations take the form

$$e^{-2\lambda}(2r\lambda' - 1) + 1 = 8\pi r^2 \rho, \tag{11.1}$$

$$e^{-2\lambda}(2r\mu' + 1) - 1 = 8\pi r^2 p, \tag{11.2}$$

$$\dot{\lambda} = -4\pi r e^{\lambda+\mu} j, \tag{11.3}$$

$$e^{-2\lambda}\left(\mu'' + (\mu' - \lambda')\left(\mu' + \frac{1}{r}\right)\right) - e^{-2\mu}(\ddot{\lambda} + \dot{\lambda}(\dot{\lambda} - \dot{\mu})) = 4\pi q. \tag{11.4}$$

The matter quantities on the right hand side of the Einstein equations are defined by

$$\rho(t, r) = \int \sqrt{1 + |v|^2} f(t, x, v) dv, \tag{11.5}$$

$$p(t, r) = \int \left(\frac{\delta_{ij} x^i v^j}{r}\right)^2 \frac{1}{\sqrt{1 + |v|^2}} f(t, x, v) dv, \tag{11.6}$$

$$j(t, r) = \int \frac{\delta_{ij} x^i v^j}{r} f(t, x, v) dv, \tag{11.7}$$

$$q(t, r) = \int \left[|v|^2 - \left(\frac{\delta_{ij} x^i v^j}{r}\right)^2\right] \frac{1}{\sqrt{1 + |v|^2}} f(t, x, v) dv. \tag{11.8}$$

In this case it is helpful to express the Vlasov equation in Cartesian coordinates. The reason is that the global existence theorem relies on the fact that at late times the characteristics of the Vlasov equation are approximately straight lines and straight lines are more easily described in Cartesian coordinates than in polar coordinates. A simplification results from using the following identity which holds for any spherically symmetric function.

$$(r^2 v^i - \delta_{jk} x^j v^k x^i) \partial_i f = (|v|^2 - \delta_{jk} x^j v^k v^i) \partial_{v^i} f. \tag{11.9}$$

After this has been done the Vlasov equation takes the form

$$\frac{\partial f}{\partial t} + e^{\mu-\lambda}\frac{v^i}{\sqrt{1+|v|^2}}\frac{\partial f}{\partial x^i} - \left(\dot{\lambda}\frac{\delta_{ij}x^i v^j}{r} + e^{\mu-\lambda}\mu'\sqrt{1+|v|^2}\right)\frac{x^i}{r}\frac{\partial f}{\partial v^i} = 0. \tag{11.10}$$

Consider first the vacuum case. The equation for λ' can be rewritten as $(re^{-2\lambda})' = 1$ and this can be solved explicitly at any fixed time to get $e^{2\lambda} = (1 - \frac{2m}{r})^{-1}$ for a constant m. The equation for $\dot{\lambda}$ shows that this constant is not dependent on the time chosen. Adding the equations for λ' and μ' gives $\lambda' + \mu' = 1$ and hence $e^{2\mu} = A(t)(1 - \frac{2m}{r})$ for an arbitrary function $A(t)$. Imposing the boundary condition of asymptotic flatness, which implies that $\lim_{r\to\infty} \mu(t,r) = 0$ for each t, fixes $A(t)$ to be identically equal to one. In this way the Schwarzschild solution is recovered. In the case of a solution with matter where the energy–momentum tensor has compact support on each hypersurface t=const. the expression for λ derived above is valid in the region outside the matter. It follows that the constant m can be identified with the ADM mass defined in Section 7.5. It follows from the equation for λ' that

$$m_{\text{ADM}} = \int_0^\infty 4\pi r^2 \rho(t,r) dr. \tag{11.11}$$

This says that the total ADM mass is equal to the integral of the energy density over space, a relation which looks simple to understand. Notice, however, that the integral is carried out with respect to the volume form of flat space and not with respect to that of the actual spatial geometry. In any case, the fact that the ADM mass is constant in time gives rise to a conserved quantity which is useful in the analysis of the Cauchy problem.

Now the general case with collisionless matter will be considered. If the data are sufficiently small then a solution is obtained which is global in the time coordinate used here and geodesically complete. The first step in proving this is to obtain a local existence theorem with a good continuation criterion. The theorem assumes that the initial datum f_0 for the distribution function has compact support. There are no additional data for the geometry – the constraints already determine everything. From the fact that f_0 has compact support it easily follows that the restriction of the distribution function to any Cauchy hypersurface has compact support. Given a choice of time coordinate t, a function $P(t)$ is defined to be the supremum of values of $|v|$ for which there is some (x,v) with $f(t,x,v) \neq 0$. This serves as a diagnostic quantity for the continued existence of a solution. If a solution exists on a time interval $[t_0, t_1)$ with $t_1 < \infty$ and $P(t)$ is bounded on this interval then the solution extends to an interval $[t_0, t_1 + \epsilon)$ for some $\epsilon > 0$.

Thus if it is possible to show that for a prescribed initial datum the quantity $P(t)$ is bounded on any time interval on which a solution exists then global existence follows. (In making this statement it is taken for granted that a local existence theorem has been proved.) Let $F = \|f_0\|_{L^\infty}$ and let R be the supremum of $|x|$ for which there is a v with $f_0(x, v) \neq 0$. A sufficient smallness condition to imply global existence is to require a fixed bound on $P(0)$ and R and smallness of F.

The proof of small data global existence proceeds by a bootstrap method. It uses the following estimates. Consider initial data f_0 satisfying $P(0) \leq P_0$, $R \leq R_0$ and $\|f_0\|_\infty \leq \epsilon$ for some positive constants P_0, R_0 and ϵ. The aim is to show that the corresponding solution exists globally if ϵ is chosen small enough. Suppose that a solution corresponding to data of the stated type is given on a time interval $[0, T)$ and that the following estimates are satisfied for some constants $\gamma > 0$ and $\delta \in (0, 1)$:

$$|\dot\lambda(t, r)| + |\dot\mu(t, r)| + |\lambda'(t, r)| + |\mu'(t, r)| < \gamma(1+t)^{-1-\delta}, \quad (11.12)$$

$$\frac{1}{r}\left(|\dot\lambda(t, r)| + |\lambda'(t, r)| + |\mu'(t, r)|\right) + |H(t, r)| < \gamma(1+t)^{-2-\delta}, \quad (11.13)$$

where

$$H(t, r) = e^{-2\lambda}\left(\mu'' + (\mu' - \lambda')\left(\mu' + \frac{1}{r}\right)\right) - e^{-2\mu}(\ddot\lambda + \dot\lambda(\dot\lambda - \dot\mu)). \quad (11.14)$$

This last quantity is equal to the matter quantity q by one of the field equations. This condition is called the free streaming condition and is the bootstrap assumption in the global existence theorem.

If a solution satisfies the free streaming condition and γ is sufficiently small then any characteristic which starts in the support of f_0 stays within the set where $|v| \leq P_0 + 1$. Using this it can be shown that if in addition it is assumed that $|\lambda(0, r)|$ and $|\mu(0, r)|$ are bounded by some constant C_0 then there is a constant C_1 such that $\|\rho(t)\|_{L^\infty} \leq C_1 t^{-3}$. The mechanism behind this decay statement is the following. Because the fields are sufficiently small the characteristics are almost straight lines – the particles spread out linearly. This means that for any spatial point t the particles arriving at that point at time t must have their velocities contained in a set of diameter t^{-1} in order to have come from the support of f_0. Thus the volume over which it is necessary to integrate when computing ρ is of order t^{-3}. Combining this with the conservation of $\|f(t)\|_{L^\infty}$ gives the desired result. The technical difficulty in implementing this idea is in precisely formulating and proving the statement about the linear spread of the characteristics.

It has now been shown that a suitable decay of the fields leads to decay of the energy density. Next it is shown that conversely decay of the energy density implies decay of the fields. Moreover the decay obtained is stronger than that which was assumed at the beginning and this provides the basic estimate for a bootstrap argument. If a solution satisfies the estimate $\|\rho(t)\|_{L^\infty} \leq C_1 t^{-3}$ and the estimates $R \leq R_0$ and $|\lambda(0,r)| \leq C_0$ then the solution satisfies the free streaming condition with some $\gamma > 0$ and $\delta = 1$. Considering the longest interval on which the free streaming condition holds for a given initial datum proves the global existence statement. To prove that there is any interval on which the free streaming condition is satisfied, so that the longest interval is well-defined, it is necessary to use the fact that the solutions of the Einstein–Vlasov system depend continuously on the initial data in a suitable sense. The mechanism of the proof is such that it shows that the solutions constructed satisfy a number of decay assumptions. It can be concluded that the L^∞ norms satisfy

$$\|\rho(t)\| + \|p(t)\| + \|j(t)\| + \|q(t)\| \leq C(1+t)^{-3}, \tag{11.15}$$

$$\|\lambda(t)\| + \|\mu(t)\| \leq C(1+t)^{-1}, \tag{11.16}$$

$$\|\Gamma^\alpha_{\beta\gamma}(t)\| \leq C(1+t)^{-2}, \tag{11.17}$$

$$\|R^\alpha_{\beta\gamma\delta}\| \leq C(1+t)^{-3}. \tag{11.18}$$

With these estimates it can be shown that the spacetime is geodesically complete. This shows that the spacetime is not only global in a coordinate sense but also in a geometrical sense.

It is reasonable to hope that a similar result holds without the assumption of spherical symmetry but this is clearly difficult to prove because it would contain the nonlinear stability of Minkowski space as a special case. For the corresponding system of equations in the Newtonian case, the Vlasov–Poisson system, an analogous result holds and was in fact the pattern which was followed in constructing the proof for the Einstein–Vlasov system. A question which remains open for spherically symmetric solutions of the the Einstein–Vlasov system with small initial data is whether the asymptotics can be refined to show that $\|D^k \rho\|_{L^\infty} \leq C(1+t)^{-3-k}$. The corresponding result has been proved in the case of the Vlasov–Poisson system.

Global existence is known for arbitrarily large initial data in the case of the Vlasov–Poisson system and this is one of the reasons for thinking that matter modelled by the Vlasov equation is very well behaved. Global existence for the spherically symmetric Einstein–Vlasov system in Schwarzschild coordinates with general large initial data is not known. In fact it is to be expected that in general black holes are formed for large data. At the same time it is believed that the Schwarzschild time coordinate has singularity-avoiding properties which would mean that only the region outside the black hole is covered by the Schwarzschild coordinates – the Schwarzschild time t would tend to infinity before the region of trapped surfaces is reached. This type of behaviour is observed in numerical calculations of solutions of the Einstein–Vlasov system but there is no analytical proof that it always happens.

It has been shown that if any singularity occurs in a solution of the spherically symmetric Einstein–Vlasov system written in Schwarzschild coordinates then the first singularity must be at the centre of symmetry. There is a case with large data where it has been shown that the solutions are geodesically complete in the future. The conditions defining these data express the idea that the matter should originally be moving outwards at a fast enough rate. Beyond this there are numerical results where the phenomenon of critical collapse has been investigated. An initial datum f_0 is rescaled to get a one-parameter family Af_0. For A sufficiently small the solution disperses to infinity, as it must due to the analytical theorems. For f large the solution collapses to a black hole of mass $M(A)$. The Schwarzschild coordinates cannot enter the region of trapped surfaces and the black hole and its mass are identified by a step in the function μ which develops at $r = 2M(A)$ and becomes larger and larger with time. When there is no black hole formed the quantity $M(A)$ is defined to be zero. Let A_* be the critical value of the parameter A, the infimum of those values of A for which there is black hole. It is found that the function has a jump discontinuity at the value A_*. There is evidence that this is associated with unstable static solutions of the equations which act as intermediate attractors.

It is interesting to compare these results for solutions of the Einstein–Vlasov system with what happens for the Einstein–dust system. The singularities of dust, in particular the shell-crossing singularities, show that in that case global existence for small initial data fails. In other words, for the Einstein–dust system Minkowski space is unstable even within the class of spherically symmetric solutions. It is also true in the case of dust that the first singularity can occur away from the centre. Thus the results which have just been presented clearly show the fundamental difference between smooth solutions of the Vlasov equation and dust as matter models in general relativity. In the case of a perfect fluid with pressure it must also be expected that smooth solutions break down after a finite time and that this can take place away from the centre in the spherically symmetric case. This phenomenon is independent of the presence of gravity and is due to the formation of shock waves. The singularity is relatively weak compared to those encountered for dust and in the future it may be possible to handle general solutions of the Einstein–Euler equations with shocks mathematically. This seems, however, to be a very difficult problem.

In some work which created a lot of attention at the time it appeared [192] it was claimed on the basis of numerical calculations that solutions of the Einstein–Vlasov system with axial symmetry (not spherically symmetry) can form naked singularities. It seems that this was due to the fact that what was actually calculated corresponded to solving the Einstein–dust equations and not to computing solutions of the Einstein–Vlasov system with smooth initial data. Corresponding calculations in the Newtonian case also showed singularities although global existence of smooth solutions is known for the Vlasov–Poisson system.

11.2.2 Maximal-isotropic and double null coordinates

Asymptotically flat spherically symmetric solutions of the Einstein–Vlasov system have also been studied analytically in other coordinates. One of the choices of coordinates is to take the hypersurfaces t=const. to be maximal ($\text{tr} k = 0$) and the radial coordinate to be isotropic, which means that the spatial metric is manifestly conformal to a flat metric. Maximal-isotropic coordinates are similar to the Schwarzschild coordinates in that the Einstein equations take a form where hyperbolic equations are of little importance and that they are believed to be singularity avoiding. They can however enter certain regions containing trapped surfaces. This provides the possibility of posing initial data containing a trapped surface. Theorems have been proved in these coordinates which are similar to those which had been obtained previously in Schwarzschild coordinates. The theorem stating that the first singularity must be in the centre is technically easier in maximal-isotropic coordinates.

In maximal-isotropic coordinates the spatial metric takes the conformally flat form $A^2(t, R)\delta_{ab}dx^a dx^b$ where the radial coordinate $R = \sqrt{\delta_{ab}x^a x^b}$ does not agree with the area radius r in general. Spherical symmetry implies that the shift vector takes the form $\beta^a = \beta x^a/R$ for some function $\beta(t, R)$. The Einstein equations are:

$$\alpha'' + +2\alpha'/R + A^{-1}A'\alpha' = \alpha A^2\left[\frac{3}{2}K^2 + 4\pi(\rho + \text{tr} S)\right], \quad (11.19)$$

$$(R^2(A^{1/2})')' = -\frac{1}{8}A^{5/2}R^2\left(\frac{3}{2}K^2 + 16\pi\rho\right), \quad (11.20)$$

$$K' + 3(A^{-1}A' + 1/R)K = 8\pi Aj, \quad (11.21)$$

$$\beta' - R^{-1}\beta = \frac{3}{2}\alpha K, \quad (11.22)$$

$$\partial_t A = -\alpha KA + (\beta A)', \quad (11.23)$$

$$\partial_t K = -A^{-2}\alpha'' + A^{-3}A'\alpha' + \alpha[-2A^{-3}A'' + 2A^{-4}A'^2 - 2A^{-3}A'/R \\ - 8\pi S_R + 4\pi \text{tr} S - 4\pi\rho] + \beta K'. \quad (11.24)$$

11.2 Small data global existence

The quantity K is obtained by contracting the second fundamental form on both indices with the unit vector in the time slices orthogonal to the orbits. S_R is obtained from the spatial projection of the energy–momentum tensor in the same way. A prime denotes $\partial/\partial R$. The Vlasov equation is

$$\frac{\partial f}{\partial t} + \left(\alpha A^{-1}\frac{v}{\sqrt{1+|v|^2}} - \beta\frac{x}{R}\right)\cdot\frac{\partial f}{\partial x}$$
$$+ \left[-A^{-1}\alpha'\sqrt{1+|v|^2}\frac{x}{R} - \frac{1}{2}\alpha K\left(v - 3v_r\frac{x}{R}\right)\right.$$
$$\left. - \alpha A^{-2}A'\left(vv_r - |v|^2\frac{x}{R}\right)\frac{1}{\sqrt{1+|v|^2}}\right]\cdot\frac{\partial f}{\partial v} = 0 \qquad (11.25)$$

where $v_r = \delta_{ij}v^i x^j/R$. The mass shell is parametrized using an orthonormal frame and so the energy–momentum tensor takes exactly the same form as in the case of Schwarzschild coordinates. The boundary conditions are that A and α tend to one as $R \to \infty$ for each fixed t.

Consider now initial data for this system where the matter has compact support. In these coordinates it is not so easy to obtain an explicit expression for the ADM mass as it is in Schwarzschild coordinates. It is worth to do so since this provides a conservation law for use in the analysis of the evolution equations. It will now be explained how this can be done. In the vacuum case $(R^3 A^3 K)' = 0$ so that $K = K_0(t)R^{-3}A^{-3}$ for some function $K_0(t)$. Putting this back into (11.20) and using the vacuum condition again gives $(R^2(A^{1/2})')' = O(R^{-4})$ for $R \to \infty$. This can be integrated to give

$$A(t, R) = (1 + A_0(t)R^{-1})^2 + O(R^{-4}). \qquad (11.26)$$

It follows from (11.22) that $(R^{-1}\beta)' = O(R^{-4})$. Hence $\beta = O(R^{-2})$ and $\beta' = O(R^{-3})$. It then follows from (11.23) that $\partial_t A = O(R^{-3})$. Integrating this last equation in time from 0 to t shows that $A(t, R) = A(0, R) + O(R^{-3})$, so that $A_0(t)$ is in fact independent of t. The ADM mass is given by $\lim_{t\to\infty}(-R^2 A')$ and so is seen to be time independent. From (11.20) it can be calculated that the ADM mass is equal to

$$m_{\mathrm{ADM}} = \frac{1}{8}\int_0^\infty A^{5/2}R^2\left(\frac{3}{2}K^2 + 16\pi\rho\right)dR. \qquad (11.27)$$

This formula gives a simple proof that a regular solution of the equations above with ADM mass zero must be flat. For in that case $\rho = 0$ and $K = 0$. The dominant energy condition then shows that the spacetime is vacuum. With this information the other field equations can be integrated to show that $A = 1$, $\alpha = 1$ and $\beta = 0$.

11: The Einstein–Vlasov system

Yet another approach uses double null coordinates where the metric takes the form

$$-\Omega^2 du\, dv + r^2 \gamma_{AB} dx^A dx^B, \tag{11.28}$$

with Ω and r being functions of u and v. Here x^A are local coordinates on the spheres of symmetry and γ_{AB} is the standard metric of the unit sphere. In this form of the metric the expression $du\, dv$ should be interpreted as $\frac{1}{2}(du \otimes dv + dv \otimes du)$. This means that $g_{01} = -\frac{1}{2}\Omega^2$ and $g^{01} = -2\Omega^{-2}$. Every spacetime with surface symmetry belonging to the important class of globally hyperbolic spacetimes can be covered by a coordinate system of this type. Techniques using double null coordinates for spherically symmetric spacetimes with rather general matter models were developed by Dafermos. He proved theorems about global structure subject to a continuation criterion of the following type. Suppose that a solution of the Einstein–matter equations is given on a 'diamond' of the form

$$\{(u, v) : U_0 \le u \le U_1, V_0 \le v \le V_1, (u, v) \ne (U_1, V_1)\}, \tag{11.29}$$

which is in the closure of the non-trapped region where $2m/r < 1$. (Here m is the Hawking mass which is defined below.) Then the solution extends smoothly to an open neighbourhood of the point (U_1, V_1). Whether this criterion holds or not depends on the particular matter model. It has been shown to be true in the case that the matter is described by the Vlasov equation. It cannot be expected to be true for dust or for generic perfect fluids with pressure.

In double null coordinates the Einstein equations take the form

$$\partial_u \partial_v r = -\frac{\Omega^2}{4r} - \frac{1}{r}\partial_u r \partial_v r + 4\pi r T_{uv}, \tag{11.30}$$

$$\partial_u \partial_v (\log \Omega) = -4\pi T_{uv} + \frac{\Omega^2}{4r^2} + \frac{1}{r^2}\partial_u r \partial_v r - \frac{\pi \Omega^2}{r^2}\gamma^{AB} T_{AB}, \tag{11.31}$$

$$\partial_u (\Omega^{-2}\partial_u r) = -4\pi r T_{uu} \Omega^{-2}, \tag{11.32}$$

$$\partial_v (\Omega^{-2}\partial_v r) = -4\pi r T_{vv} \Omega^{-2}. \tag{11.33}$$

The Vlasov equation is

$$p^u \frac{\partial f}{\partial u} + p^v \frac{\partial f}{\partial v} = (\partial_u(\log \Omega^2)(p^u)^2 + 2\Omega^{-2}r\partial_v r \gamma_{AB} p^A p^B)\frac{\partial f}{\partial p^u}$$

$$+ 2r^{-1}(p^u \partial_u r + p^v \partial_v r)p^A \frac{\partial f}{\partial p^A} \tag{11.34}$$

and the energy–momentum tensor is

$$T_{\alpha\beta} = \int_0^\infty \int_{-\infty}^\infty \int_{-\infty}^\infty r^2 p_\alpha p_\beta f(p^u)^{-1} \sqrt{\gamma}\, dp^u dp^A dp^B. \qquad (11.35)$$

The double null form is maintained if u is replaced by a function of u and v by a function of v. This can be useful for simplifying certain problems. It can in particular be used to ensure that the range of the coordinates u and v becomes finite so that the spacetime can be represented by a subset of the (u,v)-plane with compact closure. A representation of a spherically symmetric spacetime in the (u,v) plane is sometimes called a Penrose diagram.

A useful quantity is the Hawking mass which is defined in the spherically symmetric case by

$$m = \frac{r}{2}(1 - \nabla_a r \nabla^a r) = \frac{r}{2}(1 + 4\Omega^{-2}\partial_u r \partial_v r). \qquad (11.36)$$

Notice that the first expression on the right hand side of this equation has a coordinate-independent meaning. The Hawking mass satisfies the equations

$$\partial_u m = 8\pi r^2 \Omega^{-2}(T_{uv}\partial_u r - T_{uu}\partial_v r), \qquad (11.37)$$

$$\partial_v m = 8\pi r^2 \Omega^{-2}(T_{uv}\partial_v r - T_{vv}\partial_u r). \qquad (11.38)$$

In the vacuum case this implies that m is a constant. If the gradient of r is spacelike at some point then the metric can be written in the form (4.26) in neighbourhood of that point. Computing the mass shows that $1 - 2m/r = e^{-2\lambda}$. Putting this into the field equation for μ shows that $\mu = C(1 - 2m/r)$ for a constant C. This proves the special case of Birkhoff's theorem where the gradient of r is spacelike. Suppose that initial data are prescribed for which $\partial_u r < 0$. Then it follows from the field equations that if the null energy condition is satisfied $\partial_u r$ remains negative in the future of the initial hypersurface. If the dominant energy condition holds then the evolution equations for m imply that it decreases to the future along ingoing radial null geodesics and increases to the future along outgoing null geodesics as long as $\partial_v r > 0$. In the situation under consideration the last condition is equivalent to $2m/r < 1$.

11.3 Cosmological solutions

This section is concerned with solutions of the Einstein–Vlasov system which admit a compact Cauchy hypersurface. Almost all the results which have been proved up to the present require the assumption of a symmetry group which is at least two-dimensional. Furthermore the isotropy group is assumed to be of constant dimension. This means that the only possible symmetry types are surface symmetry, T^2-symmetry or specializations of these. Cases where the isotropy group has variable dimension are more difficult but may become tractable in the near future. These are Gowdy symmetry on S^3 or $S^2 \times S^1$, which each have two axes of symmetry, and spherical symmetry on S^3 which has two centres. The latter case is particularly interesting because it could be used to model the formation of black holes in a compact universe. There is one interesting result for this class of spacetimes which says that assuming the DEC and non-negative pressures condition (i.e $T_{\alpha\beta}x^\alpha x^\beta \geq 0$ for all spacelike vectors x^α) the lifetime of any solution is finite. In other words there is a finite upper bound for the lengths of all timelike geodesics in the spacetime. This result is in particular applicable to the Einstein–Vlasov system. The symmetries where the isotropy group has constant dimension are not compatible with the presence of black holes unless the cosmological constant is non-vanishing. For $\Lambda > 0$ there is a spherically symmetric solution on $S^2 \times S^1$, the Schwarzschild–de Sitter solution which does model black holes in a spacetime with a compact Cauchy hypersurface.

11.3.1 Einstein–Vlasov solutions with T^2 symmetry

Consider now spacetimes with T^2 symmetry on a three-torus. The special case of plane symmetry was discussed in Section 5.10. The metric for T^2 symmetry was written in a certain coordinate system (areal coordinates) in Section 4.4. A useful orthonormal frame $\{e_\kappa\}$ can be defined by

$$\alpha^{-1/2} e^{U-\eta} \frac{\partial}{\partial t}, \; e^{U-\eta}\left(\frac{\partial}{\partial \theta} - G\frac{\partial}{\partial x} - H\frac{\partial}{\partial y}\right), \; e^{-U}\frac{\partial}{\partial x}, \; e^{U}t^{-1}\left(\frac{\partial}{\partial y} - A\frac{\partial}{\partial x}\right). \tag{11.39}$$

The momenta can then be parametrized as in Section 5.10. The Einstein equations can be divided into the constraints

$$\frac{\partial_t \eta}{t} = (\partial_t U)^2 + \alpha(\partial_\theta U)^2 + \frac{e^{4U}}{4t^2}((\partial_t A)^2 + \alpha(\partial_\theta A)^2)$$
$$+ \frac{e^{-2\eta}}{4}(e^{4U}\Gamma^2 + t^2(\partial_t H)^2) + e^{2(\eta-U)}\alpha\rho, \tag{11.40}$$

$$\frac{\partial_\theta \eta}{t} = 2\partial_t U \partial_\theta U + \frac{e^{4U}}{2t^2}\partial_t A \partial_\theta A - \frac{\partial_\theta \alpha}{2t\alpha} - e^{2(\eta-U)}\sqrt{\alpha}J_1, \tag{11.41}$$

$$\partial_t \alpha = 2t\alpha^2 e^{2(\eta-U)}(P_1 - \rho) - \alpha t e^{-2\eta}(e^{4U}\Gamma^2 + t^2(\partial_t H)^2) \tag{11.42}$$

11.3 Cosmological solutions

and the evolution equations

$$\partial_t^2 \eta - \alpha \partial_\theta^2 \eta = \frac{\partial_\theta \eta \partial_\theta \alpha}{2} + \frac{\partial_t \eta \partial_t \alpha}{2\alpha} - \frac{(\partial_\theta \alpha)^2}{4\alpha} + \frac{\partial_\theta^2 \alpha}{2} - (\partial_t U)^2 + \alpha(\partial_\theta U)^2$$
$$+ \frac{e^{4U}}{4t^2}((\partial_t A)^2 - \alpha(\partial_\theta A)^2) - e^{-2\eta}t^2(\partial_t H)^2$$
$$- \frac{1}{2}e^{-2\eta}e^{4U}\Gamma^2 - \alpha e^{2(\eta-U)}P_3, \qquad (11.43)$$

$$\partial_t^2 U - \alpha \partial_\theta^2 U = -\frac{\partial_t U}{t} + \frac{\partial_\theta U \partial_\theta \alpha}{2} + \frac{\partial_t U \partial_t \alpha}{2\alpha} + \frac{e^{4U}}{2t^2}((\partial_t A)^2 - \alpha(\partial_\theta A)^2)$$
$$+ \frac{e^{-2\eta}e^{4U}}{2}\Gamma^2 + \frac{e^{2(\eta-U)}\alpha}{2}(\rho - P_1 + P_2 - P_3), \qquad (11.44)$$

$$\partial_t^2 A - \alpha \partial_\theta^2 A = \frac{\partial_t A}{t} + \frac{\partial_\theta \alpha \partial_\theta A}{2} + \frac{\partial_t \alpha \partial_t A}{2\alpha} - 4\partial_t A \partial_t U + 4\alpha \partial_\theta A \partial_\theta U$$
$$+ t^2 e^{-2\eta}\Gamma \partial_t H + 2t\alpha e^{2(\eta-2U)}S_{23}. \qquad (11.45)$$

There are also some auxiliary equations

$$\partial_\theta [e^{-2\eta}\alpha^{-1/2}e^{4U}\Gamma] = -2e^\eta J_2, \qquad (11.46)$$

$$\partial_t [e^{-2\eta}t\alpha^{-1/2}e^{4U}\Gamma] = 2t\alpha^{1/2}e^\eta S_{12}, \qquad (11.47)$$

$$\partial_\theta [e^{-2\eta}\alpha^{-1/2}(Ae^{4U}\Gamma + t^2 \partial_t H)] = -2e^\eta A J_2 - 2te^{\eta-2U}J_3, \qquad (11.48)$$

$$\partial_t [e^{-2\eta}t\alpha^{-1/2}(Ae^{4U}\Gamma + t^2 \partial_t H)] = 2t\alpha^{1/2}e^\eta (AS_{12} + te^{-2U}S_{13}). \qquad (11.49)$$

This is the form of the Einstein equations in a spacetime of this type with any matter model. The matter quantities in the equations are defined by $\rho = T(e_0, e_0)$, $J_i = -T(e_0, e_i)$, $P_1 = T(e_i, e_i)$, $S_{ij} = T(e_i, e_j)$.

The Vlasov equation is

$$\frac{\partial f}{\partial t} + \frac{\sqrt{\alpha}v^1}{v^0}\frac{\partial f}{\partial \theta} - \left[\left(\partial_\theta \eta - \partial_\theta U + \frac{\partial_\theta \alpha}{2\alpha}\right)\sqrt{\alpha}v^0 + (\partial_t \eta - \partial_t U)v^1 \right.$$
$$\left. - \frac{\sqrt{\alpha}e^{2U}\partial_\theta A\, v^2 v^3}{t} + \frac{\sqrt{\alpha}\partial_\theta U}{v^0}((v^3)^2 - (v^2)^2) + e^{-\eta}(e^{2U}\Gamma v^2 + t\partial_t H v^3)\right]$$
$$\times \frac{\partial f}{\partial v^1} - \left[\partial_t U v^2 + \sqrt{\alpha}\partial_\theta U \frac{v^1 v^2}{v^0}\right]\frac{\partial f}{\partial v^2}$$
$$- \left[\left(\frac{1}{t} - \partial_t U\right)v^3 - \sqrt{\alpha}\partial_\theta U \frac{v^1 v^3}{v^0} + \frac{e^{2U}v^2}{t}\left(\partial_t A + \sqrt{\alpha}\partial_\theta A\frac{v^1}{v^0}\right)\right]\frac{\partial f}{\partial v^3} = 0.$$
$$(11.50)$$

The energy–momentum tensor of the collisionless matter has the standard form for an orthonormal frame.

These equations can in principle be computed by making suitable specializations in the general 3+1 equations presented in Section 2.3. It should, however, be noted that in practice this is an arduous task. The equations just presented for the general case of T^2 symmetry were obtained with the help of computer algebra.

The first step in analysing the solutions of these equations is to obtain a local existence theorem. This could be done by constructing an iteration and proving its convergence as in the case of plane symmetry. Here another strategy will be presented. There are several steps and it is inconvenient to keep track of the differentiability in the process. For this reason only the case of C^∞ initial data will be discussed. The first step is to forget the symmetry and to apply a general local existence theorem for the Einstein–Vlasov system. Next it can be concluded from another general result that there is a Cauchy evolution of the prescribed data which has the symmetry of the initial data. Then it can be proved that the coordinates which are to be used can be introduced in that spacetime. The conclusion is a local existence and uniqueness theorem for the system of evolution equations in T^2 symmetry. The solution, like the initial datum, is C^∞. This proof does not provide a continuation criterion. A weak one can be obtained as follows. Consider a solution on a time interval $[t_0, T)$. If it can be shown that the unknowns and their derivatives of all orders with respect to all variables together with the maximum velocity $P(t)$ in the support of f at time t are uniformly bounded then the solution can be extended to a longer time interval. For all quantities are uniformly continuous and thus converge uniformly together with their derivatives to smooth quantities on the hypersurface $t = T$. These can be taken as new initial data which determines a local evolution on an interval $[T, T + \epsilon)$. The solutions on the intervals $[t_0, T)$ and $[T, T + \epsilon)$ match smoothly to give a solution on the interval $[t_0, T + \epsilon)$. To get a stronger and more useful continuation criterion it is necessary to start with a bound on a small number of quantities and derive estimates which imply the bounds of all higher order derivatives. It is typical for problems like this one that once bounds have been obtained for $P(t)$, the C^1 norm of f and the C^2 norm of the metric coefficients, it is routine if tiresome to bound the higher derivatives. The reason is that in this situation the higher derivatives solve linear systems with bounded coefficients. This localizes the estimates which are specific to the particular problem under consideration.

The following discussion concerns the equations written in an areal time coordinate and the initial data are given on a hypersurface of constant areal time. Define $P(t)$ to be the maximum momentum of any particle at time t. As in other cases of Vlasov systems it is an important diagnostic quantity. A

11.3 Cosmological solutions

global existence theorem in the future for the T^2-symmetric Einstein–Vlasov system will now be sketched. A basic tool in the proof is an estimate for the following energy functional which is similar to the one used in Section 5.10. Its is defined as follows:

$$E(t) = \int_{S^1} \left[\alpha^{-\frac{1}{2}}(\partial_t U)^2 + \sqrt{\alpha}(\partial_\theta U)^2 + \frac{e^{4U}}{4t^2}(\alpha^{-\frac{1}{2}}(\partial_t A)^2 + \sqrt{\alpha}(\partial_\theta A)^2) \right.$$

$$\left. + \frac{e^{-2\eta}\alpha^{-1/2}}{4}(e^{4U}\Gamma^2 + t^2(\partial_t H)^2) + \sqrt{\alpha}e^{2(\eta-U)}\rho \right] d\theta. \quad (11.51)$$

This quantity is monotone non-increasing and satisfies

$$\frac{d}{dt}E(t) = -\frac{2}{t}\int_{S^1}\left[\frac{(\partial_t U)^2}{\sqrt{\alpha}} + \frac{e^{4U}}{4t^2}\sqrt{\alpha}(\partial_\theta A)^2 + \frac{e^{-2\eta}}{4\sqrt{\alpha}}(e^{4U}\Gamma^2 + 2t^2(\partial_t H)^2)\right.$$

$$\left. + \frac{\sqrt{\alpha}}{2}e^{2(\eta-U)}(\rho + P_3)\right]d\theta \leq 0. \quad (11.52)$$

It is convenient to introduce the quantity $\tilde{\eta} = \eta + \frac{1}{2}\log\alpha$. The difference between $\tilde{\eta}$ at two values of θ at a given time t can be estimated by integrating the field equation for $\partial_\theta\eta$ in θ and using the energy bound. Next the integral of $\tilde{\eta}(t,\theta)$ with respect to θ is estimated using the field equation for $\partial_t\eta$. The boundedness of the energy quantity is applied once again. The fact that α is non-increasing in t is also used. The bounds just obtained for the oscillation of $\tilde{\eta}$ (difference between its values at two different points) and its integral give a bound for $\tilde{\eta}$ itself on any finite time interval (t_0, T). Next U and A are bounded successively. Again the basic idea is to bound the oscillation and the integral in space of the quantity in question. After this the quantities G and H are bounded using the auxiliary equations. The solutions of the characteristic system for the Vlasov equation satisfy an angular momentum conservation law following directly from the symmetry. It says that the quantities $v^2 e^U$ and $v^2 A e^U + v^3 t e^{-U}$ are bounded. With the bounds for the metric quantities already stated this gives the boundedness of v^2 and v^3 along the characteristics. Thus controlling the support of the distribution function f reduces to controlling its diameter in the v^1 direction.

Having estimated these basic quantities the next step is to estimate their first order derivatives. This is the stage at which the hyperbolic nature of the Einstein equations plays a role. The derivatives in null directions satisfy propagation equations along null directions. These imply integral inequalities for the size of these quantities where the quantity $P(t)$ occurs. Using the Vlasov equation an integral inequality can be derived for $P(t)$ where a quantity measuring the size of the first order derivatives of the metric components occurs. In this way a system of two integral inequalities is derived which is sufficient to ensure the boundedness of the quantities concerned on any finite time interval. It is of the form

$$F(t) \leq C \int_{t_0}^{t} (1 + F(s)) \log F(s) ds, \tag{11.53}$$

for a non-negative function F. This can be compared with a solution of the differential equation $\dot{u} = C(1 + u) \log u$. To show that a solution of this equation is bounded on any interval of the form $[t_0, T]$ it is enough to consider an interval where $u \geq 1$ and thus to consider the simplified ODE $\dot{u} = Cu \log u$. The explicit solution of the latter equation is $u(t) = \exp(\exp(C(t - t_1)))$ for a constant t_1. Thus the solution of a differential inequality of the type under consideration can show at most double exponential growth which is faster than the exponential growth obtained from Gronwall's inequality. It nevertheless suffices to imply the boundedness of the solution on any finite time interval. In fact this argument does not bound all first derivatives of the metric coefficients. It remains to bound $\partial_\theta \alpha$ and $\partial_\theta \eta$. This is done together with bounding the first order derivatives of the distribution function f. This requires making use of the fact that the propagation equations for the first derivatives of the characteristic system are essentially the geodesic deviation equation or the equation for Jacobi fields, which leads to certain cancellations as in Section 11.2. Once the first derivatives of f have been bounded the estimates for higher order derivatives are relatively routine and lead to a proof of global existence.

This proof gives no information at all about the asymptotics of the solution in the limit $t \to \infty$. In view of the results in the plane symmetric case it would seem reasonable to hope to control the asymptotic behaviour in the case that a positive cosmological constant is present. A global existence result like that just discussed has been proved for vacuum spacetimes with positive cosmological constant [63]. For data close to de Sitter data the general theorem of [82] applies to provide detailed asymptotics. Translating the general results into the coordinates used above gives the following leading terms:

$$\alpha(t, \theta) = \frac{1}{4H^2} e^{-2\eta_0(\theta)} t^{-3}(1 + o(1)), \tag{11.54}$$

$$\eta(t, \theta) = \log t + \eta_0(\theta) + o(1), \tag{11.55}$$

$$U(t, \theta) = \frac{1}{2} \log t + U_0(\theta) + o(1), \tag{11.56}$$

$$A(t, \theta) = A_0(\theta) + o(1), \tag{11.57}$$

$$G(t, \theta) = G_0(\theta) + o(1), \tag{11.58}$$

$$H(t, \theta) = H_0(\theta) + o(1). \tag{11.59}$$

For a given solution these asymptotic expansions could be simplified by a coordinate transformation. If, however, it is only required that the general conditions defining areal coordinates are satisfied the full complexity is required. For the case with collisionless matter and positive Λ no results are available yet beyond plane symmetry. It may be conjectured that the leading order asymptotics are as in the vacuum case but there is no proof of this yet. As is discussed in Section 11.3.5 global existence in an areal time coordinate does itself imply one important geometrical fact, namely that the solution cannot be extended further to the future. Thus it is enough to show the inextendibility of the maximal Cauchy development which is desired for cosmic censorship to hold in the future. In other words, it reduces the task of proving strong cosmic censorship to understanding the structure of the initial singularity [68].

The results proved directly using an areal time coordinate above only apply to the case in which initial data is prescribed on a hypersurface where the area radius is constant. Quite general initial hypersurfaces (sharing the symmetry of the spacetime) can be treated using conformal coordinates. These can be adapted so as to make any given Cauchy hypersurface take the form $t = t_0$. In conformal coordinates the metric is

$$g = e^{2(\eta-U)}(-dt^2 + d\theta^2) + e^{2U}[dx + Ady + (G+AH)d\theta]^2 + e^{-2U}R^2[dy + Hd\theta]^2. \tag{11.60}$$

An adapted orthonormal frame is given by

$$e^{U-\eta}\frac{\partial}{\partial t}, e^{U-\eta}\left(\frac{\partial}{\partial \theta} - G\frac{\partial}{\partial x} - H\frac{\partial}{\partial y}\right), e^{-U}\frac{\partial}{\partial x}, e^{U}R^{-1}\left(\frac{\partial}{\partial y} - A\frac{\partial}{\partial x}\right). \tag{11.61}$$

The Einstein equations can be divided into the constraints

$$(\partial_t U)^2 + (\partial_\theta U)^2 + \frac{e^{4U}}{4R^2}((\partial_t A)^2 + (\partial_\theta A)^2) + \frac{\partial_\theta^2 R}{R} - \frac{\partial_t \eta \partial_t R}{R} - \frac{\partial_\theta \eta \partial_\theta R}{R}$$
$$= -\frac{e^{-2\eta+4U}}{4}\Gamma^2 - \frac{R^2 e^{-2\eta}}{4}(\partial_t H)^2 - e^{2(\eta-U)}\rho, \tag{11.62}$$

$$2\partial_t U \partial_\theta U + \frac{e^{4U}}{2R^2}\partial_t A \partial_\theta A + \frac{\partial_t \partial_\theta R}{R} - \frac{\partial_t \eta \partial_\theta R}{R} - \frac{\partial_\theta \eta \partial_t R}{R} = e^{2(\eta-U)}J_1 \tag{11.63}$$

and the evolution equations

$$\partial_t^2 U - \partial_\theta^2 U = \frac{\partial_\theta U \partial_\theta R}{R} - \frac{\partial_t U \partial_t R}{R} + \frac{e^{4U}}{2R^2}((\partial_t A)^2 - (\partial_\theta A)^2) + \frac{e^{-2\eta+4U}}{2}\Gamma^2$$
$$+ \frac{1}{2}e^{2(\eta-U)}(\rho - P_1 + P_2 - P_3), \tag{11.64}$$

$$\partial_t^2 A - \partial_\theta^2 A = \frac{\partial_t R \partial_t A}{R} - \frac{\partial_\theta R \partial_\theta A}{R} + 4(\partial_\theta A \partial_\theta U - \partial_t A \partial_t U) + R^2 e^{-2\eta}\Gamma \partial_t H$$
$$+ 2Re^{2(\eta-2U)}S_{23}, \tag{11.65}$$

$$\partial_t^2 R - \partial_\theta^2 R = Re^{2(\eta-U)}(\rho - P_1) + \frac{Re^{-2\eta+4U}}{2}\Gamma^2 + \frac{R^3 e^{-2\eta}}{2}(\partial_t H)^2, \tag{11.66}$$

$$\partial_t^2 \eta - \partial_\theta^2 \eta = (\partial_\theta U)^2 - (\partial_t U)^2 + \frac{e^{4U}}{4R^2}((\partial_t A)^2 - (\partial_\theta A)^2) - \frac{e^{-2\eta+4U}}{4}\Gamma^2$$
$$- \frac{3R^2 e^{-2\eta}}{4}(\partial_t H)^2 - e^{2(\eta-U)}P_3. \tag{11.67}$$

As in the case of areal coordinates there are some auxiliary equations

$$\partial_\theta [Re^{-2\eta+4U}\Gamma] = -2Re^\eta J_2, \tag{11.68}$$

$$\partial_t [Re^{-2\eta+4U}\Gamma] = 2Re^\eta S_{12}, \tag{11.69}$$

$$\partial_\theta (R^3 e^{-2\eta}\partial_t H) + Re^{-2\eta+4U}\partial_\theta A\Gamma = -2R^2 e^{\eta-2U}J_3, \tag{11.70}$$

$$\partial_t (R^3 e^{-2\eta}\partial_t H) + Re^{-2\eta+4U}\partial_t A\Gamma = 2R^2 e^{\eta-2U}S_{13}. \tag{11.71}$$

The Vlasov equation is

$$\frac{\partial f}{\partial t} + \frac{v^1}{v^0}\frac{\partial f}{\partial \theta} - \left[(\partial_\theta \eta - \partial_\theta U)v^0 + (\partial_t \eta - \partial_t U)v^1 - \partial_\theta U\frac{(v^2)^2}{v^0}\right.$$
$$+ \left(\partial_\theta U - \frac{\partial_\theta R}{R}\right)\frac{(v^3)^2}{v^0} - \frac{\partial_\theta A}{R}e^{2U}\frac{v^2 v^3}{v^0} + e^{-\eta}(e^{2U}\Gamma v^2 + R\partial_t H v^3)\right]\frac{\partial f}{\partial v^1}$$
$$- \left[\partial_t U v^2 + \partial_\theta U\frac{v^1 v^2}{v^0}\right]\frac{\partial f}{\partial v^2} - \left[\left(\frac{\partial_t R}{R} - \partial_t U\right)v^3 - \left(\partial_\theta U - \frac{\partial_\theta R}{R}\right)\frac{v^1 v^3}{v^0}\right.$$
$$+ \left.\frac{e^{2U}v^2}{R}\left(\partial_t A + \partial_\theta A\frac{v^1}{v^0}\right)\right]\frac{\partial f}{\partial v^3} = 0. \tag{11.72}$$

Local existence for these equations can be obtained by a procedure analogous to that used in the case of areal coordinates. The equation (11.66) can be used to throw some light on the compatibility condition for an areal time coordinate with conformal coordinates mentioned in Section 4.4. This equation implies that if $R = t$ it must be the case that $\rho = P_1$ and that certain quantities related to the twist must vanish. The remarks in Section 4.4 concerned only the case of Gowdy symmetry and in that case the only condition is $\rho = P_1$ which is equivalent to what is stated there.

In conformal coordinates the conformal structure is that of a subset of flat space. In fact there is little difference between these coordinates and double null coordinates. It is possible to identify the past Cauchy development of initial data with a subset of the half cylinder $(-\infty, t_0] \times S^1$. There are two possibilities. Either the subset is the whole of the half cylinder or it is not. In the second case the area function R converges to a constant as the boundary of the subset is approached. Since it can be shown that the gradient of R is always spacelike it follows that the boundary is compact. Further estimates show that the limiting value of R on the boundary is zero. In the case where the past Cauchy development extends to $t = -\infty$ the function R also has a limit which is zero except in the case of the flat Kasner solution. In either case the existence of a Cauchy hypersurface of constant R can be concluded. These results using conformal coordinates apply to plane symmetry which is a special case of T^2 symmetry. They also apply in part to the case of hyperbolic symmetry. The existence of a Cauchy hypersurface of constant areal time and the fact that R converges to a constant value hold also in that case. It is not, however, known how generally the limiting value of R in the past is zero. There is a set of solutions, the pseudo-Schwarzschild solutions, where it is not always zero. The metric of those solutions is

$$-(1 - 2m/t)^{-1}dt^2 + (1 - 2m/t)dr^2 + t^2(d\theta^2 + \sinh^2\theta d\phi^2). \quad (11.73)$$

They satisfy the vacuum Einstein equations with zero cosmological constant. For $m > 0$ there is a curvature singularity at $t = 0$ and the limiting value of R in the past is zero. For $m < 0$ the limiting value is strictly positive and the solution can be extended smoothly to the past through a smooth Cauchy horizon. For $m = 0$ it is a flat spacetime of Bianchi type III. It is isometric to the product of S^1 with a Lorentz manifold whose universal cover is isometric to the interior of the light cone in three-dimensional Minkowski space. The universal cover can be extended smoothly to the past but this is not true if the spatial topology is chosen to be compact. The group which is used to do the identification acts on the null cone in a way which destroys its smooth structure by the identification.

It is interesting to note that there are plane symmetric solutions of the Einstein–Euler equations with linear equations of state where R does not go to zero in the past. These are some of the homogeneous solutions. They are LRS Bianchi type I. As mentioned in Section 5.5.1 there are solutions where the generalized Kasner exponents converge to $(1, 0, 0)$ as the singularity is approached. A closer examination of these leads to the conclusion that while the Kretschmann scalar blows up the metric can be extended so as to be continuous and non-degenerate. This is a weak null singularity, a concept discussed further in Section 12.2.

11.3.2 T^2 symmetry and CMC time

There is yet another coordinate system which has been used to study T^2-symmetric solutions of the Einstein–Vlasov system. This is the closest equivalent to the maximal-isotropic coordinates used in the asymptotically flat spherically symmetric case. The time coordinate is defined by hypersurfaces of constant mean curvature. The mean curvature varies in a monotone way from one of these hypersurfaces to the next and so can itself be used as a time coordinate. This is referred to as a constant mean curvature (CMC) time coordinate. The spatial metric cannot be made entirely isotropic. There remains an extra constant on each time slice which cannot be set to a standard value. This gives rise to the function $a(t)$ in the metric given below. The form of the metric is:

$$-\alpha^2 dt^2 + A^2[(dx + \beta^1 dt)^2 + a^2 \tilde{g}_{AB}(dy^A + \beta^A dt)(dy^B + \beta^B dt)]. \quad (11.74)$$

The metric $\tilde{g}_{AB}$ is parametrized in terms of two functions W and V in the following way:

$$\tilde{g}_{22} = e^W \cosh V, \quad \tilde{g}_{33} = e^{-W} \cosh V, \quad \tilde{g}_{23} = \sinh V. \quad (11.75)$$

This corresponds to a coordinate system on the hyperbolic plane different from that defined by the variables P and Q which come up in the context of the usual parametrization of Gowdy spacetimes. Some of the the Einstein equations are

$$\partial_x^2(A^{1/2}) = -\frac{1}{8}A^{5/2}\left[\frac{3}{2}\left(K - \frac{1}{3}t\right)^2 - \frac{2}{3}t^2\right.$$
$$\left. + 2\eta_A\eta^A + \tilde{\kappa}^{AB}\tilde{\kappa}_{AB} + \tilde{\lambda}^{AB}\tilde{\lambda}_{AB} + 16\pi\rho\right], \quad (11.76)$$

$$\partial_x^2\alpha + A^{-1}\partial_x A \partial_x \alpha = \alpha A^2\left[\frac{3}{2}\left(K - \frac{1}{3}t\right)^2 + \frac{1}{3}t^2 + 2\eta_A\eta^A\right.$$
$$\left. + \tilde{\kappa}_{AB}\tilde{\kappa}^{AB} + 4\pi(\rho + \mathrm{tr}S)\right] - A^2, \quad (11.77)$$

$$\partial_x K + 3A^{-1}\partial_x AK - A^{-1}\partial_x At - \tilde{\kappa}^{AB}\tilde{\lambda}_{AB} = 8\pi JA, \quad (11.78)$$

$$\partial_x \beta^1 = -a^{-1}\partial_t a + \frac{1}{2}\alpha(3K - t), \tag{11.79}$$

$$\partial_t a = a[-\partial_x \beta^1 + \frac{1}{2}\alpha(3K - t)], \tag{11.80}$$

$$\partial_t A = -\alpha K A + \partial_x(\beta^1 A). \tag{11.81}$$

Here κ_{AB} is the second fundamental form of the surfaces of symmetry corresponding to the normal to the hypersurface of constant t, λ_{AB} is the second fundamental form of the surfaces of symmetry corresponding to the normal within the hypersurfaces of constant t and η^A is a one-form representing the normal connection. It can be defined in the following way:

$$\eta_a = k_{cb}(\delta_a^c + N^c N_a)N^b. \tag{11.82}$$

Here N^a is the unit normal vector to an orbit within a hypersurface of constant time. There seems to be no advantage in expressing the quantities κ_{AB}, λ_{AB} and η_A in terms of more elementary quantities – in the original existence proof it was never done. The tracefree parts of κ_{AB} and λ_{AB} are denoted by $\tilde{\kappa}_{AB}$ and $\tilde{\lambda}_{AB}$ respectively. In the case of plane symmetry these are all the Einstein equations. Those which have not been included in the general case are those which describe the hyperbolic aspects of the propagation of the gravitational field and those which describe the behaviour of η_A. The latter correspond to those equations which were called auxiliary equations in areal and conformal coordinates. The hyperbolic equations are

$$\nabla^a(r^2\nabla_a W) = -2r^2 \tanh V \nabla^a W \nabla_a V - r^2(\cosh V)^{-1}[e^{-W}T_{22} - e^W T_{33}$$
$$- \frac{1}{2}(e^{-W}(\eta_2)^2 - e^W(\eta_3)^2)], \tag{11.83}$$

$$\nabla^a(r^2\nabla_a V) = r^2 \cosh V \sinh V \nabla^a W \nabla_a W$$
$$- 2r^2(\cosh V)^{-1}\left[\left(T_{23} - \frac{1}{2}\tilde{h}^{AB}T_{AB}\tilde{g}_{23}\right)\right.$$
$$\left. - \frac{1}{2}\left(\eta_2\eta_3 - \frac{1}{2}(\tilde{h}^{AB}\eta_A\eta_B)\tilde{g}_{23}\right)\right]. \tag{11.84}$$

Here the lower case Latin indices refer to objects which live on the quotient of the spacetime by the symmetry group.

The maximal range of a CMC time coordinate in a spacetime with local T^2 symmetry is $(-\infty, 0)$. This is because a maximal hypersurface (trk = 0) is ruled out by an argument already seen in a special case in Section 4.2. Consider a maximal hypersurface in a spacetime satisfying the weak energy condition. Then the Hamiltonian constraint implies that its scalar curvature is non-negative. If the hypersurface is compact it follows by general results about scalar curvature that the manifold is Yamabe positive or zero. If it is Yamabe zero then the metric must in fact be flat. In that case the second fundamental form and the energy density both vanish on the hypersurface. If the Einstein–matter equations are satisfied for a matter model satisfying the dominant energy condition then it can be concluded that the spacetime is flat. Thus non-trivial solutions with a maximal hypersurface are only possible on manifolds which are Yamabe positive. Neither the torus nor the twisted topologies with local T^2 symmetry have this property and so maximal hypersurfaces are ruled out in the case of present interest.

For the Einstein–Vlasov system data with T^2 symmetry on a hypersurface of constant mean curvature $t_0 < 0$ give rise to a spacetime which exists on the interval $(-\infty, t_0)$. The CMC hypersurfaces cover the entire past of the initial hypersurface in the maximal Cauchy development. These facts imply in particular that the mean curvature tends to $-\infty$ in the past so that the past singularity is crushing. This theorem does not come with much more information about the asymptotics of the solutions in the past. A local existence theorem is obtained by using the method already described of forgetting the symmetry, using a general existence theorem and then showing that the coordinates of interest can be introduced in a neighbourhood of the initial hypersurface. This uses a general theorem on the existence of CMC hypersurfaces close to a given one. Global existence in the past is then shown by proving the boundedness of all geometric and matter quantities together with their derivatives of all orders on any interval where a solution exists. A pleasant property of these coordinates is that many quantities can be bounded on the basis of energy conditions alone, without using information about the particular type of matter. Some of these general estimates will now be presented.

In the class of spacetimes with T^2 symmetry it is possible to define an area radius r to be the square root of the area of the group orbits. Furthermore the Hawking mass can be defined in this case by $m = -\frac{1}{2}\nabla_\alpha r \nabla^\alpha r$. It is also of interest to consider the expansions θ and θ' of the null geodesics starting orthogonal to the orbits. It can be shown using estimates of Malec and O Murchadha [142] that θ and θ' can be bounded by $4|\mathrm{tr} k|$. It can also be shown assuming only the dominant energy condition that the Hawking mass is everywhere non-negative and is only zero if the spacetime is flat. This means that excluding the flat case the gradient of r is always timelike. It can be concluded that r increases to the future along any timelike curve. This implies that it is bounded on the past of the initial hypersurface. Another useful estimate is one which holds for any CMC time coordinate with compact level hypersurfaces in a spacetime satisfying the strong energy condition. The lapse function of a foliation of this kind satisfies the

equation:
$$-\Delta\alpha + \alpha[k_{ab}k^{ab} + 4\pi(\rho + \mathrm{tr}S)] = 1. \tag{11.85}$$

At a point where α attains its maximum $\Delta\alpha \leq 0$ and so $\frac{1}{3}\alpha(\mathrm{tr}k)^2 \leq 1$ and $\alpha \leq \frac{3}{(\mathrm{tr}k)^2}$. The bounds which have been obtained up to now can be used in conjunction with the field equations to show that on any time interval of the form $(t_1, t_2]$ the following quantities are bounded:

$$\alpha, \partial_x\alpha, A, A^{-1}, \partial_x A, K, \beta^1, a, a^{-1}, \partial_t a, \partial_t A, \partial_x \beta^1. \tag{11.86}$$

One important step in proving this is to integrate the equation (11.76) in space. This leads to bounds on the integrals $\int_0^{2\pi} \rho$ and $\int_0^{2\pi} (2\eta_A \eta^A + \tilde{\kappa}^{AB}\tilde{\kappa}_{AB} + \tilde{\lambda}^{AB}\tilde{\lambda}_{AB})$.

The field equations (11.83) and (11.84) which are used to control V and W are hyperbolic. These quantities may be thought of as describing gravitational waves. The fact that these equations are coupled with the matter equations and with themselves means intuitively that the waves interact with the matter and with each other. Let S_W and S_V denote the right hand sides of these equations. It will be shown that the modulus of each of these quantities can be bounded by a constant multiple of the expression

$$\rho + \eta_A \eta^A + \tilde{\kappa}_{AB}\tilde{\kappa}^{AB} + \tilde{\lambda}_{AB}\tilde{\lambda}^{AB}, \tag{11.87}$$

which implies that the L^1 norms of S_W and S_V in space are bounded by a constant which does not depend on time. For this purpose it is necessary to calculate the lengths of $\tilde{\lambda}_{AB}$ and $\tilde{\kappa}_{AB}$ explicitly in terms of W and V.

$$\tilde{\lambda}_{AB}\tilde{\lambda}^{AB} = \frac{1}{2}A^{-2}(\cosh^2 V(\partial_x W)^2 + (\partial_x V)^2), \tag{11.88}$$

$$\tilde{\kappa}_{AB}\tilde{\kappa}^{AB} = \frac{1}{2}\alpha^{-2}[\cosh^2 V(\partial_t W - \beta^1\partial_x W)^2 + (\partial_t V - \beta^1\partial_x V)^2]. \tag{11.89}$$

This shows that the first term on the right hand side of each of the equations (11.83) and (11.84) can be bounded by $\tilde{\lambda}_{AB}\tilde{\lambda}^{AB} + \tilde{\kappa}_{AB}\tilde{\kappa}^{AB}$. To bound the other terms on the right hand side of (11.83) and (11.84), define an orthonormal frame on each orbit by:

$$e_2 = (Aa)^{-1}(e^{-W/2}\cosh(V/2)\partial/\partial y^2 - e^{W/2}\sinh(V/2)\partial/\partial y^3), \tag{11.90}$$

$$e_3 = (Aa)^{-1}(-e^{-W/2}\sinh(V/2)\partial/\partial y^2 + e^{W/2}\cosh(V/2)\partial/\partial y^3). \tag{11.91}$$

Then

$$e^{-W/2}\partial/\partial y^2 = Aa[\cosh(V/2)e_2 + \sinh(V/2)e_3], \tag{11.92}$$

$$e^{W/2}\partial/\partial y^3 = Aa[\sinh(V/2)e_2 + \cosh(V/2)e_3]. \tag{11.93}$$

11 : The Einstein–Vlasov system

The components of the covector η_A expressed in an orthormal frame can be bounded in terms of $\eta^A \eta_A$. Thus if the latter expression is bounded it follows that the components of η_A expressed with respect to the basis consisting of ($e^{-W/2} \partial/\partial y^2$ and $e^{W/2} \partial/\partial y^3$) can be bounded by a constant multiple of $\cosh(V/2)$ or, equivalently, by a constant multiple of $(\cosh V)^{1/2}$. This means that $e^{-W/2}\eta_2$ and $e^{W/2}\eta_3$ can be bounded by $C\eta^A \eta_A (\cosh V)^{1/2}$ for some constant C. This allows the expressions on the right hand side of equations (11.83) and (11.84) containing η_A to be bounded in modulus by a constant multiple of $\eta_A \eta^A$. The terms involving the energy–momentum tensor can be handled in a very similar way. The dominant energy condition implies that the components of the energy–momentum tensor in an orthonormal frame are bounded in modulus by ρ and using (11.92) and (11.93) allows this to be translated into a bound on the matter terms on the right hand side of equations (11.83) and (11.84) in terms of ρ.

The next step is to bound the quantities W, V, η_A, β^A and $\partial_x \beta^A$. The main thing is to write integral formulae for the quantities $r^2 W$ and $r^2 V$. Fix a point (t, x) in the domain of the solution and let γ_1 and γ_2 be the two future directed characteristics of equation (11.83) (or (11.84)) through this point. Together with the initial hypersurface they form a triangle T. Integrate equation (11.83) over T and apply Stokes' theorem. This leads to a term on the initial surface, which can be bounded in terms of the initial data, a surface integral and terms on the characteristics. The surface integral can be bounded by the integral in time of an L^1 norm in space whose boundedness has already been discussed. The terms on the characteristics can be treated by integration by parts. This leads to the desired bound for W and V can be handled in an analogous way.

The only properties of the matter model used in the estimates up to now are the dominant and strong energy conditions. To go further the equations of motion of the matter, in the present case the Vlasov equation, must be used. The coefficients in the Vlasov equation, which are defined by rotation coefficients, are complicated and in fact the necessary estimates can be obtained without ever writing them out explicitly in detail. Knowing certain qualitative properties of the coefficients suffices. One important fact is that the existence of two local Killing vectors leads to two conserved quantities for solutions of the characteristic equations of the Vlasov equation, i.e. for geodesics. Explicitly these conserved quantities take the form

$$Aae^{W/2}[\cosh(V/2)v^2 + \sinh(V/2)v^3], \qquad (11.94)$$

$$Aae^{-W/2}[\sinh(V/2)v^2 + \cosh(V/2)v^3]. \qquad (11.95)$$

Using these and the boundedness of V and W shows that v^2 and v^3 are bounded along characteristics. Let $P(t)$ be the supremum of $|v|$ over the support of $f(t)$. Since a, A, W and V have already been controlled pointwise the components T_{AB} of the energy–momentum tensor occurring on the right hand side of (11.83) and (11.84) can be estimated in terms of the corresponding frame components. Looking at the explicit expressions for these frame components and using the boundedness of v^2 and v^3 in the support of f shows that:

$$\|T_{AB}(t)\|_\infty \le CP(t), \qquad (11.96)$$

where C is a constant which only depends on the initial data. To make use of (11.96) an estimate for v^1 must be obtained. Define:

$$Q(t) = \|\partial_x W(t)\|_\infty + \|\partial_t W(t)\|_\infty + \|\partial_x V(t)\|_\infty + \|\partial_t V(t)\|_\infty. \qquad (11.97)$$

It is possible to prove a differential inequality of the following form:

$$1 + P(t_1) \le C\left(1 + P(t) + \int_0^{t-t_1} 1 + P(t-s) + Q(t-s)ds\right). \qquad (11.98)$$

In doing this it is important to observe that terms in the characteristic system which depend on derivatives of V and W do so linearly. Furthermore, the combinations of components of the velocity which occur in the rotation coefficients are of the form $v^i v^j / v^0$. Due to the boundedness of v^2 and v^3 the only case in which this is not known to be bounded is the case $i = j = 1$. Fortunately the rotation coefficients γ^1_{i1} vanish and so the bad case makes no contribution.

In order to find a pointwise estimate for the first derivatives of W and V it is useful to rewrite (11.83) and (11.84) in a slightly different way.

$$\nabla^a \nabla_a W + 2\tanh V \nabla^a W \nabla_a V = -(2/r)\nabla^a r \nabla_a W$$

$$- (\cosh V)^{-1}\left[e^{-W}T_{22} - e^W T_{33}\frac{1}{2}(e^{-W}(\eta_2)^2 - e^W(\eta_3)^2)\right], \qquad (11.99)$$

$$\nabla^a \nabla_a V - \sinh V \cosh V \nabla^a W \nabla_a W = -(2/r)\nabla^a r \nabla_a V$$

$$- 2(\cosh V)^{-1}\left[\left(T_{23} - \frac{1}{2}\tilde{h}^{AB}T_{AB}\tilde{g}_{23}\right) - \frac{1}{2}\left(\eta_2\eta_3 - \frac{1}{2}(\tilde{h}^{AB}\eta_A\eta_B)\tilde{g}_{23}\right)\right]. \qquad (11.100)$$

The advantage of this is that if the right hand sides of (11.99) and (11.100) are replaced by zero the resulting equations are those for a wave map with target space $\mathbb{R}^2$, endowed with the metric $\cosh^2 V dW^2 + dV^2$. This is a representation of the standard metric of the hyperbolic plane in a certain coordinate system. It is natural to try to generalize estimates which have been used in the study of wave maps to the present situation. This can be done with an estimate of Gu [94], who used it to prove global existence of classical solutions in the Cauchy problem for wave maps defined on two-dimensional Minkowski space. Define two null vectors on the two-dimensional space coordinatized by t and r by

$$e_+ = \alpha^{-1}(\partial/\partial t - \beta\partial/\partial x) + A^{-1}\partial/\partial x, \qquad (11.101)$$

$$e_- = \alpha^{-1}(\partial/\partial t - \beta\partial/\partial x) - A^{-1}\partial/\partial x. \qquad (11.102)$$

Computing the derivatives $\nabla_{e_-} e_+$ and $\nabla_{e_+} e_-$ shows that they are given by linear expressions in e_+ and e_- with coefficients which have already been bounded plus a term coming from the matter fields. Integrating these terms along characteristics gives an estimate for $Q(t)$. Combining this with the integral inequality for $P(t)$ obtained previously gives a linear integral inequality for $P(t) + Q(t)$. Applying Gronwall's inequality to this proves the boundedness of P and Q and hence that of $\partial_t W$, $\partial_x W$, $\partial_t V$ and $\partial_x V$.

If f were zero (the vacuum case) it would now be easy to bound second order and all higher order derivatives of V and W since the equations for each successive derivative are linear with coefficients depending on lower order derivatives. When f is non-zero things are not so simple. When the Vlasov equation is differentiated once with respect to x terms arise which involve second derivatives of V and W multiplied by first derivatives of f and so Gronwall's inequality cannot be applied directly. The problem can be solved by a simple variant of an idea which was used by Glassey and Strauss in the study of the Vlasov–Maxwell system. This will not be written out in detail here but the main idea will be explained. Differentiating the basic equations (11.83) and (11.84) with respect to x gives evolution equations for second derivatives of W and V along the characteristics of the wave equations. The type of term which is difficult to estimate is an integral of $\partial_x f$ with some weight factor with respect to t and v. Consider the case where the vector e_+ points along the characteristic under consideration. Then there is no problem estimating the derivative $e_+ f$ by integrating along the characteristic. The idea to estimate $\partial_x f$ is to express ∂_x as a linear combination of e_+ and the vector

$$m = \partial/\partial t + (\alpha A^{-1}(v^1/v^0) - \beta^1)\partial/\partial x. \qquad (11.103)$$

The result is

$$\partial/\partial x = \alpha^{-1}A(1 - v^1/v^0)(e_+ - m). \qquad (11.104)$$

This allows the integral of interest to be replaced by a corresponding integral of mf in the estimate. Using the Vlasov equation mf may be replaced by an expression involving only derivatives of f with respect to the velocity variables. The latter can be estimated by integrating by parts in the velocity variables. These estimates lead to a differential inequality which, using Gronwall's inequality, shows that the second derivatives of W and V and the first derivatives of f are bounded.

Higher order derivatives of all quantities of interest can now be bounded by induction. Differentiating the equations repeatedly leads to equations which are linear in the highest order derivatives while the coefficients, which depend on lower order derivatives, are bounded as a consequence of the inductive hypothesis. Using this fact it is relatively straightforward to bound all derivatives by Gronwall's inequality. It follows that a solution evolving from initial data on a CMC hypersurface extends to the whole interval $(-\infty, t_0]$ of CMC time. Moreover, similar arguments show that the solution can be extended to all negative real values. The argument does not show that the whole future of the initial hypersurface is covered by the CMC foliation. If the cosmological constant is chosen negative similar arguments can be used to prove a stronger theorem. With $\Lambda < 0$ the result is that the whole spacetime can be covered by a CMC foliation with the mean curvature taking all real values. The reason for this difference can be traced to the estimate for α following from the lapse equation which in general reads $\alpha \leq (\frac{1}{3}t^2 - \Lambda)^{-1}$.

If the Vlasov equation is replaced by dust as a matter model the analogous result does not hold. Instead it can be shown that there are initial data for dust with the property that the time of existence of the corresponding solution in CMC time is arbitrarily short. It can also be shown that there are examples where the CMC foliation does not cover the whole maximal Cauchy development. The examples which lead to these conclusions have plane symmetry. The intuitive interpretation is that they form shell-crossing singularities. This explanation has not been made rigorous but it did serve a guide in constructing the non-existence theorems. In the case of a perfect fluid with pressure similar results should hold, with shocks replacing shell crossing. There is as yet no published proof of this.

Returning to the Einstein–Vlasov system, a question left open by the results presented so far is whether a spacetime with T^2 symmetry evolving from data on a general compact Cauchy hypersurface contains even one CMC hypersurface. This question can be answered in the affirmative using the concept of a prescribed mean curvature (PMC) foliation introduced and analysed in the work of Henkel. The definition is as follows. Let S be a compact Cauchy surface in a spacetime and let S_t be a foliation with $S = S_{t_0}$ for some t_0. Let $\mathrm{tr} k$ be the mean curvature of this foliation which is a function on spacetime. The integral curves of the normal vector field orthogonal to the leaves of the foliation form a congruence of timelike curves. Parametrize them by the parameter t labelling the leaves of the foliation and denote the corresponding tangent vector by t^α. The foliation is said to be PMC if the relation $t^\alpha \nabla_\alpha(\mathrm{tr}k) = 1$ holds. In other words the mean curvature grows at unit rate along the congruence. The concept generalizes that of a CMC foliation since if the Cauchy hypersurface S is CMC and has mean curvature t_0 a PMC foliation based on S is CMC. The equation satisfied by the leaves of a CMC foliation, expressing them as graphs in Gaussian coordinates based on one hypersurface, is elliptic. In the PMC case the corresponding equations form a coupled elliptic–hyperbolic system. For this system it is possible to prove a local existence theorem without symmetry assumptions. The proof is similar to that for hyperbolic systems and the solution belongs to a suitable Sobolev space of L^2 type or the class of C^∞ functions.

The arguments presented above for CMC hypersurfaces can be adapted to the case of PMC hypersurfaces. The result is that the past of any Cauchy hypersurface S in a solution of the Einstein–Vlasov system with T^2 symmetry can be covered by a PMC foliation based on S whose mean curvature tends uniformly to $-\infty$ in the past. This shows that the past singularity is crushing and ensures that there is at least one CMC hypersurface. This fact can then be put into the previous results to obtain a full foliation by CMC hypersurfaces. It is possible to show that the CMC foliation covers the entire spacetime by an additional argument. It has already been shown that the past of any CMC hypersurface is covered by CMC hypersurfaces. It remains to show that if p is any point of the spacetime there exists a CMC hypersurface with p in its past. The point p lies on a hypersurface S_1 of constant areal time and its mean curvature is given (in the variables related to areal coordinates) by

$$\mathrm{tr}k = -e^{-\eta+U}\alpha^{-1/2}(\partial_t \eta - \partial_t U + t^{-1}). \tag{11.105}$$

From the field equations it follows that

$$\partial_t \eta - \partial_t U + t^{-1} \geq t(\partial_t U)^2 - \partial_t U + t^{-1} = \frac{3}{4}(\partial_t U)^2 + t\left(\frac{1}{2}\partial_t U - t^{-1}\right)^2. \tag{11.106}$$

This shows that the mean curvature of S_1 is strictly negative. Hence it has a maximal value $-3H_1 < 0$. Let S_2 be the compact CMC hypersurface with mean curvature $-3H_1/2$. Then the infimum of the mean curvature of S_2 is greater than the supremum of the mean curvature of S_1 and a standard argument [143] shows that S_2 is to the future of S_1. Hence p is in the past of S_2 as required.

11.3.3 Einstein–Vlasov solutions with surface symmetry

Consider now surface symmetric solutions of the Einstein–Vlasov system. The plane symmetric solutions constitute a subset of the T^2 symmetric spacetimes and thus the results just listed for the case of T^2 symmetry apply to them. The case of hyperbolic symmetry will be discussed next. The field equations can be written in areal coordinates where they take a form similar to those for plane symmetry discussed in Section 5.10. A global existence theorem in the future in the areal time coordinate can be proved in a similar way. As mentioned above, conformal coordinates can also be used to give information about the structure of the solution in the past. This leads to the conclusion that any solution of the Einstein–Vlasov system with hyperbolic symmetry which is a maximal globally hyperbolic development admits a Cauchy hypersurface of constant areal time and that the whole spacetime can be covered by an areal time coordinate with range (R, ∞) where $R \geq 0$. That the case $R > 0$ can occur is shown by the pseudo-Schwarzschild solution. It is not known whether this is the only example. For $\Lambda = 0$ future geodesic completeness is known for a certain open set of initial data but not in general. For $\Lambda > 0$ it is known in general. In the latter case the asymptotics is very similar to that in the plane symmetric case.

The surface symmetric spacetimes can also be expressed with respect to a CMC time coordinate. The metric is

$$-\alpha^2 dt^2 + A^2[(dx + \beta dt)^2 + a^2 d\Sigma^2]. \tag{11.107}$$

The Einstein equations, with $\Lambda = 0$, are

$$(A^{1/2})'' = -\frac{1}{8} A^{5/2} \left[\frac{3}{2} \left(K - \frac{1}{3} t \right)^2 - \frac{2}{3} t^2 + 16\pi\rho \right] + \frac{1}{4} \epsilon A^{-1/2} a^{-2}, \tag{11.108}$$

$$\alpha'' + A^{-1} A' \alpha' = \alpha A^2 \left[\frac{3}{2} \left(K - \frac{1}{3} t \right)^2 + \frac{1}{3} t^2 + 4\pi(\rho + \mathrm{tr} S) \right] - A^2, \tag{11.109}$$

$$K' + 3A^{-1} A' K - A^{-1} A' t = 8\pi j A, \tag{11.110}$$

$$\beta' = -a^{-1} \partial_t a + \frac{1}{2} \alpha (3K - t), \tag{11.111}$$

$$\partial_t a = a \left[-\beta' + \frac{1}{2} \alpha (3K - t) \right], \tag{11.112}$$

$$\partial_t A = -\alpha K A + (\beta A)', \tag{11.113}$$

$$\partial_t K = \beta K' - A^{-2}\alpha'' + A^{-3}A'\alpha',$$
$$+ \alpha[-2A^{-3}A'' + 2A^{-4}A'^2 + Kt - 8\pi S_1^1 + 4\pi \mathrm{tr} S - 4\pi\rho]. \quad (11.114)$$

The Vlasov equation is

$$\frac{\partial f}{\partial t} + \left(\alpha A^{-1}\frac{v^1}{v^0} - \beta\right)\frac{\partial f}{\partial x} + \left[-A^{-1}\alpha' v^0 + \alpha K v^1 + \alpha A^{-2} A' \frac{(v^2)^2 + (v^3)^2}{v^0}\right]$$
$$\times \frac{\partial f}{\partial v^1} - \alpha\left[A^{-2}A'\frac{v^1}{v^0} + \frac{1}{2}(K-t)\right]v^B\frac{\partial f}{\partial v^B} = 0. \quad (11.115)$$

It will now be shown that vacuum solutions of these equations are spatially homogeneous. For this purpose the following lemma will be used.

Lemma 11.1 *Consider the ordinary differential equation $d^2u/dx^2 = f(u)$, where $f : (0,\infty) \to \mathbb{R}$ is Lipschitz. Suppose that $f(u_0) = 0$ for some u_0, $f(u) < 0$ for $0 < u < u_0$ and $f(u) > 0$ for $u > u_0$. Then any periodic solution is constant.*

Proof Let u be a periodic solution. By periodicity there exists a point x_0 where d^2u/dx^2 vanishes. At that point $u = u_0$. If $du/dx(x_0)$ is positive then it is easy to show that du/dx remains positive for $x > x_0$, contradicting periodicity. Similarly the assumption $du/dx(x_0) < 0$ leads to a contradiction. Hence in fact $du/dx(x_0) = 0$. By uniqueness for solutions of the ordinary differential equation it follows that u is constant. □

Consider now vacuum solutions of equations (11.108)–(11.114). The momentum constraint can be solved explicitly, giving $K - \frac{1}{3}t = CA^{-3}$ for some constant C. Substituting this into the Hamiltonian constraint gives:

$$(A^{1/2})'' = -\frac{3}{16}C^2 A^{-7/2} + \frac{1}{12}t^2 A^{5/2} + \frac{1}{4}\epsilon a^{-2} A^{-1/2}. \quad (11.116)$$

It can be checked straightforwardly that this ordinary differential equation for $A^{1/2}$ satisfies the hypotheses of the lemma and so A is constant. Then the same lemma may be applied to the lapse equation to show that α is constant. The constancy of A implies that of K and the equation for β then gives $\beta = 0$. Hence every vacuum solution of equations (11.108)–(11.114) is spatially homogeneous. These solutions will now be identified with known exact solutions. This will be done by examining the Cauchy data on one spacelike hypersurface. Suppose that constants t, a and K are

11.3 Cosmological solutions

given and satisfy the following sign condition, which is necessary for the constraints to have a solution:

1. if $\epsilon = 1$ then $\frac{3}{2}(K - \frac{1}{3}t)^2 - \frac{2}{3}t^2 > 0$;
2. if $\epsilon = 0$ then $\frac{3}{2}(K - \frac{1}{3}t)^2 - \frac{2}{3}t^2 = 0$;
3. if $\epsilon = -1$ then $\frac{3}{2}(K - \frac{1}{3}t)^2 - \frac{2}{3}t^2 < 0$.

Suppose that $t = 0$. Then the sign condition is incompatible with $\epsilon = -1$. It is only compatible with $\epsilon = 0$ if $K = 0$. In that case the data give rise to flat space, identified in a simple way. If $\epsilon = 1$ then the Hamiltonian constraint can be solved for A in terms of a and K. For $t \ne 0$ the sign condition can readily be analysed by dividing the expression of interest by t^2 and studying the resulting quadratic expression in K/t.

The case $\epsilon = 0$ is the simplest. There are two possible values for K/t, namely $-1/3$ and 1. These solutions of the constraints can be realized by the $\tau = $ const. hypersurfaces in the Kasner solution

$$-d\tau^2 + b^2\tau^{2p}dx^2 + \tau^{1-p}(dy^2 + dz^2), \tag{11.117}$$

where $p = -1/3$ or $p = 1$ and b is a positive constant. In the case $\epsilon = 1$ the quantity K/t takes all values in the intervals $(-\infty, -1/3)$ and $(1, \infty)$ and these solutions of the constraints can be realized by the $\tau = $ const. hypersurfaces in the following metric, which is obtained by identifying the part of the Schwarzschild solution inside the horizon:

$$-(2m/\tau - 1)^{-1}d\tau^2 + b^2(2m/\tau - 1)dx^2 + \tau^2 d\Sigma^2. \tag{11.118}$$

Here $d\Sigma^2$ is the standard metric on the sphere. Similarly, the solutions with $\epsilon = -1$ produce all values of K/t in the interval $(-1/3, 1)$ and these solutions of the constraints can be realized by the $\tau = $ const. hypersurfaces in the pseudo-Schwarzschild metric:

$$-(2m/\tau + 1)^{-1}d\tau^2 + b^2(2m/\tau + 1)dx^2 + \tau^2 d\Sigma^2. \tag{11.119}$$

In this case $d\Sigma^2$ is a metric of constant negative curvature on a compact manifold obtained by identifying the hyperbolic plane by means of a discrete group of isometries. For $m > 0$ the initial singularity in this solution, which occurs at $t = 0$, is a crushing singularity.

The method using PMC hypersurfaces to show that a solution of the Einstein–Vlasov system with T^2 symmetry contains a CMC hypersurface also applies to the case of hyperbolic symmetry. In that case too a compact CMC hypersurface always exists. It can then be shown that there is a unique compact CMC hypersurface for each real number in the interval $(-\infty, 0)$. It is also true that the CMC hypersurfaces cover the entire maximal Cauchy development, with the same method of proof as in the case of T^2 symmetry.

In some classes of surface symmetric spacetimes it has been possible to show directly that the initial singularity is a curvature singularity by looking at the Kretschmann scalar. It is given by the expression

$$R^{\alpha\beta\gamma\delta}R_{\alpha\beta\gamma\delta} = 4K^2 + 4r^{-4}(k - \nabla^a r \nabla_a r)^2 + 12r^{-2}\nabla_a\nabla_b r \nabla^a \nabla^b r, \quad (11.120)$$

where K is the Gaussian curvature of the quotient of the spacetime by the symmetry group and is given explicitly in double null coordinates by

$$K = 4\Omega^{-2}(\Omega^{-1}\partial_u\Omega\partial_v\Omega - \Omega^{-2}\partial_u\Omega\partial_v\Omega). \quad (11.121)$$

The definition of the Hawking mass m in a general surface symmetric spacetime is such that the second term on the right hand side of eqn (11.120) is equal to $16m^2/r^6$. If the gradient of the area function is timelike near the singularity, which is automatic provided $k \leq 0$, then the Hawking mass is non-decreasing along any past-directed causal curve and thus m is bounded away from zero near the past singularity. If in addition the radius function converges to zero as the singularity is approached then the Kretschmann scalar diverges uniformly in that limit.

There is a restricted class of initial data for the Einstein–Vlasov system for which it can be shown that the radius function tends to zero in the past and thus that there is an initial curvature singularity. The restriction on the data is rather complicated and here only some aspects of it will be described. The result is based on a small data restriction. In the simplest case, $\Lambda = k = 0$ it reads

$$c = \frac{1}{2} - 10\pi^2 w_0 F_0 \sqrt{1 + w_0^2 + F_0^2} \|e^{2\mu_0}\|_{L^\infty}\|f_0\|_{L^\infty}\| > 0. \quad (11.122)$$

Here μ_0 and f_0 are the initial data for μ and f respectively on a hypersurface of constant areal time which has been chosen to be $t = 1$. A similar but slightly more complicated condition can be written down for an arbitrary value of the initial time. The quantity w_0 is the maximum value of v^1 on the support of f_0 while F is the maximum value of $(v^2)^2 + (v^3)^2$ there. Under the smallness condition much more can be said about the structure of the initial singularity beyond the fact that it is curvature singularity. In particular, for any characteristic of the Vlasov equation in the support of f the quantity v^i decays like t^c as $t \to 0$. Furthermore the generalized Kasner exponents converge uniformly to the values $(-1/3, 2/3, 2/3)$. What does this smallness condition mean intuitively? In an FLRW model with $k = 0$ the generalized Kasner exponents have the constant values $(1/3, 1/3, 1/3)$. Thus the smallness condition must necessarily exclude the FLRW case. With this in mind it can be seen that in a sense the condition says that the initial data are such that the contribution of the matter in the Hamiltonian constraint is not too large.

11.3.4 Spherical symmetry and CMC time

In spherically symmetric solutions with topology $S^2 \times S^1$ it cannot necessarily be expected that a global areal time coordinate exists, which is due intuitively to the fact that these models recollapse. They do, however, admit a CMC foliation which covers the whole spacetime and where the mean curvature of the CMC hypersurfaces takes all real values. In particular, there is a maximal hypersurface where tr$k = 0$. This is related to a general conjecture about spatially compact spacetimes. It has already been shown that spacetimes with a Cauchy hypersurface which is Yamabe negative cannot contain a maximal Cauchy hypersurface so that in this sense they cannot recollapse. If the Cauchy hypersurface is Yamabe zero then recollapse is only possible under very special conditions. Thus essentially recollapse implies that the spatial topology is Yamabe positive. It has been conjectured that the converse is true.

The closed universe recollapse conjecture says roughly that any spacetime whose topology is Yamabe positive must recollapse. One way to make this precise would be to interpret it as saying for example that the maximal Cauchy development of a spacetime with data on a manifold with this kind of topology must contain a maximal hypersurface. This cannot be expected to be true without restricting the matter model. For instance it must be expected to be false in the case of dust due to shell-crossing singularities. In vacuum or with a better behaved matter model such as collisionless matter it may be true. No counter-examples are known and the result for spherical symmetry just presented confirms it in one special case. There is also a more robust version of the conjecture which might be true for any matter model, even dust. This says that if the topology of an initial hypersurface is of this type there is a finite upper bound on the length of all timelike curves in the maximal Cauchy development. Here matter-generated singularities, if they occur, can only help. There is no proof of this conjecture in any class of spacetimes more general than the spherically symmetric ones.

11.3.5 Strong cosmic censorship without full asymptotics

The straightforward strategy for proving cosmic censorship is to show that for spacetimes evolving from generic initial data some curvature invariant blows up along any incomplete causal geodesic. This ensures that, under the genericity assumption, no causal geodesic can leave the maximal Cauchy development in any extension. At least this is true if 'extension' means 'C^2 extension' and this is the usage in this subsection. This straightforward strategy may be very hard to carry out in practice – it means actually proving much more than strong cosmic censorship. Now two alternative strategies will be sketched. The first is not dependent on the particular choice of matter model while the second is specific to the Vlasov equation (or possibly other kinetic equations).

Let K^α be a Killing vector on a globally hyperbolic spacetime. Taking a further covariant derivative in the Killing equation and commuting derivatives leads to the equation

$$\nabla_\beta \nabla_\gamma K_\delta = R^\alpha{}_{\beta\gamma\delta} K_\alpha. \tag{11.123}$$

A consequence of this is that Killing vector fields cannot blow up in the approach to the boundary of their domain of definition. The equation leads to a statement about boundedness of the second derivative of the Killing vector field and using this information in the mean value theorem this leads to uniform continuity of the first derivatives. In this way it can be shown that any Killing vector on a maximal globally hyperbolic development extends continuously to a Cauchy horizon, if one exists.

Consider now the case of T^2 symmetry. There are two Killing vectors K_1 and K_2. The square root of the determinant of the 2×2 matrix of inner products $g(K_i, K_j)$ is proportional to the areal time. As a consequence, if the metric and the Killing vectors extend to a Cauchy horizon the same is true of the areal time. In a situation like that of the Einstein–Vlasov system with T^2 symmetry where global existence in an areal time coordinate is known, it follows that no extension of the maximal Cauchy development to the future is possible. There are other applications of this idea. In the global existence theorem for certain vacuum spacetimes with $U(1)$ symmetry the time coordinate is related to the length of the orbits of the Killing vector field. This geometrical quantity has to extend to any Cauchy horizon so that a statement about cosmic censorship in the future is obtained. The case of plane-symmetric solutions of the Einstein–Vlasov–scalar field system is also covered. In the case of hyperbolic symmetry similar techniques can be applied but there are some extra technical difficulties due to the fact that the areal time must be read off from the Killing vectors which are only locally defined on the orbits and form a three-dimensional space. It can, however, be done. To sum up, this strategy is useful for investigating cosmic censorship in an expanding phase for a variety of spacetimes with at least some symmetry.

The second strategy can sometimes be used to prove the part of cosmic censorship relating to the past time direction. The basic idea is to follow the trajectory of a particle which crosses the Cauchy horizon and show that under certain circumstances the conservation laws for the particle motion associated to the symmetries of the spacetime lead to a contradiction. It is essential to take account of non-radial trajectories. In some cases the success of this strategy is dependent on the assumption that the initial data for the Vlasov equation has non-compact support in momentum space.

The results discussed up to now, including local existence for the Einstein–Vlasov system, were for compactly supported initial data but in relevant cases they can be extended to the case where only power law decay of f as $|v| \to \infty$ is assumed for a sufficiently negative power. Before coming to some details of this approach it will be pointed out that these techniques lead to a proof of strong cosmic censorship for solutions of the Einstein–Vlasov system with T^2 symmetry and $\Lambda = 0$ assuming that enough particles are present. More precisely, the condition is that there is a positive constant δ such that on the set where the angular momentum is less than δ the function f does not vanish identically on any open set. These techniques do not apply to the vacuum case. The vacuum case itself is still open except for the Gowdy solutions and even for Gowdy it is a lot more difficult to prove cosmic censorship than when Vlasov matter is present.

For solutions of the Einstein–Vlasov system with T^2 symmetry, $\Lambda = 0$ and f not vanishing identically the area function converges to zero as the initial singularity is approached. Supposing that there is an extension of the maximal global hyperbolic development to the past the structure of the past Cauchy horizon $\mathcal{H}_-$ is examined. As already mentioned the Killing vectors extend continuously to the horizon. It can be shown that there is an open dense set of points of $\mathcal{H}_-$ at which the span of the Killing vectors contains a null vector K^α. It can then be shown that $R_{\alpha\beta} K^\alpha K^\beta \leq 0$. Combining this with the null convergence condition gives $R_{\alpha\beta} K^\alpha K^\beta = 0$. At the same time, using the assumption on the support of the initial data for f and following particles through the horizon shows that $R_{\alpha\beta} K^\alpha K^\beta > 0$. This contradiction implies that there can be no such extension.

The results for T^2 symmetry and $\Lambda = 0$ just described can be extended to the case of hyperbolic symmetry and to the case of plane symmetry with $\Lambda > 0$. In the case of hyperbolic symmetry the proofs are technically more complicated due, among other things, to the facts that the Killing vector fields are not globally defined and that it cannot be ruled out a priori that the area function tends to a strictly positive limit on the Cauchy horizon.

The case of spherical symmetry with $\Lambda > 0$ introduces a whole new class of phenomena. In that case black holes can exist in cosmological spacetimes. The fundamental example is the Schwarzschild–de Sitter spacetime. The metric takes the form

$$-\left(1 - \frac{2m}{r} - \frac{\Lambda r^2}{3}\right) dt^2 + \left(1 - \frac{2m}{r} - \frac{\Lambda r^2}{3}\right)^{-1} dr^2 + r^2 d\Sigma^2. \quad (11.124)$$

This solution can be defined in such a way that it has a Cauchy hypersurface which is diffeomorphic to $S^2 \times \mathbb{R}$ and the initial data are periodic. Identifying this solution with different periods leads to spacetimes with Cauchy hypersurface diffeomorphic to $S^2 \times S^1$ containing different numbers of black holes. In a spacetime of this type there are asymptotic regions with late-time behaviour similar to that of the de Sitter solution and black hole regions containing singularities resembling the singularity in the Schwarzschild spacetime. These different regions are separated by event horizons. The Schwarzschild–de Sitter solution depends on a mass parameter m as well as the cosmological constant. In the black hole solutions these parameters satisfy the inequality $m < \frac{1}{\sqrt{3\Lambda}}$. For $m = \frac{1}{\sqrt{3\Lambda}}$ there is an extreme solution with a different causal structure. The greatest difficulties in analysing the global structure of spherically symmetric solutions of the Einstein–Vlasov system on $S^2 \times S^1$ with $\Lambda > 0$ are associated to the possible formation of degenerate horizons resembling those in the extreme vacuum solution. It seems reasonable to hope that this process only occurs for exceptional initial data but there is no proof of this. At the moment inextendibility of the maximal Cauchy development has only been proved in cases where extreme horizons do not occur.

11.4 Isotropic singularities

A spacetime (M, g) is said to have an isotropic singularity if there is a function Ω such that the conformally rescaled metric $\Omega^{-2}g$ can be extended through a hypersurface where Ω vanishes in a regular way. A simple example is the Einstein–de Sitter model where the metric can be written in the form

$$T^4(-dT^2 + dx^2 + dy^2 + dz^2). \tag{11.125}$$

Another example is provided by a spatially flat FLRW model with radiation fluid where the metric can be written in the form

$$T^2(-dT^2 + dx^2 + dy^2 + dz^2). \tag{11.126}$$

In these solutions the conformally rescaled metric is flat but this is in general not the case. It is discussed in this section how many solutions of the Einstein–matter system there are with isotropic singularities.

The vacuum case is not of much interest since in that case there are almost no examples. A more interesting case is that of solutions of the Einstein equations coupled to a perfect fluid with linear equation of state. The first general results on this concerned the radiation equation of state with $w = 1/3$ [62]. It was shown how to prescribe asymptotic data at the singularity. The number of functions which could be given was half that defining the most general initial data on a regular Cauchy hypersurface. The conformal metric $\tilde{g}$ and the conformal factor Ω are both smooth if the initial data are smooth. Later the case of a more general linear equation of state with $0 \leq w \leq 1$ was handled, with similar results [13]. The main difference is that it is not possible to make both $\tilde{g}$ and Ω smooth. However it is possible to make $\tilde{g}$ and a suitable power of Ω, depending on γ, smooth. The freedom in the data is to prescribe a Riemannian metric g_{ab} without any constraints. There is no independent matter data. These results are proved by methods related to those described in Section 8.11.

The corresponding problem for the Vlasov equation with massless particles has been solved with results which are significantly different [12]. Subject to a finite number of constraints an entire distribution function at a fixed time can be given. It follows that there is much more free data for solutions of the Einstein–Vlasov system with isotropic singularities than there is in the case of a perfect fluid. A case which might be expected to be intermediate between these two cases is that of the Einstein–Boltzmann system. An investigation of this question on the level of formal series tends to confirm this idea [200]. If the collision kernel grows fast enough at high momenta the result resembles that in the Euler case. The distribution function is forced to be in local equilibrium. For slow enough growth the result is similar to that found for the Vlasov equation. In the spatially homogeneous case the existence results for spacetimes with isotropic singularities have been extended to the Vlasov equation with massive particles and to include a non-vanishing cosmological constant [201].

The motivation for studying isotropic singularities is the Weyl curvature hypothesis of Penrose. He suggested to explain the observed increase of entropy in the universe by an initial state of low gravitational entropy and that the entropy of the gravitational field should in some sense be measured by the Weyl tensor. Thus it was suggested to impose the condition that the Weyl tensor tends to zero, or at least is bounded, at an initial physical singularity. It is not immediately obvious how to express this mathematically but in the end the concept of isotropic singularity was found to fulfil this task.

11.5 Weak cosmic censorship and internal structure of black holes

Consider a spherically symmetric globally hyperbolic solution of the Einstein–Vlasov system with asymptotically flat initial data. If this solution contains a trapped surface then it follows from the Penrose singularity theorem that it must be causally geodesically incomplete in the future. In this case it can be shown that weak cosmic censorship holds for this solution in a sense which is made precise below. Note that this is a stronger statement than is available in the absence of a trapped surface, in which case weak cosmic censorship is still open. The precise statement is as follows. Let γ be a future-complete outgoing radial null geodesic. Let l be an ingoing future-directed radial null vector field which is parallelly transported along γ. Finally let γ' be an ingoing affinely parametrized null geodesic starting at a point of γ with initial tangent vector l. Then the desired condition is that as the starting point of the geodesic tends to infinity along γ the parameter length of γ' tends to infinity. This has implications for the structure of the Penrose diagram of the spacetime. It is possible to define a horizon such that any inextendible causal geodesic which remains outside the horizon must be complete. In other words, all singularities are covered by the horizon, which is the statement of weak cosmic censorship.

Another question of great interest is that of the internal structure of black holes. It is known that under some circumstances the presence of matter can change the Schwarzschild singularity into something else, a weak null singularity, immediately after the horizon. For more on weak null singularities see Section 12.1. It is not known whether weak null singularities can occur in the case of the Einstein–Vlasov system but there is at least a sufficient condition ensuring their absence. To explain this two quantities must be defined. The Hawking mass is non-decreasing along outgoing radial null geodesics and is bounded by the ADM mass of the initial data. Hence it tends to a limit $M(u)$, the Bondi mass. It may well be that in the case of the spherically symmetric Einstein–Vlasov system the Bondi mass is constant but for other matter models this will in general not be the case. Since the Hawking mass is also non-increasing along ingoing radial null geodesics it follows that the Bondi mass is non-increasing with increasing u. Since the Bondi mass is non-negative it tends to a non-negative limit, the final Bondi mass M_f. The radius function r is non-decreasing along the event horizon and it can be shown that it is bounded above by $2M_f$. In particular it tends to a limit r_+ in the future. There is a universal constant $\delta_0 < 1$ such that if in a solution of the Einstein–Vlasov system $r_+ \leq 2\delta_0 M_f$ then there is no weak null singularity immediately following the event horizon.

11.6 Further reading

The two global existence proofs for the Vlasov–Poisson system, both of which build on work of [106], are in [159] and [138]. Global existence for the Vlasov–Nordström system was proved in [41]. The small data global existence theorem for the spherically symmetric Einstein–Vlasov system was first proved in [164]. The result for the Vlasov–Poisson system which it built

11.6 Further reading

on is contained in [17] while the improved decay estimates for that system are in [110]. That the first singularity if any of a solution of the spherically symmetric Einstein–Vlasov system must be at the centre is proved in [165]. The global solutions with outgoing matter are constructed in [9]. The issue of the alleged violation of cosmic censorship in axially symmetric solutions of the Einstein–Vlasov system is discussed in [167]. Global existence in maximal-isotropic coordinates was proved in [171]. The continuation criterion in double null coordinates is the subject of [66]. Its verification for collisionless matter was carried out in [67]. Singularities in solutions of the spherically symmetric Einstein–dust system are analysed in [51] and interesting explicit examples can be found in [209]. The finiteness of the lifetime of spherically symmetric spacetimes on S^3 is a result of [40]. The theorem on global existence in the future for T^2 symmetry in areal coordinates is a result of [11]. The technique of Glassey and Strauss occurring in the global existence proof for the Einstein–Vlasov system with T^2 symmetry is in [93] and a simplified form sufficient for the application here is in [92]. For the non-existence theorems for solutions with dust see [114]. The cosmic censorship results in Section 11.5 are proved in [67], [69] and [70].

12 The Einstein–scalar field system

The field theoretic matter model for which the most mathematical results are known for the Einstein–matter system is a massless linear scalar field solving the wave equation. The available results on this subject are surveyed in this chapter.

12.1 Asymptotically flat solutions

This section is concerned with spherically symmetric solutions of the Einstein equations with vanishing cosmological constant coupled to a scalar field. This is the subject of a series of papers by Christodoulou whose motivation was to find a model for gravitational collapse which was as simple as possible. To study gravitational collapse in the spherically symmetric case it is necessary to add some matter field to get around Birkhoff's theorem. The scalar field is an obvious choice to start with. The first paper of the series contains a global existence theorem for small initial data with initial data prescribed on a future light cone. It is not hard to relate this characteristic initial value problem to the corresponding problem with compactly supported data on a Cauchy hypersurface. By Cauchy stability, if the initial datum is small there exists a solution on a long but finite interval to the past and to the future. Moreover, the solution is close to flat space. Thus there exists a point at $r = 0$ whose future light cone reaches the initial hypersurface $t = 0$ at a point outside the support of the initial data. From there it can be continued to infinity within the Schwarzschild solution. Restricting the solution to that light cone gives the desired initial datum for the characteristic initial value problem. The proof of global existence is carried out in Bondi coordinates which generalize the outgoing Eddington–Finkelstein coordinates in the Schwarzschild metric. This result has been generalized [44] to the case of the Maxwell–Higgs system. This includes the charged scalar field and nonlinear wave equations with potentials which are sufficiently small near zero. It includes in particular the nonlinear wave equations with polynomial nonlinearity mentioned in Section 8.9. Note that the method used for proving small data global existence in the case of the Einstein–Vlasov system would fail in the case of a scalar field. The reason is that the Vlasov proof relies on the L^∞ norm of the energy–momentum tensor decaying strictly faster than t^{-2} and that the L^∞ norm of the energy–momentum tensor of a scalar field decays exactly like t^{-2}. This is a problem which is likely to affect any field theoretic matter model which allows radiation in spherical symmetry. The proof of the stability of Minkowski space due to Lindblad and Rodnianski [137] extends rather directly to the Einstein–scalar field system. Thus there is now a global existence theorem for asymptotically flat solutions of the Einstein–scalar field system with small initial data which does not require any symmetry assumption.

In later papers of Christodoulou the existence of a global weak solution for arbitrary initial data was shown and for any such solution a quantity $M_f \geq 0$, the final Bondi mass, was defined. For more information on this quantity see Section 11.5. It was proved that if $M_f > 0$ the solution converges to a Schwarzschild solution of mass M_f as $t \to \infty$ in the region $r > 2M_f$. These results concerned only the outside of any black hole formed since this is the maximal region which can be covered by Bondi coordinates. To investigate what happens inside the black hole it is necessary to change to another coordinate system, for instance double null coordinates. In this context a global existence theorem has been proved for data which are allowed to be rather rough. They belong to a space of functions of bounded variation. These BV functions are a little more general than $W^{1,1}$. The advantage in using them is that the size of a solution of the Einstein–scalar field system can be measured in a norm which is additive and diffeomorphism invariant. The first property means that the norm of a function on $A \cup B$ for disjoint sets A and B is equal to the sum of the norms of the restrictions of the function to A and B. The second means that the norm of f on A is equal to the norm of $f \circ \psi$ on $\psi(A)$ for any diffeomorphism ψ. In the context of BV solutions general statements about the global structure of the spacetime can be made.

There is a sufficient condition known on initial data for $M_f > 0$, i.e. the formation of a black hole, which applies to initial data not containing a trapped surface. (Until very recently there was no comparable criterion known for any other matter model but now progress has been made [10]). It has been proved that there are solutions which describe the formation of a globally naked singularity from regular initial data. This is a counterexample to the simplest formulation of weak cosmic censorship. It has also been proved that the initial data with this unpleasant property can be changed into data which do not have it by an arbitrarily small perturbation. This shows the validity of a somewhat more sophisticated formulation of weak cosmic censorship (including a genericity assumption) for this model.

The solutions containing naked singularities which have been constructed are self-similar in a certain sense. They have the property that there is a vector field X such that $\mathcal{L}_X g = \alpha g$ and $\mathcal{L}_X \phi = \beta$ for constants α and β. The solutions with naked singularities are only obtained for $\beta \neq 0$.

In the case of a charged scalar field things look very different. Note first that the Reissner–Nordström solution has a Cauchy horizon and is thus, in itself, a counterexample to a formulation of strong cosmic censorship without a genericity requirement. There is an old idea that a generic perturbation of the initial data for the Reissner–Nordström solution should lead to a solution where the Cauchy horizon is replaced by a singularity like that in the Schwarzschild solution. The suggested mechanism, known as the 'blue sheet instability' was that the frequency of radiation propagating towards the Cauchy horizon would be increased (blue-shifted) without limit resulting in a blow-up of the energy density. A later heuristic analysis painted a different picture and suggested that the Cauchy horizon should be replaced by something called a 'weak null singularity'. At a singularity like this curvature invariants may blow up but the metric itself can be extended through the singularity in such a way that it is continuous and non-degenerate. The picture was later established rigorously. This shows that the meaning of strong cosmic censorship can change if the degree of differentiability of extensions used in the definition is switched between C^2 and C^0. This is discussed further in Section 12.2.

These results may have a significance going beyond spherical symmetry. Charge may serve as a kind of simplified analogue of angular momentum. The Reissner–Nordström solution shows some evident similarities with the Kerr solution, which also has a Cauchy horizon. There is heuristic work indicating that a kind of weak null singularity may occur in vacuum spacetimes which are perturbations of the Kerr solution. Unfortunately this seems to be far beyond the reach of rigorous results at the moment.

The subject of critical collapse has already been mentioned in Chapter 11. The origin of the subject lies in the study of the Einstein–scalar field system. When the data is scaled in that case and the black hole mass M is plotted as a function of the scaling parameter then it is found that M approaches zero when the parameter tends to the critical value which marks the transition from dispersion to black hole formation. The behaviour of the evolution for parameters close to the critical value seems to be controlled by a discretely self-similar solution. There is no proof of the existence of a solution of this kind although there is a direct numerical construction. Many of the phenomena seen in critical collapse for the Einstein–matter equations are also seen in singularity formation in solutions of various types of partial differential equations. What is remarkable is that the occurrence of a discretely self-similar solution with a central importance for the dynamics has been found in no system of PDE other than the Einstein–matter system. This is an obstacle to obtaining a better mathematical understanding of the phenomenon since it would be much more favourable to have a simpler system showing analogous behaviour. Critical collapse has also been observed numerically in solutions of the vacuum Einstein equations in higher dimensions. Many features are found which recall those known for the collapse of a spherically symmetric scalar field in four dimensions. Some things are easier to see numerically in the higher dimensional case due to values of certain constants which occur in the problem. It is interesting to note that the solutions of the Einstein–scalar field containing naked singularities seem to have no analogue in higher dimensional vacuum solutions. The symmetry condition with $\beta \neq 0$ does not seem to generalize.

12.2 Weak null singularities

The rigorous results on weak null singularities inside black holes concern the Einstein equations coupled to an uncharged scalar field and a Maxwell field, all in spherical symmetry. Since the Maxwell field has no source it would have to vanish if there existed a complete Cauchy hypersurface with the topology of Euclidean space. The way around this is to choose a different topology similar to that of a Cauchy hypersurface in the Reissner–Nordström solution. In that case a Cauchy hypersurface has topology $S^2 \times \mathbb{R}$ but for the present purposes it suffices to consider an exterior region. Data is given on the event horizon and on an ingoing characteristic. In the original version of this work the data on the horizon was exactly that of Reissner–Nordström. This was later extended to the case where the scalar field decays like some negative power of an affine parameter along the horizon. The result is the same in both cases. It can be shown that there is a finite interval of u, the ingoing null coordinate, for which the solution exists up to the value of v corresponding to the Cauchy horizon in the Reissner–Nordström solution. The metric can be extended further in v in such a way as to be continuous and non-degenerate but not so as to be C^1. In the usual discussion in the literature the occurrence of this weak null singularity is related to the rate of fall-off of ϕ along the event horizon. The rigorous result shows that this is not relevant. A weak null singularity is created even when the scalar field is identically zero on the horizon. In the usual discussion it is backscatter of waves outside the black hole which is supposed to be important but the complete analysis can be interpreted as showing that backscattering inside the black hole is enough. In the traditional approach the statement about the decay of the scalar field on the horizon comes from Price's law which is discussed in the next section. It should be noted that this result as it stands does not apply to spacetimes evolving from data on a regular Cauchy hypersurface with Euclidean topology. For this it would be necessary to use a different matter model such as a charged scalar field.

Another aspect of the heuristic analysis which is captured by the rigorous results is that of *mass inflation*. This means by definition that the Hawking mass tends to infinity as the weak null singularity is approached.

In the proof of the theorem the solution is followed through several regions where different effects are important and different techniques are used to deal with them. The details will not be presented here but the result will be considered in a wider perspective. The first work on the instability of the Cauchy horizon in the Reissner–Nordström solution was based on considering a test matter field on a background spacetime. It was found that the energy density of the test field became unbounded. For want of more information it was assumed that this singularity would not be smoothed out by taking into account nonlinearity. Since the sources in the Einstein equations are getting very large it is in principle problematic to ignore them. Of course it later turned out that the test field analysis led to a misleading picture of what happens. On the other hand the idea underlying the analysis with a test field does enter the proof of the theorem in an important way in analysing the behaviour of solutions in a regime occurring in a certain part of spacetime.

It was once suggested that the instability of the Reissner–Nordström Cauchy horizon is suppressed by the presence of a positive cosmological constant. Thus it was concluded that strong cosmic censorship was violated in the collapse of charged matter with $\Lambda > 0$. Later these heuristic arguments were replaced by others supporting strong cosmic censorship. The presence of a positive cosmological constant may suppress one mechanism leading to the instability of the Cauchy horizon in the Reissner–Nordström solution but even when that mechanism fails there is another which can take over.

As mentioned in Subsection 11.3.1 there are LRS Bianchi type I solutions of the Einstein–Euler equations with weak null singularities. This will now be examined in some detail. The metric of Bianchi I perfect fluid spacetimes with linear equation of state can be written in the form (see [203], p. 199):

$$ds^2 = -A^{2w} dt^2 + t^{2p_1} A^{2q_1} dx^2 + t^{2p_2} A^{2q_2} dy^2 + t^{2p_3} A^{2q_3} dz^2. \quad (12.1)$$

It is supposed that $0 \leq w < 1$. Here $p_1 + p_2 + p_3 = p_1^2 + p_2^2 + p_3^2 = 1$ and $q_i = \frac{2}{3} - p_i$ for $i = 1, 2, 3$. The function A is defined by $A^{1-w} = \alpha + m^2 t^{1-w}$ for positive constants α and m. The case of interest here is where $t = 0$ is a

12.2 Weak null singularities

weak null singularity. If we single out the translations in y and z as a group of isometries then the area radius of the orbits is proportional to $t^{2(p_2+p_3)}$. In order to have a weak null singularity this must tend to a non-zero constant for $t \to 0$. This is only possible if $p_2 = p_3 = 0$ and $p_1 = 1$. In this case the metric reduces to

$$ds^2 = -A^{2w}dt^2 + t^2 A^{-2/3}dx^2 + A^{4/3}dy^2 + A^{4/3}dz^2, \quad (12.2)$$

which is LRS.

In order to see the structure of the singularity at $t = 0$ it is useful to begin by transforming to Gauss coordinates. This is achieved by defining

$$\tau = \int_0^t (\alpha + m^2 s^{1-w})^{\frac{w}{1-w}} ds. \quad (12.3)$$

Note that the integral exists for w in the range of interest. It follows that

$$\tau = \alpha^{\frac{w}{1-w}} t \left[1 + \frac{w}{(1-w)(2-w)} \frac{m^2}{\alpha} t^{1-w} + O(t^{2-2w}) \right]. \quad (12.4)$$

Inverting this leads to

$$t = \alpha^{-\frac{w}{1-w}} \tau \left[1 - \frac{w}{(1-w)(2-w)} m^2 \alpha^{-1-w} \tau^{1-w} + O(t^{2-2w}) \right]. \quad (12.5)$$

This allows A to be expressed in terms of τ up to a remainder of higher order. Let $\tilde{\alpha} = \alpha^{1/(1-w)}$. Then the metric is of the form

$$ds^2 = -d\tau^2 + \tau^2 \tilde{\alpha}^{-\frac{2}{3}-2w}(1+r_1(\tau))^2 dx^2 + \tilde{\alpha}^{\frac{4}{3}}(1+r_2(\tau))^2(dy^2+dz^2), \quad (12.6)$$

where the remainders $r_1(\tau)$ and $r_2(\tau)$ are $O(\tau^{1-w})$. Next introduce a new spatial coordinate by

$$x' = x + \tilde{\alpha}^{\frac{1}{3}+w} \left(\log \tau - \int_0^\tau \sigma^{-1} r_1(\sigma)(1+r_1(\sigma))^{-1} d\sigma \right). \quad (12.7)$$

Finally, let $\tau' = \tau^2$. The result is that in the coordinates (τ', x') the metric takes the form:

$$-\tilde{\alpha}^{-\frac{1}{3}-w}(1+\tilde{r}_1(\tau'))d\tau' dx' + \tau'\tilde{\alpha}^{-\frac{2}{3}-2w}(1+\tilde{r}_2(\tau'))(dx')^2$$
$$+\tilde{\alpha}^{\frac{4}{3}}(1+\tilde{r}_3(\tau'))(dy^2+dz^2), \quad (12.8)$$

where $r_1(\tau')$, $r_2(\tau')$ and $r_3(\tau')$ are $O((\tau')^{\frac{1-w}{2}})$. The metric extends continuously to $\tau' = 0$ while remaining non-degenerate. In fact the extension is Hölder continuous with exponent $\frac{1-w}{2}$.

In the present case the Hawking mass is given by:

$$m_H = -\frac{1}{2}rg^{00}(\partial_t r)^2 = \frac{2}{9}A^{-2w}(\partial_t A)^2. \qquad (12.9)$$

The time derivative of A is given by:

$$\partial_t A = m^2(\alpha + m^2 t^{1-w})^{\frac{w}{1-w}} t^{-w}. \qquad (12.10)$$

As $t \to 0$ the quantity A tends to a non-zero constant and so the mass behaves like a non-zero constant times t^{-2w}. This blows up at the singularity when $w > 0$ and is constant when $w = 0$. Thus mass inflation is observed in all cases except that of dust, where the Hawking mass tends to a finite value at the singularity. As a consequence of equation (11.120) the Kretschmann scalar blows up at the singularity in the case $w > 0$ but not manifestly in the dust case. Nevertheless, the singularity is a curvature singularity even in the dust case since the invariant $R^{\alpha\beta}R_{\alpha\beta}$ blows up there. This follows from general results in [169] or can be checked by direct calculation.

In the case of black hole interiors the mass inflation singularity can be seen as arising from a perturbation of the Cauchy horizon in the Reissner–Nordström solution. A similar interpretation is possible in the cosmological models considered here. The analogue of the Reissner–Nordström solution in the present context is the Rosen magnetovacuum solution which can be written in the following form [128]:

$$-(1+k^2 t^2)^2 dt^2 + t^2(1+k^2 t^2)^{-2} dx^2 + (1+k^2 t^2)^2(dy^2 + dz^2), \quad (12.11)$$

where $k \geq 0$ is a constant. Similar manipulations to those done above show that the metric can be extended through $t = 0$. However in this case the components of the metric in the coordinates (τ', x') are analytic functions and so the boundary which has been added at $t = 0$ is in fact a Cauchy horizon through which the spacetime can be extended analytically. The limiting case $k = 0$ of this solution has vanishing magnetic field and is the flat Kasner spacetime. The fluid solutions without electromagnetic field discussed above can be thought of as perturbations of the flat Kasner solution which turn the Cauchy horizon into a weak null singularity. Note for comparison that in solutions of the Einstein–Vlasov system with LRS Bianchi I symmetry there are no Cauchy horizons or weak null singularities.

12.3 Price's law

An important open question in general relativity is that of the nonlinear stability of the Schwarzschild solution. A proof of this seems a rather distant prospect at the moment. This is true even if only the exterior of the black hole is to be treated. One difficulty is that small perturbations of initial data for the Schwarzschild solution include data for the Kerr solution. The hope that the static Killing vector field in the exterior Schwarzschild solution might lead to some useful energy estimates for the perturbed solution is endangered by the occurrence of the ergosphere. Furthermore it does not seem to be possible to single out a large family of nonlinear perturbations of the Schwarzschild solution which rules out the Kerr solution. Thus it seems that it is not possible to prove the nonlinear stability of the Schwarzschild solution without proving the nonlinear stability of the Kerr solution at least for some range of parameters.

Given that the full stability for the Schwarzschild solution appears so difficult it is natural to look for some simplified problems. One approach is to look at the linearized problem. The linearized equations can be thought of as defining a linear field theory on the fixed background of the Schwarzschild solution. The aim is then to study the stability of the zero solution. A further simplification is to consider a simpler linear field theory, for instance the linear scalar field. A different type of simplification is to restrict to the spherically symmetric case. Of course this option is not useful for the case of the vacuum Einstein equations, due to Birkhoff's theorem.

A heuristic analysis of a linear massless scalar field on the Schwarzschild spacetime was given by Price [162]. The problem is invariant under the rotation group and so it is possible to decompose into spherical harmonics. In other words, for each point in the quotient of spacetime by the symmetry group the solution is expanded in a basis of eigenfunctions of the Laplacian on the unit sphere. The eigenvalues are conventionally indexed by integers l and m where l indexes the eigenvalues and m indexes the different eigenfunctions for a given eigenvalue. The wave equation decomposes into a sequence of wave equations in one space dimension indexed by l. The case of spherically symmetric solutions of the wave equation corresponds to $l = 0$. Let $\psi = r\phi$. In terms of the tortoise coordinate $r^* = r + 2m \log(r-2m)$ the wave equation takes the form

$$\frac{\partial^2 \psi}{\partial t^2} - \frac{\partial^2 \psi}{\partial (r^*)^2} = F_l(f^*)\psi, \qquad (12.12)$$

where

$$F_l(r^*) = (1 - 2m/r)(2m/r^3 + l(l+1)/r^2). \qquad (12.13)$$

The coordinates here are conformal coordinates and lead to an equation which has the form of a wave equation on flat space with a certain potential.

The heuristic analysis indicates that field perturbations with spatial dependence corresponding to the eigenvalue l decay like $u^{-(2l+3)}$ along the event horizon, where u is an affine parameter. In particular this gives the decay rate u^{-3} for $l = 0$. It has been proved in [71] that the field decays at a rate which is at least $u^{-3+\epsilon}$ for any $\epsilon > 0$. This has been done both for a spherically symmetric scalar field on a Schwarzschild background and for spherically symmetric solutions of the coupled Einstein–scalar field system. The heuristics indicate that vacuum perturbations of a black hole should, when expressed in suitable variables, have the same decay properties as the scalar field. Beyond the spherically symmetric case the decay embodied in Price's law has not yet been proved rigorously. The strongest available result at least shows that the field decays.

12.4 Cosmological solutions

The theorem about the finite lifetimes of spherically symmetric spacetimes on S^3 mentioned in the last chapter does not directly apply to the case of a scalar field since a spherically symmetric scalar field need not have nonnegative pressures. However there is an analogous result which does apply to the case of a scalar field [39].

The equations (11.40)–(11.49) which are the Einstein–matter equations in areal coordinates for a spacetime with T^2 symmetry apply also to the scalar field. Only the explicit form of the matter terms is different. These equations for the Einstein–scalar field system have never been applied in the general case of a T^2-symmetric spacetime. Accordingly no global existence theorem for the coupled system in these coordinates has been proved although it is presumably possible by analogy with known results. Local in time existence for the system can be proved just as described for the Einstein–Vlasov case. The monotonicity formula for the energy-like quantity also holds since it is only dependent on the fact that the energy–momentum tensor is divergence free. Similar comments apply to the equations in conformal coordinates. Again they have not been used in the case of general T^2-symmetric solutions of the Einstein–scalar field system.

In contrast to these negative statements, CMC coordinates have been applied to T^2-symmetric solutions of the Einstein–scalar field system with results similar to those found in the Einstein–Vlasov case. In particular a global existence theorem in CMC time has been proved. In fact a more general case has been treated. The proof works for any wave map whose target is a complete Riemannian manifold in place of the scalar field. The key observation is that in this situation the hyperbolic degrees of freedom of the gravitational field themselves satisfy equations which are similar to a wave map. Hence the techniques used to handle those equations can be taken over directly and from most points of view things only get simpler. The only difficulty is the bookkeeping problem of working with several charts when the target manifold is topologically non-trivial.

Under the assumption of plane symmetry the Einstein–scalar field system can be put in a very simple form. The only essential field equation is exactly the same linear wave equation as that occurring in the polarized Gowdy class, although the unknown has a different interpretation. Known results for the polarized Gowdy equations can be applied directly to the plane symmetric Einstein–scalar field system. In particular this provides a proof of future geodesic completeness for these spacetimes. It also allows the structure of the singularity to be analysed in detail. This results in a proof of strong cosmic censorship for this class of spacetimes. In the case of hyperbolic symmetry global existence in the future has been proved but geodesic completeness remains an open question.

A scalar field can lead to a dramatic simplification in the structure of spacetime singularities. The BKL heuristics suggests that there should be an open set of initial data having simple asymptotics. This has not been proved but it is known, as an application of Fuchsian techniques, that there is a family of solutions having this asymptotics depending on the same number of free functions as the general solution. This result assumes analyticity of the free functions but does not require any symmetry assumptions. The BKL picture suggests that no corresponding result should hold for the vacuum Einstein equations in four dimensions. However it does hold for the vacuum Einstein equations in dimension at least eleven.

The asymptotics obtained in these results using Fuchsian techniques are such that the generalized Kasner exponents converge uniformly to limiting functions as the singularity is approached. The limiting values are not too far away from $(1/3, 1/3, 1/3)$. In other words this represents a certain kind of stability statement for the singularities in FLRW models in the presence of a scalar field. In these solutions it can be shown that the Kretschmann scalar blows up uniformly in the approach to the singularity and thus these solutions provide support for strong cosmic censorship. This result may usefully be compared with what is known in the spatially homogeneous case (cf. Subsection 5.7.1).

For the Einstein–Vlasov–scalar field system global existence in the future is known in the cases of plane and hyperbolic symmetry but there is no information available on the asymptotic behaviour.

12.5 Further reading

Small data global existence for the spherically symmetric Einstein–scalar field system was proved in [52]. The paper which initiated the research area of critical collapse in general relativity was [45]. For the claims about the effects of a positive cosmological constant on the formation of a weak null singularity see [145] and [34]. The available rigorous results relating to Price's law are in the papers [71] and [72]. Results on cosmological solutions of the Einstein–scalar field and Einstein–Vlasov–scalar field system can be found in [199]. For the results based on Fuchsian methods see [7] and [74].

References

[1] Adams, R. A. (1975) *Sobolev spaces*. Academic Press, New York.
[2] Alinhac, S. and Gérard, P. (1991) *Opérateurs pseudo-différentiels et théorème de Nash-Moser*. InterEditions, Paris.
[3] Amann, H. (1990) *Ordinary differential equations*. De Gruyter, Berlin.
[4] Anderson, M. T. (2005) Existence and stability of even dimensional asymptotically de Sitter spacetimes. *Ann. H. Poincaré* **6**, 801–820.
[5] Andersson, L. and Moncrief, V. (2004) Future complete vacuum spacetimes. In: Chruściel, P. T. and Friedrich, H. (eds.) *The Einstein equations and the large scale behavior of gravitational fields*. Birkhäuser, Basel.
[6] Andersson, L., Moncrief, V. and Tromba, A. J. (1997) On the global evolution problem in 2+1 gravity. *J. Geom. Phys.* **23**, 191–205.
[7] Andersson, L. and Rendall, A. D. (2001) Quiescent cosmological singularities. *Commun. Math. Phys.* **218**, 479–511.
[8] Andréasson, H. (2005) The Einstein–Vlasov system/kinetic theory. *Liv. Rev. Relativity*. lrr-2005-2.
[9] Andréasson, H., Kunze, M. and Rein, G. (2006) Global existence for the spherically symmetric Einstein–Vlasov system with outgoing matter. Preprint gr-qc/0611115.
[10] Andréasson, H., Kunze, M. and Rein, G. (2007) The formation of black holes in spherically symmetric gravitational collapse. Preprint arXiv:0706.3787.
[11] Andréasson, H., Rendall, A. D. and Weaver, M. (2004) Existence of CMC and constant areal time foliations in T^2 symmetric spacetimes with Vlasov matter. *Commun. PDE* **29**, 237–262.
[12] Anguige, K. (2000) Isotropic cosmological singularities III: the Cauchy problem for the inhomogeneous conformal Einstein–Vlasov equations. *Ann. Phys. (NY)* **282**, 395–419.
[13] Anguige, K. and Tod, K. P. (1999) Isotropic cosmological singularities I: polytropic perfect fluid spacetimes. *Ann. Phys (NY)* **276**, 257–293.
[14] Arnold, V. I. and Ilyashenko, Yu. S. (1988) Ordinary differential equations. In: Anosov, D. V. and Arnold, V. I. (eds.)

Dynamical systems I: Ordinary differential equations and smooth dynamical systems. Springer, Berlin.

[15] Aubin, T. (1982) *Non-linear analysis on manifolds. Monge–Ampère equations.* Springer, Berlin.

[16] Bancel, D. and Choquet-Bruhat, Y. (1973) Existence, uniqueness and local stability for the Einstein–Maxwell–Boltzmann system. *Commun. Math. Phys.* **33**, 83–96.

[17] Bardos, C. and Degond, P. (1985) Global existence for the Vlasov–Poisson system in 3 space variables with small initial data. *Ann. Inst. H. Poincaré (Anal. Non Linéaire)* **2**, 101–118.

[18] Bartnik, R. (1986) The mass of an asymptotically flat manifold. *Commun. Pure Appl. Math.* **39**, 661–693.

[19] Bartnik, R. and Isenberg, J. (2004) The constraint equations. In: Chruściel, P. T. and Friedrich, H. (eds.) *The Einstein equations and the large scale behavior of gravitational fields.* Birkhäuser, Basel.

[20] Bauer, S., Kunze, M., Rein, G. and Rendall, A. D. (2006) Multipole radiation in a collisionless gas coupled to electromagnetism or scalar gravitation. *Commun. Math. Phys.* **266**, 267–288.

[21] Baumgarte, T. W. and Shapiro, S. L. (2003) General relativistic magnetohydrodynamics for the numerical construction of dynamical spacetimes. *Astrophys. J.* **585**, 921–929.

[22] Beig, R. and Schmidt, B. G. (2003) Relativistic elasticity. *Class. Quantum Grav.* **20**, 889–904.

[23] Belinskii, V. A., Grishchuk, L. P., Zeldovich, Ya. B. and Khalatnikov, I. M. (1986) Inflationary stages in cosmological models with scalar fields. *Sov. Phys. JETP* **62**, 195–203.

[24] Belinskii, V. A., Khalatnikov, I. M. and Lifshitz, E. M. (1970) Oscillatory approach to a singular point in the relativistic cosmology. *Adv. Phys.* **19**, 525–573.

[25] Belinskii, V. A., Khalatnikov, I. M. and Lifshitz, E. M. (1982) A general solution of the Einstein equations with a time singularity. *Adv. Phys.* **31**, 639–637.

[26] Bérard Bergery, L. (1978) Sur la courbure des métriques riemanniennes invariantes des groupes de Lie et des espaces homogènes. *Ann. Sci. Ecole Norm. Sup.* **11**, 543–576.

[27] Berger, B., Chruściel, P. T. and Moncrief, V. (1995) On 'asymptotically flat' space-times with G_2-invariant Cauchy surfaces. *Ann. Phys. (NY)* **237**, 322–354.

[28] Bergh, J. and Löfström, J. (1976) *Interpolation spaces.* Springer, Berlin.

[29] Bieli, R. (2005) Algebraic expansions for curvature coupled scalar field models. *Class. Quantum Grav.* **22**, 4363–4375.

[30] Bieli, R. (2006) Coupled quintessence and curvature-assisted acceleration. *Class. Quantum Grav.* **23**, 5983–5995.

[31] Binney, J. and Tremaine, S. (1987) *Galactic dynamics*. Princeton University Press, Princeton.

[32] Bizoń, P., Chmaj, T. and Rostworowski, A. 2007 On asymptotic stability of the skyrmion. *Phys. Rev. D* **75**, 121702.

[33] Bizoń, P. and Schmidt, B. G. (2006) How to bypass Birkhoff through extra dimensions (a simple framework for investigating the gravitational collapse in vacuum). *Int. J. Mod. Phys. D* **15**, 2217–2222.

[34] Brady, P. R., Moss, I. G. and Myers, R. C. (1998) Cosmic censorship: as strong as ever. *Phys. Rev. Lett.* **80**, 3432–3435.

[35] Brauer, U., Rendall, A. D. and Reula, O. (1994) The cosmic no-hair theorem and the nonlinear stability of homogeneous Newtonian cosmological models. *Class. Quantum Grav.* **11**, 2283–2296.

[36] Brauer, U. (1995) *Singularitäten in relativistischen Materiemodellen*. PhD thesis, University of Potsdam.

[37] Bray, H. L. and Lee, D. A. (2007) On the Riemannian Penrose inequality in dimensions less than eight. Preprint arXiv:0705.1128.

[38] Burnett, G. A. (1991) Incompleteness theorems for the spherically symmetric spacetimes. *Phys. Rev. D* **43**, 1143–1149.

[39] Burnett, G. A. (1994) Closed spherically symmetric massless scalar field spacetimes have finite lifetimes. *Phys. Rev. D* **50**, 6158–6164.

[40] Burnett, G. A. (1995) Lifetimes of spherically symmetric closed universes. *Phys. Rev. D* **51**, 1621–1631.

[41] Calogero, S. (2006) Global classical solutions to the 3D Nordström–Vlasov system. *Commun. Math. Phys.* **266**, 343–353.

[42] Carmeli, M., Charach, Ch. and Malin, S. (1981) Survey of cosmological models with gravitational, scalar and electromagnetic waves. *Phys. Rep.* **76**, 79–156.

[43] Carr, J. (1981) *Applications of centre manifold theory*. Springer, Berlin.

[44] Chae, D. (2003) Global existence of solutions to the coupled Einstein and Maxwell–Higgs system in the spherical symmetry. *Ann. H. Poincaré* **4**, 35–62.

[45] Choptuik, M. W. (1993) Universality and scaling in the gravitational collapse of a massless scalar field. *Phys. Rev. Lett.* **70**, 9–12.

[46] Choquet-Bruhat, Y. (1958) Théorèmes d'existence en mécanique des fluides relativistes. *Bull. Soc. Math. France* **86**, 155–175.

[47] Choquet-Bruhat, Y. (2004) Future complete $U(1)$ symmetric Einsteinian spacetimes, the unpolarized case. In: Chruściel, P. T. and Friedrich, H. (eds.) *The Einstein equations and the large scale behavior of gravitational fields*. Birkhäuser, Basel.

[48] Choquet-Bruhat, Y., Chruściel, P. T. and Loizelet, J. (2006) Global solutions of the Einstein–Maxwell equations in higher dimensions. *Class. Quantum Grav.* **23**, 7383–7394.

[49] Choquet-Bruhat, Y. and Geroch, R. (1969) Global aspects of the Cauchy problem in general relativity. *Commun. Math. Phys.* **14**, 329–335.

[50] Choquet-Bruhat, Y. and Moncrief, V. (2001) Future global in time Einsteinian spacetimes with $U(1)$ isometry group. *Ann. H. Poincaré* **2**, 1007–1064.

[51] Christodoulou, D. (1984) Violation of cosmic censorship in the gravitational collapse of a dust cloud. *Commun. Math. Phys.* **93**, 171–195.

[52] Christodoulou, D. (1986) The problem of a self-gravitating scalar field. *Commun. Math. Phys.* **105**, 337–361.

[53] Christodoulou, D. (1987) A mathematical theory of gravitational collapse. *Commun. Math. Phys.* **109**, 613–647.

[54] Christodoulou, D. (1991) The formation of black holes and singularities in spherically symmetric gravitational collapse. *Commun. Pure Appl. Math.* **44**, 339–373.

[55] Christodoulou, D. (1993) Bounded variation solutions of the spherically symmetric Einstein–scalar field equations. *Commun. Pure Appl. Math.* **46**, 1131–1220.

[56] Christodoulou, D. (1994) Examples of naked singularity formation in the gravitational collapse of a scalar field. *Ann. Math.* **140**, 607–653.

[57] Christodoulou, D. (1999) The instability of naked singularities in the gravitational collapse of a scalar field. *Ann. Math.* **149**, 183–217.

[58] Christodoulou, D. (2007) *The formation of shocks in 3-dimensional fluids*. EMS Publishing, Zürich.

[59] Christodoulou, D. and Klainerman, S. (1993) *The global nonlinear stability of the Minkowski space*. Princeton University Press, Princeton.

[60] Christodoulou, D. and Tahvildar-Zadeh, A. S. (1993) On the regularity of spherically symmetric wave maps. *Commun. Pure Appl. Math.* **46**, 1041–1091.

[61] Christodoulou, D. and Tahvildar-Zadeh, A. S. (1993) On the asymptotic behavior of spherically symmetric wave maps. *Duke Math. J.* **71**, 31–69.

[62] Claudel, C. M. and Newman, K. P. (1998) The Cauchy problem for quasi-linear hyperbolic evolution problems with a singularity in the time. *Proc. R. Soc. Lond. A* **454**, 1073–1107.

[63] Clausen, A. and Isenberg, J. (2007) Areal foliation and AVTD behaviour in T^2 symmetric spacetimes with positive cosmological constant. Preprint gr-qc/0701054.

[64] Corvino, J. (2000) Scalar curvature deformation and a gluing construction for the Einstein constraint equations. *Comun. Math. Phys.* **214**, 137–189.

[65] Dafermos, M. (2003) Stability and instability of the Cauchy horizon for the spherically symmetric Einstein–Maxwell–scalar field equations. *Ann. Math.* **158**, 875–928.

[66] Dafermos, M. (2005) Spherically symmetric spacetimes with a trapped surface. *Class. Quantum Grav.* **22**, 2221–2232.

[67] Dafermos, M. and Rendall, A. D. (2005) An extension principle for the Einstein–Vlasov system in spherical symmetry. *Ann. H. Poincaré* **6**, 1137–1155.

[68] Dafermos, M. and Rendall, A. D. (2005) Inextendibility of expanding cosmological models with symmetry. *Class. Quantum Grav.* **22**, L143–L147.

[69] Dafermos, M. and Rendall, A. D. (2006) Strong cosmic censorship for T^2 symmetric cosmological spacetimes with collisionless matter. Preprint gr-qc/0610075.

[70] Dafermos, M. and Rendall, A. D. (2007) Strong cosmic censorship for surface-symmetric cosmological spacetimes with collisionless matter. Preprint gr-qc/0701034.

[71] Dafermos, M. and Rodnianski, I. (2005) A proof of Price's law for the collapse of a self-gravitating scalar field. *Invent. Math.* **162**, 381–457.

[72] Dafermos, M. and Rodnianski, I. (2005) The red-shift effect and radiation decay on black hole spacetimes. Preprint gr-qc/0512119.

[73] Dain, S. (2006) Elliptic systems. In Frauendiener, J., Giulini, D, and Perlick, V. (eds.) *Analytical and numerical approaches to mathematical relativity*. Springer, Berlin.

[74] Damour, T., Henneaux, M., Rendall, A. D. and Weaver, M. (2002) Kasner-like behaviour for subcritical Einstein–matter systems. *Ann. H. Poincaré* **3**, 1049–1111.

[75] Douglis, A. and Nirenberg, L. (1955) Interior estimates for elliptic systems of partial differential equations. *Commun. Pure Appl. Math.* **8**, 503–538.

[76] Ehlers, J. (1991) The Newtonian limit of general relativity. In Ferrarese, G. (ed.) Classical mechanics and relativity: relationship and consistency. Bibliopolis, Naples.

[77] Erdelyi, A. (1956) *Asymptotic expansions*. Dover, New York.

[78] Escobedo, M., Mischler, S. and Valle, M. A. (2003) *Homogeneous Boltzmann equation in quantum relativistic kinetic theory.* Electronic J. Diff. Equations Monograph 04. EJDE, San Marcos.
[79] Faraoni, V. (2004) *Cosmology in scalar-tensor gravity.* Kluwer, Dordrecht.
[80] Fischer, A. E. and Moncrief, V. (1999) The Einstein flow, the σ-constant and the geometrization of 3-manifolds. *Class. Quantum Grav.* **16**, L79–L87.
[81] Fourès-Bruhat, Y. (1952) Théorème d'existence pour certains systèmes d'équations aux dérivées partielles non linéaires. *Acta Math.* **88**, 141–225.
[82] Friedrich, H. (1986) Existence and structure of past asymptotically simple solutions of Einstein's field equations with positive cosmological constant. *J. Geom. Phys.* **3**, 101–117.
[83] Friedrich, H. (1995) Einstein equations and conformal structure: existence of anti-de Sitter type spacetimes. *J. Geom. Phys.* **17**, 125–184.
[84] Friedrich, H. and Nagy, G. (1999) The initial boundary value problem for Einstein's vacuum field equations. *Commun. Math. Phys.* **201**, 619–655.
[85] Friedrich, H. and Rendall, A. D. (2000) The Cauchy problem for the Einstein equations. In Schmidt, B. G. (ed.) *Einstein's field equations and their physical implications.* Springer, Berlin.
[86] Fujii, Y. and Maeda, K. (2003) *The scalar-tensor theory of gravitation.* Cambridge University Press, Cambridge.
[87] Gantmacher, F. R. (2000) *Matrix theory.* American Mathematical Society, Providence.
[88] Garfinkle, D. (2004) Numerical simulations of generic singularities. *Phys. Rev. Lett.* **93**, 161101.
[89] Gerhard, C. (1983) H-surfaces in Lorentzian manifolds. *Commun. Math. Phys.* **89**, 523–553.
[90] Gilbarg, D. and Trudinger, N. S. (1983) *Elliptic partial differential equations of second order.* Springer, Berlin.
[91] Glassey, R. T. (1996) *The Cauchy problem in kinetic theory.* SIAM, Philadelphia.
[92] Glassey, R. T. and Schaeffer, J. (1990) On the 'one and one half dimensional' relativistic Vlasov–Maxwell system. *Math. Meth. Appl. Sci.* **13**, 169–179.
[93] Glassey, R. T. and Strauss, W. (1986) Singularity formation in a collisionless plasma could only occur at large velocities. *Arch. Rat. Mech. Anal.* **92**, 59–90.

[94] Gu, C.-H. (1980) On the Cauchy problem for harmonic maps defined on two-dimensional Minkowski space. *Commun. Pure Appl. Math.* **33**, 727–737.

[95] Guckenheimer, J. and Holmes, P. (1983) *Nonlinear oscillations, dynamical systems, and bifurcations of vector fields.* Springer, Berlin.

[96] Hall, G. S. (2004) *Symmetries and curvature structure in general relativity.* World Scientific, Singapore.

[97] Hamilton, R. (1982) The inverse function theorem of Nash and Moser. *Bull. Amer. Math. Soc.* **7**, 65–222.

[98] Hartman, P. (1982) *Ordinary differential equations.* Birkhäuser, Boston.

[99] Hawking, S. W. and Ellis, G. F. R. (1973) *The large scale structure of space-time.* Cambridge University Press. Cambridge.

[100] Heinzle, J. M. and Rendall, A. D. (2007) Power-law inflation in spacetimes without symmetry. *Commun. Math. Phys.* **269**, 1–15.

[101] Henkel, O. (2002) Global prescribed mean curvature foliations in cosmological spacetimes with matter. I. *J. Math. Phys.* **43**, 2439–2465.

[102] Henkel, O. (2002) Global prescribed mean curvature foliations in cosmological spacetimes with matter. II. *J. Math. Phys.* **43**, 2466–2485.

[103] Henkel, O. (2003) Local prescribed mean curvature foliations in cosmological spacetimes. *Math. Proc. Camb. Phil. Soc.* **134**, 551–571.

[104] Hervik, S. (2002) Multidimensional cosmology: spatially homogeneous models of dimension 4+1. *Class. Quantum Grav.* **19**, 5409–5427.

[105] Hervik, S., Van den Hoogen, R. J., Lim, W. C. and Coley, A. A. (2007) Late-time behaviour of the tilted Bianchi type VI_h models. *Class. Quantum Grav.* **24**, 3859–3895.

[106] Horst, E. (1982) On the classical solutions of the initial value problem for the unmodified non-linear Vlasov equation II. *Math. Meth. Appl. Sci.* **4**, 19–32.

[107] Hubbard, J. H. and West, B. H. (1991) *Differential equations: a dynamical systems approach. Ordinary differential equations.* Springer, Berlin.

[108] Hughes, T. J. R., Kato, T. and Marsden, J. E. (1977) Well-posed quasilinear second-order hyperbolic systems with applications to nonlinear elastodynamics and general relativity. *Arch. Rat. Mech. Anal* **63**, 273–294.

[109] Huisken, G. and Ilmanen, T. (2001) The inverse mean curvature flow and the Riemannian Penrose inequality. *J. Diff. Geom.* **59**, 353–437.

[110] Hwang, H. J., Rendall, A. D. and Velázquez, J. J. L. (2006) Optimal gradient estimates and asymptotic behaviour for the Vlasov–Poisson system with small initial data. Preprint Math.AP/0606389.

[111] Isenberg, J. and Kichenassamy, S. (1999) Asymptotic behavior in polarized T^2-symmetric vacuum spacetimes. *J. Math. Phys.* **40**, 340–352.

[112] Isenberg, J. and Moncrief, V. (1992) On spacetimes containing Killing vector fields with non-closed orbits. *Class. Quantum Grav.* **9**, 1683–1691.

[113] Isenberg, J. and Moncrief, V. (2002) Asymptotic behaviour in polarized and half-polarized $U(1)$-symmetric vacuum spacetimes. *Class. Quantum Grav.* **19**, 5361–5386.

[114] Isenberg, J. and Rendall, A. D. (1998) Cosmological spacetimes not covered by a constant mean curvature slicing. *Class. Quantum Grav.* **15**, 3679–3688.

[115] John, F. (1984) *Partial differential equations.* Springer, Berlin.

[116] John, F. (1990) *Nonlinear wave equations, formation of singularities.* AMS, Providence.

[117] Kichenassamy, S. (1996) Fuchsian equations in Sobolev spaces and blow-up. *J. Diff. Eq.* **125**, 299–327.

[118] Kichenassamy, S. (1996) The blow-up problem for exponential nonlinearities. *Commun. PDE* **21**, 125–162.

[119] Kichenassamy, S. and Rendall, A. D. (1998) Analytic description of singularities in Gowdy spacetimes. *Class. Quantum Grav.* **15**, 1339–1355.

[120] Kirchgraber, U. and Palmer, K. J. (1990) *Geometry in the neighbourhood of invariant manifolds of maps and flows and linearization.* Longman, New York.

[121] Klainerman, S. and Nicolò, F. (2003) *The evolution problem in general relativity.* Birkhäuser, Boston.

[122] Klainerman, S. and Nicolò, F. (2003) Peeling properties of asymptotically flat solutions to the Einstein vacuum equations. *Class. Quantum Grav.* **20**, 3215–3257.

[123] Klainerman, S. and Rodnianski, I. (2005) Causal geometry of Einstein vacuum spacetimes with finite curvature flux. *Invent. Math.* **159**, 437–529.

[124] Klainerman, S. and Rodnianski, I. (2006) A geometric approach to the Littlewood–Paley theory. *Geom. Funct. Anal.* **16**, 126–163.

[125] Kobayashi, S. (1972) *Transformation groups in differential geometry.* Springer, Berlin.

[126] Kreiss, H.-O. and Lorenz, J. (1989) *Initial-boundary value problems and the Navier–Stokes equations.* Academic Press, New York.

[127] Krieger, J., Schlag, W. and Tataru, D. (2006) Renormalization and blow up for charge one equivariant critical wave maps. Preprint math.AP/0610248
[128] LeBlanc, V. G., Kerr, D. and Wainwright, J. (1995) Asymptotic states of magnetic Bianchi VI$_0$ cosmologies. Class. *Quantum Grav.* **12**, 513–541.
[129] Lee, H. (2004) Asymptotic behaviour of the Einstein–Vlasov system with a positive cosmological constant. *Math. Proc. Camb. Phil. Soc.* **137**, 495–509.
[130] Lee, H. (2005) The Einstein–Vlasov equation with a scalar field. *Ann. H. Poincaré* **6**, 697–723.
[131] Lemou, M., Méhats, F. and Raphael, F. (2005) On the orbital stability and the singularity formation for the gravitational Vlasov–Poisson system. Preprint.
[132] Leray, J. (1953) *Hyperbolic differential equations.* Lecture notes, Princeton.
[133] Lichnerowicz, A. (1967) *Relativistic hydrodynamics and magnetohydrodynamics. Lectures on the existence of solutions.* Benjamin, New York.
[134] Lifshitz, E. M. and Khalatnikov, I. M. (1963) Investigations in relativistic cosmology. *Adv. Phys.* **12**, 185–249.
[135] Lin, X. F. and Wald, R. M. (1990) Proof of the closed universe recollapse conjecture for general Bianchi type IX cosmologies. *Phys. Rev. D* **41**, 2444–2448.
[136] Lindblad, H. and Rodnianski, I. (2003) The weak null condition for Einstein's equations. *C. R. Acad. Sci. Paris* **336**, 901–906.
[137] Lindblad, H. and Rodnianski, I. (2005) Global existence for the Einstein vacuum equations in wave coordinates. *Commun. Math. Phys.* **256**, 43–110.
[138] Lions, P.-L. and Perthame, B. (1991) Propagation of moments and regularity for the 3-dimensional Vlasov–Poisson system. *Invent. Math.* **105**, 415–430.
[139] Lojasiewicz, S. (1988) *The theory of real functions.* Wiley, New York.
[140] Majda, A. (1984) *Compressible fluid flow and systems of conservation laws in several space variables.* Springer, Berlin.
[141] Makino, T. (1986) On a local existence theorem for the evolution equation of gaseous stars. In: Nishida, T. (ed.) *Patterns and Waves.* North Holland, Amsterdam.
[142] Malec, E. and O Murchadha, N. (1994) Optical scalars and singularity avoidance in spherical spacetimes. *Phys. Rev. D* **50**, 6033–6036.

[143] Marsden, J. E. and Tipler, F. J. (1980) Maximal hypersurfaces and foliations of constant mean curvature in general relativity. *Phys. Rep.* **66**, 109–139.
[144] McOwen, R. C. (1996) *Partial differential equations*. Prentice Hall, London.
[145] Mellor, F. and Moss, I. G. (1990) Stability of black holes in de Sitter space. *Phys. Rev. D* **41**, 403–409.
[146] Moncrief, V. (1981) Global properties of Gowdy spacetimes with $T^3 \times R$ topology. *Ann. Phys. (NY)* **132**, 87–107.
[147] Moncrief, V. (1982) Neighbourhoods of Cauchy horizons in cosmological spacetimes with one Killing field. *Ann. Phys. (NY)* **141**, 83–103.
[148] Moncrief, V. (1984) The space of generalized Taub–NUT spacetimes. *J. Geom. Phys.* **1**, 107–130.
[149] Mukhanov, V. (2005) *Physical foundations of cosmology*. Cambridge University Press, Cambridge.
[150] Narita, M. (2002) On the existence of global solutions for T^3-Gowdy spacetimes with stringy matter. *Class. Quantum Grav.* **19**, 6279–6288.
[151] Narita, M. (2003) Global existence problem in T^3 Gowdy symmetric IIB superstring cosmology. *Class. Quantum Grav.* **20**, 4983–4994.
[152] Narita, M. (2004) Global properties of higher-dimensional cosmological spacetimes. *Class. Quantum Grav.* **21**, 2071–2087.
[153] Narita, M., Torii, T. and Maeda, K. (2000) Asymptotic singular behaviour of Gowdy spacetimes in string theory. *Class. Quantum Grav.* **17**, 4597–4613.
[154] Noakes, D. R. (1983) The initial value formulation of higher derivative gravity. *J. Math. Phys.* **24**, 1846–1850.
[155] Oliynyk, T. (2007) The Newtonian limit for perfect fluids. *Commun. Math. Phys.* **276**, 131–188.
[156] Oppenheimer, J. R. and Snyder, H. (1939) On continued gravitational contraction. *Phys. Rev.* **56**, 455–459.
[157] Ori, A. (1999) Oscillatory null singularity inside realistic spinning black holes. *Phys. Rev. Lett.* **83**, 5423–5426.
[158] Palais, R. (1968) *Foundations of global nonlinear analysis*. Benjamin, New York.
[159] Pfaffelmoser, K. (1992) Global classical solutions of the Vlasov–Poisson system in three dimensions for general initial data. *J. Diff. Eq.* **95**, 281–303.
[160] Poisson, E. and Israel, W. (1990) Internal structure of black holes. *Phys. Rev. D* **41**, 1796–1809.

[161] Pretorius, F. (2005) Evolution of binary black hole spacetimes. *Phys. Rev. Lett.* **95**, 121101.

[162] Price, R. H. (1972) Nonspherical perturbations of relativistic gravitational collapse I. Scalar and gravitational perturbations. *Phys. Rev. D* **5**, 2419–2438.

[163] Rein, G. (1996) Cosmological solutions of the Vlasov–Einstein system with spherical, plane, or hyperbolic symmetry. *Math. Proc. Camb. Phil. Soc.* **119**, 739–762.

[164] Rein, G. and Rendall, A. D. (1992) Global existence of solutions of the spherically symmetric Vlasov–Einstein system with small initial data. *Commun. Math. Phys.* **150**, 561–583.

[165] Rein, G., Rendall, A. D. and Schaeffer, J. (1995) A regularity theorem for solutions of the spherically symmetric Vlasov–Einstein system. *Commun. Math. Phys.* **168**, 467–478.

[166] Rendall, A. D. (1992) The initial value problem for a class of general relativistic fluid bodies. *J. Math. Phys.* **33**, 1047–1053.

[167] Rendall, A. D. (1992) Cosmic censorship and the Vlasov equation. *Class. Quantum Grav.* **9**, L99–L104.

[168] Rendall, A. D. (1994) The Newtonian limit for asymptotically flat solutions of the Vlasov–Einstein system. *Commun. Math. Phys.* **163**, 89–112.

[169] Rendall, A. D. (1995) Global properties of locally spatially homogeneous cosmological models with matter. *Math. Proc. Camb. Phil. Soc.* **118**, 511–526.

[170] Rendall, A. D. (1997) Existence of constant mean curvature foliations in spacetimes with two-dimensional local symmetry. *Commun. Math. Phys.* **189**, 145–164.

[171] Rendall, A. D. (1997) An introduction to the Einstein–Vlasov system. *Banach Centre Publications* **41**, 35–68.

[172] Rendall, A. D. (2000) Fuchsian analysis of singularities in Gowdy spacetimes beyond analyticity. *Class. Quantum Grav.* **17**, 3305–3316.

[173] Rendall, A. D. (2002) Cosmological models and centre manifold theory. *Gen. Rel. Grav.* **34**, 1277–1294.

[174] Rendall, A. D. (2004) Asymptotics of solutions of the Einstein equations with positive cosmological constant. *Ann. Inst. H. Poincaré* **5**, 1041–1064.

[175] Rendall, A. D. (2004) Accelerated cosmological expansion due to a scalar field whose potential has a positive lower bound. *Class. Quantum Grav.* **21**, 2445–2454.

[176] Rendall, A. D. (2005) Theorems on existence and global dynamics for the Einstein equations. *Liv. Rev. Relativity* lrr-2005-6.
[177] Rendall, A. D. (2005) The nature of spacetime singularities. In: Ashtekar, A. (ed.) *100 years of relativity. Spacetime structure: Einstein and beyond.* World Scientific, Singapore.
[178] Rendall, A. D. (2005) Intermediate inflation and the slow-roll approximation. *Class. Quantum Grav.* **22**, 1655–1666.
[179] Rendall, A. D. (2006) Dynamics of k-essence. *Class. Quantum Grav.* **23**, 1557–1569.
[180] Rendall, A. D. (2007) Late-time oscillatory behaviour for self-gravitating scalar fields. *Class. Quantum Grav.* **24**, 667–677.
[181] Rendall, A. D. and Weaver, M. (2001) Manufacture of Gowdy spacetimes with spikes. *Class. Quantum Grav.* **18**, 2959–2975.
[182] Reula, O. 2004 Strongly hyperbolic systems in general relativity. Preprint gr-qc/0403007.
[183] Ringström, H. (2000) Curvature blow up in Bianchi VIII and IX vacuum solutions. *Class. Quantum Grav.* **17**, 713–731.
[184] Ringström, H. (2001) The Bianchi IX attractor. *Ann. H. Poincaré* **2**, 405–500.
[185] Ringström, H. (2004) On a wave map equation arising in general relativity. *Commun. Pure Appl. Math.* **57**, 657–703.
[186] Ringström, H. (2006) Existence of an asymptotic velocity and implications for the asymptotic behavior in the direction of the singularity in T^3-Gowdy. *Commun. Pure Appl. Math.* **59**, 977–1041.
[187] Ringström, H. (2006) On the T^3-Gowdy symmetric Einstein–Maxwell equations. *Ann. H. Poincaré* **7**, 1–20.
[188] Rodnianski, I. and Sterbenz, J. (2006) On the formation of singularities in the critical O(3) sigma model. Preprint math.AP/0605023.
[189] Rudin, W. (1987) *Real and complex analysis*. McGraw-Hill, New York.
[190] Rudin, W. (1991) *Functional analysis*. McGraw-Hill, New York.
[191] Sakai, N. and Shibata, S. (2003) General relativistic electromagnetism and particle acceleration in a pulsar polar cap. *Astrophys. J.* **584**, 427–432.
[192] Shapiro, S. L. and Teukolsky, S. A. (1991) Formation of naked singularities: the violation of cosmic censorship. *Phys. Rev. Lett.* **66**, 994–997.
[193] Shapiro, S. L. and Teukolsky, S. A. (1993) Scalar gravitation: a laboratory for numerical relativity. *Phys. Rev. D* **47**, 1529–1540.
[194] Sogge, C. (1993) *Fourier integrals in classical analysis*. Cambridge University Press, Cambridge.

[195] Ståhl, F. (2002) Fuchsian analysis of $S^2 \times S^1$ and S^3 Gowdy spacetimes. *Class. Quantum Grav.* **19**, 4483–4504.
[196] Tao, T. (2004) Geometric renormalization of large energy wave maps. Preprint math.AP/0411354.
[197] Taylor, M. E. (1996) *Partial differential equations.* Springer, Berlin.
[198] Tchapnda, S. B. and Rendall, A. D. (2003) Global existence and asymptotic behaviour in the future for the Einstein–Vlasov system with positive cosmological constant. *Class. Quantum Grav.* **20**, 3037–3049.
[199] Tegankong, D. (2005) Global existence and future asymptotic behaviour for solutions of the Einstein–Vlasov–scalar field system with surface symmetry. *Class. Quantum Grav.* **22**, 2381–2391.
[200] Tod, K. P. (2003) Isotropic cosmological singularities: other matter models. *Class. Quantum Grav.* **20**, 521–534.
[201] Tod, K. P. (2007) Isotropic cosmological singularities in spatially homogeneous models with a cosmological constant. *Class. Quantum Grav.* **24**, 2415–2432.
[202] Varadarajan, V. S. (1984) *Lie groups, Lie algebras and their representations.* Springer, Berlin.
[203] Wainwright, J. and Ellis, G. F. R. (eds.) (1997) *Dynamical systems in cosmology.* Cambridge University Press, Cambridge.
[204] Wainwright, J. and Hsu, L. (1989) A dynamical systems approach to Bianchi cosmologies: orthogonal models of class A. *Class. Quantum Grav.* **6**, 1409–1431.
[205] Wald, R. M. (1983) Asymptotic behaviour of homogeneous cosmological models in the presence of a positive cosmological constant. *Phys. Rev. D* **28**, 2118–2120.
[206] Wald, R. M. (1984) *General relativity.* Chicago University Press, Chicago.
[207] Wasserman, R. H. (1992) *Tensors and manifolds.* Oxford University Press, Oxford.
[208] Weaver, M. (2000) Dynamics of magnetic Bianchi type VI_0 cosmologies. *Class. Quantum Grav.* **17**, 421–434.
[209] Yodzis, P., Seifert, H.-J. and Müller zum Hagen, H. (1973) On the occurrence of naked singularities in general relativity. *Commun. Math. Phys.* **34**, 135–148.

Index

k-essence, 35

ADM mass, 129, 216, 221
anti-de Sitter space, 199
approximation, 4
　slow roll, 198
asymptotic expansion, 76
asymptotic velocity, 189

Banach space, 104
Bel–Robinson tensor, 201
Besov space, 116
Bianchi type, 58
big bang, 2
BKL picture, 182, 261
black hole, 2, 52, 219, 250
Boltzmann equation, 45, 249
bootstrap argument, 201, 217
Brans–Dicke theory, 37, 48, 214

Casimir invariant, 188
Cauchy development, 172
　maximal, 174
Cauchy horizon, 180
　compact, 207
Cauchy hypersurface, 18
Cauchy stability, 175
Christoffel symbols, 12
closed universe recollapse conjecture, 245
CMC time coordinate, 125, 232, 245
collapsing of Riemannian manifolds, 206
comparison principle, 18, 72, 148
conformal transformation, 50
constraint
　Hamiltonian, 24
　momentum, 24
convergence
　Gromov–Hausdorff, 206
coordinates
　areal, 224, 241
　conformal, 65, 229
　double null, 62, 222, 253

Eddington–Finkelstein, 52
harmonic, 171, 200
Kruskal, 53
maximal-isotropic, 220
Schwarzschild, 215, 218
corrected energy, 205
cosmic acceleration, 2
cosmic censorship, 261
　strong, 181, 245
　weak, 180, 250
cosmic expansion, 2
cosmic microwave background, 3
cosmological constant, 21
cosmology, 2
critical, 162
critical collapse, 219, 254

dark energy, 196
de Sitter space, 57, 102
　stability of, 193
derivative
　covariant, 12
　Lie, 13, 51
distribution, 106
　tempered, 107
domain of dependence, 144
domain of influence, 144
dust, 42, 44
dynamical system, 75

Einstein equations, 21
　reduced, 171
Einstein–de Sitter model, 57, 63, 183
elliptic, 120
　Douglis–Nirenberg, 123
　strongly, 120
energy condition
　dominant, 129
energy functional, 188, 205
energy–momentum tensor, 21
ergosphere, 54
Euler equations, 41, 140
event horizon, 52

foliation, 52
formal power series, 77
formula
   dipole, 209
   quadrupole, 208
Fréchet space, 108
Fredholm operator, 123
free streaming, 217
Fuchsian methods, 261
Fuchsian system, 164

galaxies, 54, 58
generalized Kasner exponents, 82
geodesic, 13
geodesic completeness, 218
geometrization conjecture, 59
Gevrey space, 114, 154
globally hyperbolic, 18
Gowdy spacetime, 65, 165, 186
gravitational collapse, 2, 55
gravitational waves, 3, 65, 98
group action, 51

Hölder space, 109
Hawking mass, 223
heat equation, 163
Hilbert space, 103
Hubble parameter, 60
hyperbolic, 34, 132
   Leray, 152
   strongly, 150
   symmetric, 137
hyperbolic plane, 35, 66, 69, 186
   complex, 68

ill-posed, 134
initial boundary value problem, 155
isometry, 50

Kaluza–Klein reduction, 48, 197
Kasner circle, 82
Kasner relations, 82
Kasner solutions, 82
Kerr solution, 54
Kretschmann scalar, 53, 244, 261

Laplace equation, 120
lapse function, 24
Lie algebra, 51
Lie group, 51, 59

linearization
   ordinary differential equation, 79
   partial differential equation, 119
Lipschitz condition, 18, 72, 180
Littlewood–Paley theory, 115
locally rotationally symmetric, 82
Lopatinskii–Shapiro conditions, 126
Lorentzian metric, 11
Lyapunov function, 76

manifold
   centre, 86
   stable and unstable, 80
mass inflation, 256
maximally dissipative, 157
Maxwell equations, 39, 139
mean curvature, 23
Minkowski space, 12
   stability of, 199
Moser estimates, 113
multiindex, 106

neutron star, 3, 55
Newton–Cartan theory, 209
Newtonian gravity, 1, 207
Newtonian limit, 208
no hair theorem, 53
   cosmic, 193
null condition, 159, 200
   weak, 160
numerical methods, 5

ordinary differential equation, 4

parabolic, 162
partial differential equation, 4
peeling theorem, 200
PMC time coordinate, 239
Poincaré–Bendixson theory, 75
post-Newtonian approximation, 208
Price's law, 258
principal symbol, 120
pseudodifferential operator, 117

qualitative analysis, 5

reduction theorem, 86
regularity, 6
Ricci flow, 163

scalar field, 33, 252
second fundamental form, 22
shift vector, 25
shock wave, 161
singularity, 2, 14
  coordinate, 12
  crushing, 32
  isotropic, 248
  weak null, 232, 250, 254, 255
singularity theorem
  Hawking, 19
  Penrose, 19
Sobolev embedding theorem, 111
Sobolev space, 109
  weighted, 111
solution
  Bertotti–Robinson, 54
  Oppenheimer–Snyder, 62
  pseudo-Schwarzschild, 206, 231, 243
  Reissner–Nordström, 54, 254
  Schwarzschild, 52
  Schwarzschild–de Sitter, 247
spacetime, 1
special relativity, 12
spike, 190
star, 2
static spacetime, 52
stationary spacetime, 52
supernova, 2, 196
support, 106
symmetry
  T^2, 64, 224
  $U(1)$, 68
  cylindrical, 68
  Gowdy, 65
  hyperbolic, 61
  Kantowski–Sachs, 58
  local T^2, 64
  plane, 61, 64
  polarized Gowdy, 65
  spherical, 61, 245
  surface, 61, 241

Teichmüller space, 204
theorem
  Arzela–Ascoli, 109
  Birkhoff's, 53
  Cauchy–Kovalevskaya, 155
  Frobenius', 52
  Hartman–Grobman, 79
  Nash–Moser, 109
  Wald's, 197
  Yamabe, 128
topologically equivalent, 79
trapped surface, 19, 219

vector field
  conformal Killing, 51
  Killing, 51
Vlasov equation, 44
Vlasov–Maxwell system, 214
Vlasov–Nordström system, 214
Vlasov–Poisson system, 213, 218
  relativistic, 214

Wainwright–Hsu system, 87
wave equation, 34
wave map, 186, 237
weak convergence, 105
well posed, 122, 134
Weyl tensor, 15

Yang–Mills equations, 139